Stability-Constrained Optimization for Modern Power System Operation and Planning

Stability-Constrained Optimization for Modern Power System Operation and Planning

Yan Xu
Nanyang Technological University
Singapore

Yuan Chi
Chongqing University
China

Heling Yuan
Nanyang Technological University
Singapore

IEEE Press Series on Power and Energy Systems

IEEE PRESS

WILEY

Published by John Wiley & Sons, Inc., Hoboken, New Jersey.
Published simultaneously in Canada.

For general information on our other products and services or for technical support, please contact our Customer Care Department within the United States at (800) 762-2974, outside the United States at (317) 572-3993 or fax (317) 572-4002.

Wiley also publishes its books in a variety of electronic formats. Some content that appears in print may not be available in electronic formats. For more information about Wiley products, visit our web site at www.wiley.com.

Library of Congress Cataloging-in-Publication Data:
Names: Xu, Yan (Associate professor), author.
Title: Stability-constrained optimization for modern power system operation and planning / Yan Xu, Nanyang Technological University, Singapore, Yuan Chi, Chongqing University, China, Heling Yuan, Nanyang Technological University, Singapore.
Description: Hoboken, NJ : Wiley-IEEE Press, [2023] | Series: IEEE Press series on power and energy systems | Includes index.
Identifiers: LCCN 2023002456 (print) | LCCN 2023002457 (ebook) | ISBN 9781119848868 | ISBN 9781119848875 (adobe pdf) | ISBN 9781119848882 (epub)
Subjects: LCSH: Electric power system stability.
Classification: LCC TK1010 .X83 2023 (print) | LCC TK1010 (ebook) | DDC 621.319–dc23/eng/20230214
LC record available at https://lccn.loc.gov/2023002456
LC ebook record available at https://lccn.loc.gov/2023002457

Cover Design: Wiley
Cover Image: © Leo Pakhomov/Shutterstock

Set in 9.5/12.5pt STIXTwoText by Straive, Pondicherry, India

Contents

About the Authors

Yan Xu received the B.E. and M.E. degrees from South China University of Technology, China, and the Ph.D. degree from University of Newcastle, Australia, in 2008, 2011, and 2013, respectively. He conducted postdoctoral research with the University of Sydney Postdoctoral Fellowship, and then joined Nanyang Technological University (NTU) with the Nanyang Assistant Professorship. He is now an Associate Professor at School of Electrical and Electronic Engineering and a Cluster Director at Energy Research Institute @ NTU (ERI@N). His research interests include power system stability, microgrid, and data analytics for smart grid applications. Dr Xu's research in Singapore is funded by a range of funding agencies (including Singapore NRF, EMA, MOE, HDB, etc.) and industry partners (including Rolls-Royce Electrical, Singapore Power Group, Singtel, Infineon, EDF Lab, Lite-On, etc.). Many of his research outcomes have been practically applied/licensed to industry partners. Dr Xu has received 10 IEEE/IET paper contest and conference best paper awards, the 2022 IET Premium Award (Best Paper), the 2021 IEEE Transactions on Smart Grid Outstanding Paper Award, and the 2018 Applied Energy Highly Cited Paper Award. His professional service roles include Associate Editor for *IEEE Trans. Smart Grid* and *IEEE Trans. Power Systems*, Chairman of the IEEE Power & Energy Society (PES) Singapore Chapter (2021 to 2022) and the General Co-Chair of the 11th IEEE ISGT-Asia Conference, Nov. 2022.

Yuan Chi received the B.E. degree from Southeast University, Nanjing, China, in 2009, and the M.E. degree from Chongqing University, Chongqing, China, in 2012, and the Ph.D. degree from Nanyang Technological University, Singapore, in 2021. From 2012 to 2017, he worked as an Electrical Engineer of Power System Planning consecutively with State Grid Chongqing Electric Power Research Institute and Chongqing Economic and Technological Research Institute. He is currently a Research Associate with Chongqing University. His research interests include planning, resilience, and voltage stability of power systems. Dr Chi's research in

China is funded by a range of funding agencies and industry partners, including Ministry of Finance of PRC, China Postdoctoral Science Foundation, State Grid Corporation of China, China Southern Power Grid, etc.

Heling Yuan received the B.E., M.Sc., and Ph.D. degrees from North China Electric Power University, Beijing, China, the University of Manchester, UK, and Nanyang Technological University, Singapore, in 2016, 2017, and 2022, respectively. She is currently a Research Fellow at Rolls-Royce @ NTU Corporate Lab, Singapore. Her research interests include modeling, optimization, stability analysis and control of power systems. Dr Yuan's research in Singapore is funded by Singapore NRF, MOE and Rolls-Royce Electrical.

Foreword

The stability of a power system is defined by the IEEE as its ability "for a given initial operating condition, to regain a state of operating equilibrium after being subjected to a physical disturbance, with most system variables bounded so that practically the entire system remains intact". Recent years have seen significant integration of renewable energy resources such as wind and solar power into power grids globally. Yet, such renewable energy-based generators can significantly complicate the power system's dynamic behavior and introduce considerable operational uncertainties due to their power-electronic converter interface and variable power output.

Practically, the stability of a power system can be maintained and enhanced through three general approaches: (i) accurately modeling and analyzing the power system's dynamic characteristics, then designing and deploying real-time controllers to make the system well behaved under disturbances; (ii) dispatching the power system to a state that can better withstand the disturbances; (iii) reinforcing the power grid with fast-responding resources such as FACTS (Flexible AC Transmission Systems) to support the power system dynamics to ride through the disturbances. While the first approach involves power system dynamics modeling, stability analysis, and controller design, the latter two require advanced optimization methods to optimally operate the power system and determine the optimal size and site of such resources for maximum cost effectiveness.

While most existing books are focused on the first approach to power system dynamics modeling, stability analysis, and controller design, very few address the latter two approaches that require advanced optimization. This book fills this gap by presenting a series of stability-constrained optimization methodologies for power system operation and planning. Two major foci of the book are transient stability enhancement through optimal power system dispatch and operational control and voltage stability enhancement through optimally sizing and siting dynamic VAR resources in the power grid, respectively. The book presents a series of corresponding mathematical models and solution methods to achieve the these objectives.

This book is written by a dedicated research team with over 10 years of research experience in this field, and the presented methodologies are based on their original research outcomes and insights into these topics. With a balance between theory and practice, this book serves as a timely reference and guidebook for graduate students, researchers, and power system operation and planning engineers in this field.

David John Hill, PhD
Professor of Power and Energy Systems, Monash University, Australia
Emeritus Professor, University of Sydney, and University of Hong Kong
Fellow of IEEE, IFAC, SIAM,
Australian Academy of Science,
Australian Academy of Technological Sciences and Engineering
Hong Kong Academy of Engineering Sciences
Foreign Member of Royal Swedish Academy of Engineering Science

Preface

The electrical power system is essential to a modern society, and its stability is a fundamental requirement during both online operation and offline planning studies. In general, the stability of a power system refers to its ability to regain a state of operating equilibrium after experiencing a physical disturbance, such as a short-circuit fault. In practice, the stability of the power system mainly depends on both its inherent dynamic characteristics, i.e. how the system responds to disturbances, and its steady-state operating conditions, i.e. how the system is dispatched.

In recent years, renewable energy sources such as solar photovoltaic and wind power have rapidly penetrated modern power systems, which have inherently stochastic and intermittent power output and are connected to the grid through power electronic converters. Consequently, both the static and dynamic behaviors of the power system have become much more complex, creating a series of challenges for maintaining system stability. These challenges include long-distance power transmission from renewable power stations to load centers, reduced synchronous inertia in the power systems, complex dynamics of power electronic interfaced devices, lack of reactive power resources, and fast fluctuation of magnitude and direction of power flows through the transmission network. Recent large-scale blackout events, e.g. the September 2016 South Australia blackout and the August 2019 UK blackout, have clearly demonstrated the adverse impact of these challenges on power system stability.

This book focuses on two power system stability problems, namely transient stability and voltage stability. Transient stability, also known as large-disturbance rotor angle stability, is the most stringent requirement for a power system because instability can develop rapidly within several cycles after a disturbance. Voltage stability is becoming increasingly critical since poor dynamic voltage performance of the power system could lead to the failure of wind and solar power generators riding through disturbances. This book presents a series of optimization methodologies that we have originally proposed to (i) optimally dispatch the power system to an operating state that can maintain transient stability in the event of a large disturbance and (ii) optimally allocate dynamic VAR resources, including STAT-COM and SVC, in the power grid to reinforce the grid's capability to counteract voltage instability.

The book consists of 20 chapters, which are organized into three parts:

- Part I (Chapters 1–3) provides an overall introduction to power system stability, including fundamental concepts, definitions, mathematical models, metrics, and analysis methods, as well as presents a review of major large-scale blackout events in recent years.
- Part II (Chapters 4–12) focuses on transient stability-constrained (TSC) power system dispatch and operational control. The problems are generally modeled as TSC-optimal power flow (TSCOPF) and TSC-unit commitment (TSCUC), and a series of mathematical formulations are presented in this part. The formulations cover deterministic, stochastic, and robust optimization models with uncertainties arising from dynamic load components and renewable power generation, and two-stage optimization models for coordinating preventive and corrective control actions. To solve these problems, a set of computationally efficient methods have been correspondingly presented, which are based on quantitative transient stability assessment, trajectory sensitivity analysis, linear transient stability constraint construction, machine learning-based stability constraint extraction, hybrid computation, and decomposition-based solution frameworks. The proposed methodologies are numerically tested on IEEE benchmark testing systems, showing their effectiveness with simulation results.
- Part III (Chapters 13–20) focuses on voltage stability enhancement through dynamic VAR resource allocation in the power system. This part first introduces the general framework and mathematical models for optimal VAR resource planning in the power grid, then presents several quantitative voltage stability indices, followed by an overview of the fundamentals and the mathematical models of the dynamic VAR resources. To reduce the size of the optimization model, two methods for candidate bus selection are presented afterward. After that, a series of different mathematical formulations and solution methods for optimal dynamic VAR resource allocation are presented, including multi-objective optimization model, retirement-driven planning model, multi-stage coordinated planning model, and many-objective robust optimization model. The proposed methodologies are numerically demonstrated on IEEE benchmark testing systems with simulation results.

The book is targeted at scholars, researchers, and postgraduate students who are seeking optimization methodologies for power system stability enhancement. Additionally, it provides practical solutions to operational dispatch and network reinforcement planning for power system operators, planners, and optimization algorithm developers in the power industry.

We would like to express our sincere gratitude to the funding supports for the research presented in this book, including the Hong Kong Research Grant Council (RGC) GRF Grant, the Australia Research Council (ARC) Linkage Project, the University of Newcastle Faculty Strategic Pilot Grant, the University of Sydney Postdoctoral Fellowship (USYD-PF), the Nanyang Assistant Professorship (NAP) and PhD scholarships from Nanyang Technological University, and the Singapore Ministry of Education (MOE) Tier-1 Grants.

Yan Xu
Nanyang Technological University
Singapore

Yuan Chi
Chongqing University
China

Heling Yuan
Nanyang Technological University
Singapore

Part I

Power System Stability Preliminaries

List of Acronyms

ACC   alternative current control
CCT   critical clearing time
CIG   converter interfaced generation
CMs   critical machines
COI   center of inertia
DAE   differential algebraic equations
DFIG   doubly-fed induction generators
EEA   energy emergency alert
EEAC   extended equal-area criterion
ERCOT   electric reliability council of texas
FE   first energy
IGE   induction generator effect
LPB-G   low pressure turbine B to generator
LVRT   low voltage ride through
MISO   midwest independent system operator
NMs   non-critical machines
PLL   phase-locked loop
SA   South Australia
SIME   single machine equivalence
SOL   system operational limits
SSR   subsynchronous resonance
T&D   transmission and distribution
TDS   time domain simulation
TEF   transient energy function
TSA   transient stability assessment

Stability-Constrained Optimization for Modern Power System Operation and Planning,
First Edition. Yan Xu, Yuan Chi, and Heling Yuan.
© 2023 The Institute of Electrical and Electronics Engineers, Inc.
Published 2023 by John Wiley & Sons, Inc.

TSC   transient stability control
TSIs   transient stability indexes
TSO   transmission system operator
UCTE   Union for the Co-ordination of Transmission of Electricity

1

Power System Stability: Definition, Classification, and Phenomenon

1.1 Introduction

An electrical power system is a fundamental infrastructure of a society. As a large-scale time-varying dynamic system, maintaining its stability is a basic and essential requirement during its operation and planning decision-making process. In general, the stability of a power system refers to its ability to regain a state of operating equilibrium after being subjected to a physical disturbance (such as a short-circuit fault) [1]. In practice, the stability of the power system depends on both its dynamic characteristics, i.e. how the system would behave in response to disturbances, and its steady-state operating conditions, i.e. how the power system is dispatched.

In recent years, modern power systems have started to integrate high shares of renewable energy sources, such as solar photovoltaic and wind power, which are inherently stochastic and intermittent in their power outputs and are interfaced with the power grid through power electronic converters. While these renewable energy-based converter interfaced generators (CIGs) are environmentally beneficial, they significantly complicate the power grid's static and dynamic characteristics. As a result, the dynamic behaviors of the power system become much more complex, which introduces a series of challenges to the control, operation, and planning for maintaining system stability.

In a nutshell, this chapter gives a brief introduction to the modern power system stability, including its definition, classification, and phenomenon.

Stability-Constrained Optimization for Modern Power System Operation and Planning,
First Edition. Yan Xu, Yuan Chi, and Heling Yuan.
© 2023 The Institute of Electrical and Electronics Engineers, Inc.
Published 2023 by John Wiley & Sons, Inc.

1.2 Definition

In this book, the definition of power system stability given in [1] is adopted. Conforming to definitions from system theory, the definition is based in physics, thus easily understood and readily applied by power system engineering practitioners:

> Power system stability is the ability of an electric power system, for a given initial operating condition, to regain a state of operating equilibrium after being subjected to a physical disturbance, with most system variables bounded so that practically the entire system remains intact.

The definition is applied to interconnected power systems at large while also concerning the stability of a single generator or a group of generators. When a power system is subjected to a disturbance, its stability depends on its initial operating point and the nature of the disturbance. In practice, a power system can suffer from various disturbances regardless of small or large. A small disturbance can be a continuous load change that will not alter the system topologies. A large disturbance is one which may result in structural changes due to the isolation of the faults, such as a short circuit on the transmission line or losses of a large generator. After small or large disturbances, the system must be capable of riding through the disturbances and returning to a viable equilibrium.

It should be noted that, at an initial operating point, a power system may be stable for a given large disturbance but become unstable for another. It is in general not practical or economical to ensure stability against all possible disturbances. It is more reasonable to select contingencies that show a high probability of occurrence and criticality based on historical data and system topology information. Large-disturbance stability is generally validated under a set of specified disturbances. A stable equilibrium has a finite region of attraction; the larger the region, the more robust the system is against large disturbances. However, it should be noted that the region of attraction changes with the operating point of the power system. Though the power system continually fluctuates with small magnitudes, it is usually acceptable to assume that the system is initially in a steady-state operating point when assessing its stability after being subjected to a specific disturbance.

1.3 Classification

Stability is a condition of equilibrium between opposing forces undergoing continuous imbalance, which results in different forms of instability subject to the network topology, system operating point, and the type of disturbance. The different

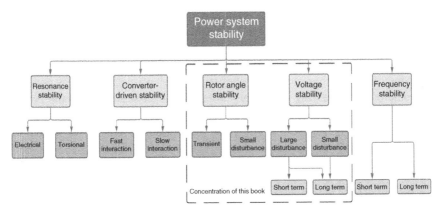

Figure 1.1 Classification of power system stability.

forms of instabilities need to be properly analyzed. To facilitate the analysis of stability problems which are in nature of high dimension and complexity, it is necessary to firstly classify the stability into appropriate categories.

Figure 1.1 shows the overall picture of the power system stability classification with its categories and subcategories. Traditionally, the power system stability was classified into three categories: rotor angle stability, voltage stability, and frequency stability [1]. The power system stability primarily dealt with fairly slow electromechanical phenomena, which are typically caused by synchronous machines and induction machines. Given the increased penetration of CIGs in modern power systems and their substantial impacts on system dynamic behavior, two new stability categories have been added [2], i.e. resonance stability and converter-driven stability, to deal with faster dynamics within electromagnetic time scales.

The focus of this book is on the rotor angle stability and voltage stability, which are the two most stringent stability requirements for the power system operation and planning. Rotor angle stability and voltage stability can be approached by optimization-based operating point dispatches and network reinforcement (dynamic reactive power device deployment).

1.4 Rotor Angle Stability

Rotor angle stability is the ability of synchronous generators to remain in synchronism after being subjected to a disturbance. It depends on the ability of each synchronous generator to maintain equilibrium between electromagnetic torque and mechanical torque. Instability will occur when the rotor angles of some generators increase continuously with regard to other generators. Namely, the generators lose synchronism with others. The loss of synchronism can occur between one machine and the rest of the system, or between groups of machines, with synchronism maintained within each group after separating from each other. The possible outcome of the instability is generator tripping and/or separation of the power systems.

As presented in Figure 1.1, the rotor angle stability can be divided into large-disturbance rotor angle stability (also called transient stability) and small-disturbance rotor angle stability, based on the severity of the disturbance.

1.4.1 Large-Disturbance Rotor Angle Stability

For transient stability, it is always relative to a severe disturbance, such as a short circuit on a transmission line, which will result in large excursions of rotor angles and is involved by the nonlinear power–angle relationship. Transient stability depends on both the initial operating point and the severity of the disturbance. Instability is usually in the form of aperiodic angular separation due to insufficient synchronizing torque, indicating *first-swing instability*. Figures 1.2 and 1.3 indicate

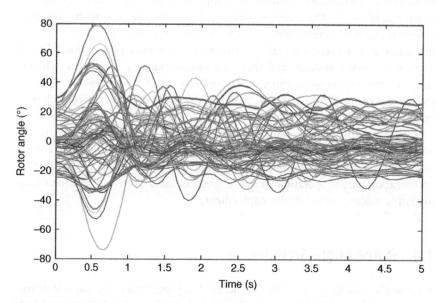

Figure 1.2 Simulated rotor angles of a transient stable case. *Source:* Xu [3].

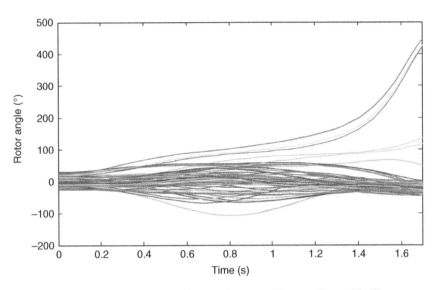

Figure 1.3 Simulated rotor angles of a transient unstable case. *Source:* Xu [3].

the simulated post-disturbance rotor angle trajectories of a real large power grid for a stable case and an unstable case, respectively. From the two figures, the stable case corresponds to keeping the synchronism of all the generators, while the unstable case corresponds to the loss of synchronism of some generators after the disturbance.

1.4.2 Small-Disturbance Rotor Angle Stability

Small-disturbance rotor angle stability refers to the ability of the power system to maintain synchronism under small disturbances. The disturbances are considered to be sufficiently small that linearization of system equations is permissible for purposes of analysis, e.g. the sudden variation of the load demands and/or the change of generation output. Small-disturbance rotor angle stability usually depends on the initial operating point. Instability is often in two forms: (i) non-oscillatory increment of rotor angle due to lack of synchronizing torque; or (ii) rotor oscillations of amplitude due to lack of damping torque. For modern power systems, small-disturbance rotor angle stability is usually associated with insufficient damping torque. The illustration of the small-disturbance stability problem due to the lack of damping torque is presented in Figure 1.4.

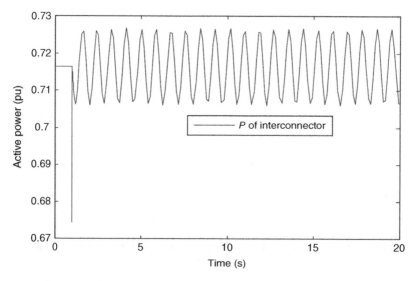

Figure 1.4 Active power oscillation due to the small-disturbance instability. *Source:* Xu [3].

1.5 Voltage Stability

Voltage stability is the ability of the power system to maintain steady voltages at all buses after being subjected to a disturbance. It depends on the ability to maintain equilibrium between load demand and load supply from the power system. Instability will occur in the form of a successive fall or rise of voltage in some buses. A possible outcome of instability is loss of load in an area or transmission lines and other elements tripping by their protective systems, leading to cascading outages. It should be noted that a progressive drop in bus voltage may be associated with rotor angle instability.

The driving force for voltage instability is usually the loads. A run-down situation causing voltage instability occurs when load dynamics attempt to restore power consumption beyond the capability of the transmission network and the connected generation.

As in the case of rotor angle stability, classifying voltage stability into large-disturbance voltage stability and small-disturbance voltage stability is applicable.

1.5.1 Large-Disturbance Voltage Stability

Large-disturbance voltage stability refers to the system's ability to maintain steady voltages following large disturbances such as system faults, loss of generation, or circuit contingencies. This ability is determined by the system and load characteristics, and the interactions of both continuous and discrete controls and protections.

The simulated post-disturbance voltage trajectories of the New England 10-machine 39-bus system are indicated in Figures 1.5 and 1.6 for a stable and unstable case, respectively. In the stable case, all voltage trajectories fluctuate and return

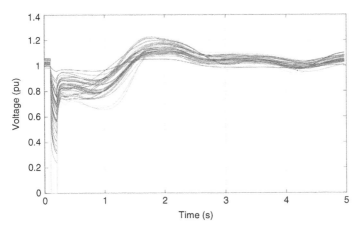

Figure 1.5 Simulated voltage trajectories of the large-disturbance voltage stable case.

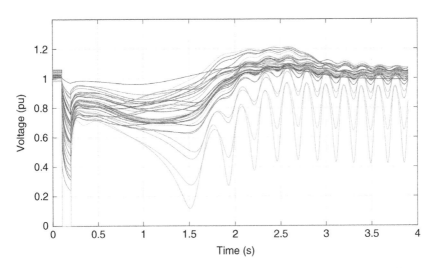

Figure 1.6 Simulated voltage trajectories of the large-disturbance voltage unstable case.

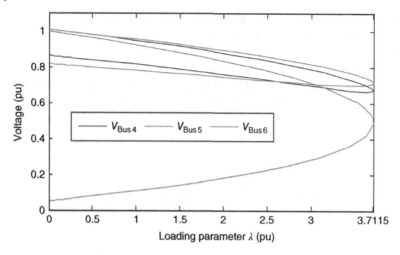

Figure 1.7 Simulated power–voltage (PV) curve. *Source:* Xu [3].

to the nominal level finally, while in the unstable case, the voltage trajectories fluctuated dramatically without recovery.

1.5.2 Small-Disturbance Voltage Stability

Small-disturbance voltage stability refers to the system's ability to maintain steady voltages when subjected to small perturbations, such as incremental changes in system load. Under this case, the system voltage will drop continuously with the load increment until a critical point is reached. After the critical point, the system voltage will drop dramatically and finally collapse.

A simulated power–voltage curve is presented in Figure 1.7, where the loading parameter means the load-increasing rate. When it increases to 3.7115 times the initial operating point, voltage instability will occur.

1.6 Frequency Stability

Frequency stability refers to the ability of a power system to maintain steady frequency following a severe system upset resulting in a significant imbalance between generation and load. It depends on the ability to maintain/restore equilibrium between system generation and load, with minimum unintentional loss of load. The instability that may result occurs in the form of sustained frequency swings, leading to the tripping of generating units and/or loads.

Figures 1.8 and 1.9 indicate the simulated post-disturbance frequency trajectory for a stable and unstable case.

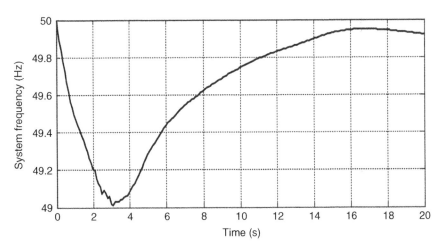

Figure 1.8 Simulated post-disturbance frequency trajectory of a frequency stable case. *Source:* Xu [3].

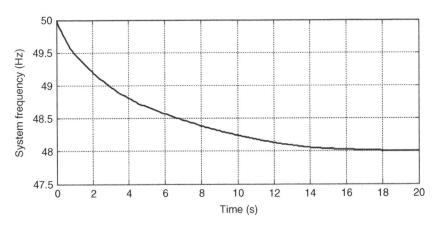

Figure 1.9 Simulated post-disturbance frequency trajectory of a frequency unstable case. *Source:* Xu [3].

1.7 Resonance Stability

Generally, resonance instability may occur when the oscillatory magnitude of voltage/current/torque exceeds specified thresholds due to insufficient dissipation of energy in the flow path. The term resonance stability encompasses subsynchronous resonance (SSR) and can be indicated in two forms: (i) due to a resonance between series compensation and the mechanical torsional frequencies of the turbine-generator shaft; (ii) due to a resonance between series compensation and the electrical characteristics of the generator. Thus, the resonance stability has been divided into two categories shown in Figure 1.1.

1.7.1 Torsional Resonance

Torsional resonance is the SSR due to torsional interactions between the series compensated line(s) and a turbine-generator mechanical shaft, particularly as it pertains to conventional synchronous generation.

Figure 1.10a indicates the torsional torque in low-pressure turbine B to generator 1 (LPB-G1) shaft section with 70% and 30% compensation levels. The dark blue shade shows the torsional torque with a 70% compensation level, and the gray shade shows the torque with a 30% compensation level. From the figure, the torque is growing slowly for the dark blue shade, which means that the 70% compensation level leads the system unstable. While for the gray shade, the torque is decaying. It represents the system becomes stable with the 30% compensation level. It can be found that the higher the compensation level, the weaker the damping.

1.7.2 Electrical Resonance

As the variable-speed DFIG generator is an induction generator connected to the grid, the electrical resonance between the generator and series compensation may exist, which would be highly susceptible to IGE-self-excitation type SSR. In this case, the self-excitation type SSR occurs when the series capacitor forms a resonant circuit, at subsynchronous frequencies, with the effective inductance of the induction generator.

The phenomena can be realized by the IGE simulation performed with the torsional system disabled. Figure 1.10b indicates the dynamic responses with various wind speeds and a constant 75% compensation level. From the figure, it can be found that the higher the wind speed, the better the SSR damping for the doubly-fed induction generator (DFIG) system.

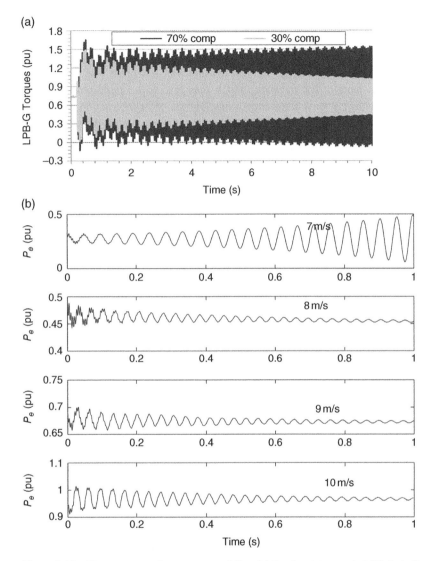

Figure 1.10 Phenomenon of resonance stability. (a) Torsional torque in LPB-G shaft section. *Source:* Adrees and Milanović [4]/(© [2014] IEEE. (b) Dynamic response of Pe under different wind speeds. Compensation level: 75%. *Source:* Fan et al. [5]/© [2010] IEEE.

1.8 Converter-Driven Stability

Since the timescale referring to the controls of CIGs is wide, the cross-coupling occurs with both the electromechanical dynamics of machines and the electromagnetic transients of the network, which may lead to unstable power system oscillations within a wide frequency range. Hence, slow-interaction converter-driven stability (less than 10 Hz) and fast-interaction converter-driven stability (tens to hundreds of Hz to kHz) are categorized.

1.8.1 Fast-Interaction Converter-Driven Stability

Fast-interaction converter-driven stability is driven by fast dynamic interactions of power electronic-based control systems, such as CIGs, HVDC, and FACTS, which are fast response elements. Instabilities produced by fast converter interactions may indicate various forms. For example, interactions of the fast inner-current loops of CIG with passive system components may cause high-frequency oscillations.

Figure 1.11a shows the harmonic instability phenomenon due to the interactions between the inner alternative current control (ACC) loops of the three paralleled VSCs. The bandwidth of the ACC loop is increased from $f_s/20$ to $f_s/15$ at the time instant of T_i. It is obvious to see that the three paralleled VSCs become unstable with the bandwidth of $f_s/15$.

1.8.2 Slow-Interaction Converter-Driven Stability

Slow-interaction converter-driven stability is driven by slow dynamic interactions of power electronic-based devices' control systems, such as the electromechanical dynamics of synchronous generators. Like voltage stability, a weak system can be the primary cause of instability. However, these two types of stability are fundamentally different since voltage instability is driven by loads, while converter-driven instability is involved by power electronic converters.

Figure 1.11b shows low-frequency oscillation and subsynchronous frequency oscillation phenomena for four phase-locked loop (PLL) parameters. For PLL1, there is a 4.5 Hz undamped oscillation, which is defined as low-frequency oscillation. For PLL2, 30 Hz oscillations are dominant. The low-frequency mode can also be found in the initial few cycles. For PLL3, both modes can be damped. For PLL4, the 40 Hz oscillation mode is dominant and the system becomes unstable.

This book mainly focuses on transient stability and large-disturbance voltage stability, as marked in the dashed square in Figure 1.1. Besides, transient stability in this book refers to first-swing stability. Multi-swing stability is not considered.

(a)

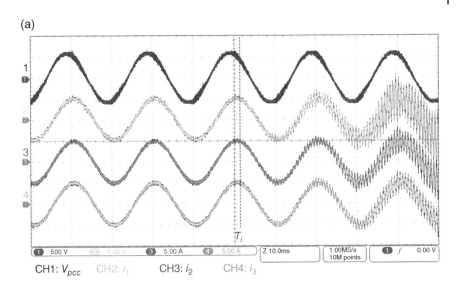

CH1: V_{pcc} CH2: i_1 CH3: i_2 CH4: i_3

(b)

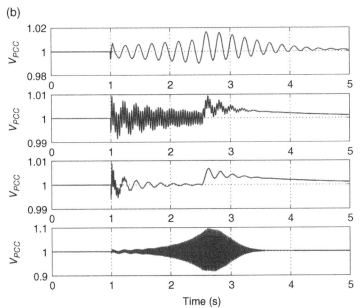

Figure 1.11 Phenomenon of converter-driven stability. (a) Measured waveforms for the harmonics instability. *Source:* Wang and F. Blaabjerg [6]/(© [2019] IEEE. (b) Response of V_{PCC} for four sets of PLL parameters, from upper to bottom: PLL1, PLL2, PLL3, and PLL4. *Source:* Fan and Miao [7]/© [2018] IEEE.

Thus, the transient stability in the subsequent chapters all refers to the first-swing stability.

References

1 Kundur, P., Paserba, J., Ajjarapu, V. et al. (2004). Definition and classification of power system stability IEEE/CIGRE joint task force on stability terms and definitions. *IEEE Transactions on Power Systems* 19 (3): 1387–1401.

2 Hatziargyriou, N., Milanovic, J., Rahmann, C. et al. (2021). Definition and classification of power system stability – revisited & extended. *IEEE Transactions on Power Systems* 36 (4): 3271–3281.

3 Xu, Y. (2013). Dynamic security assessment and control of modern power systems using intelligent system technologies. PhD thesis. The University of Newcastle, Newcastle.

4 Adrees, A. and Milanović, J.V. (2014). Methodology for evaluation of risk of subsynchronous resonance in Meshed compensated networks. *IEEE Transactions on Power Systems* 29 (2): 815–823.

5 Fan, L., Kavasseri, R., Miao, Z.L., and Zhu, C. (2010). Modeling of DFIG-based wind farms for SSR analysis. *IEEE Transactions on Power Delivery* 25 (4): 2073–2082.

6 Wang, X. and Blaabjerg, F. (2019). Harmonic stability in power electronic-based power systems: concept, modeling, and analysis. *IEEE Transactions on Smart Grid* 10 (3): 2858–2870.

7 Fan, L. and Miao, Z. (2018). Wind in weak grids: 4 Hz or 30 Hz oscillations? *IEEE Transactions on Power Systems* 33 (5): 5803–5804.

2

Mathematical Models and Analysis Methods for Power System Stability

2.1 Introduction

This chapter introduces the general mathematical models for power system stability study, the stability criteria, and several analytical methods that will be adopted in the subsequent chapters of this book.

2.2 General Mathematical Model

As a time-varying large-scale dynamic system, the general dynamic behavior of the power system can be modeled as a high-dimensional nonlinear differential-algebraic equation (DAE) set:

$$\dot{\mathbf{x}} = \mathbf{f}(\mathbf{x}, \mathbf{y}, \mathbf{k}, t) \tag{2.1}$$

$$0 = \mathbf{g}(\mathbf{x}, \mathbf{y}, \mathbf{k}, t) \tag{2.2}$$

where $\mathbf{f}(f : \mathbb{R}^{n_x} \times \mathbb{R}^{n_y} \times \mathbb{R}^{n_k} \times \mathbb{R}^\dagger \mapsto \mathbb{R}^{n_x})$ are differential equations, where synchronous generators, "dynamic" load such as induction motors, including their controls, and other devices whose dynamics are modeled; $\mathbf{g}(g : \mathbb{R}^{n_x} \times \mathbb{R}^{n_y} \times \mathbb{R}^{n_k} \times \mathbb{R}^\dagger \mapsto \mathbb{R}^{n_y})$ are algebraic equations compromising of network and "static" load; $\mathbf{x}(x \in \mathbb{R}^{n_x})$ are state variables (e.g. rotor angles and speeds of synchronous generators, dynamic load responses, and system controllers, etc.); $\mathbf{y}(y \in \mathbb{R}^{n_y})$ are algebraic variables (e.g. bus voltages magnitudes, and phases); $\mathbf{k}(k \in \mathbb{R}^{n_k})$ are discrete variables that model disturbance events, e.g. short-circuit faults and line switching, etc., and t is the time.

Imposing $\dot{\mathbf{x}} = 0$ in (2.1), yields the overall equilibrium equations, which allow computing the steady state conditions of the system in its pre-fault state or in its post-fault state.

Stability-Constrained Optimization for Modern Power System Operation and Planning,
First Edition. Yan Xu, Yuan Chi, and Heling Yuan.
© 2023 The Institute of Electrical and Electronics Engineers, Inc.
Published 2023 by John Wiley & Sons, Inc.

Assuming that the system is steady-state stable at the pre-fault condition, the stability analysis is to evaluate the system response when subjected to various disturbances. Namely, for stability assessment, it considers whether the system can reach its post-fault operating point. The dynamics in (2.1) may usually be decomposed into two subsets: one faster, generally only related to the electro-mechanical angle dynamics of the synchronous machines, i.e. transient stability; and one slower, normally only related to the stability of voltage (and load) restoration process due to the action of the slow subset of automatic voltage control devices [1]. Therefore, by solving the above DAE set, the dynamic response including, dynamic voltage response and rotor angle response, can be obtained.

2.3 Transient Stability Criteria

Transient stability is the ability of synchronous generators to remain in synchronism after being subjected to a large disturbance. Instability often occurs in the form of unlimited angular increments of some generators, which leads to the loss of synchronism with other generators.

As indicated in Section 2.2, the dynamic behavior of rotor angles in power systems can be expressed mathematically by a set of differential-algebraic equations (DAE). The dynamics of the generators (also known as swing equations) in the classical model can be formulated as follows [2]:

$$\begin{cases} M_i \dfrac{d\omega_i}{dt} = P_{mi} - P_{ei} \\ \dfrac{d\delta_i}{dt} = \omega_i \end{cases}, \quad \{i \in S_G\} \tag{2.3}$$

where M_i is the moment of inertia of i-th generator, which can be calculated as $H/\pi f$ in per-unit. H is the inertia of i-to generator. δ_i and ω_i denote the angle and the difference between the angular speed and synchronous speed, respectively. P_{mi} and P_{ei} are the mechanical power and electrical power with i-to generator, respectively.

It should be noted that, except for the swing equation, the full dynamic model of the machine contains stator transients, flux linkage, as well as damper windings. In this chapter, the details of these equations are not given here but can be readily found in the literature and relevant books. To obtain accurate analysis results, the full model should be used in time-domain simulation (TDS) studies.

Transient stability assessment (TSA) is to evaluate the stability status (and degree) of the power system under a set of selected contingencies. Conventionally, transient stability indexes (TSIs) are used to classify/quantify the stability. In general, they can be classified into the following categories:

1) Maximum rotor angle deviation during the transient period [3, 4].
2) Lyapunov-based transient energy function (TEF), which is determined by the kinetic energy and potential energy of a post-fault power system [5].
3) Stability margin η, which is calculated based on extended equal-area criterion (EEAC) theory, also known as single machine equivalence (SIME). A positive value of the stability margin means stable, while a negative value means unstable [1, 6, 7].

The first TSI is the most traditional, which needs to run the full TDS to obtain the rotor angle trajectories. To make the value practically meaningful, it is commonly expressed in the center of inertia (COI) framework:

$$\delta_i^{COI}(t) = \delta_i(t) - \frac{\sum\limits_{i=1}^{S_G} M_i \cdot \delta_i(t)}{\sum\limits_{i=1}^{S_G} M_i} \tag{2.4}$$

where for generator i, δ_i is the angle, and δ_i^{COI} is the COI angle. In general, the stability threshold of the δ_i^{COI} ranges from π to 3π according to different stability requirements of utilities [8]. However, a proper stability threshold is heuristic and system dependent, which is difficult to determine.

The second TSI can calculate a numerical stability degree, but it often results in a conservative result due to the limitations of the TEF method, e.g. difficulty in constructing a suitable Lyapunov function.

The third TSI is a hybrid method that combines full TDS and EAC. Its principle is to transform the multi-machine trajectories (which are obtained from TDS) to an equivalent one-machine-infinite-bus (OMIB) trajectory and apply the EAC to the equivalent OMIB. In such a way, it provides a good engineering approximation for investigating the stability characteristics of the original multi-machine system. In the subsequent chapters of this book, we mainly use EEAC for quantitative transient stability assessment. The details of the EEAC method will be introduced in Section 2.5.

2.4 Time-Domain Simulation

As the most basic and conventional stability analysis approach, the TDS assesses the system's dynamic behavior subjected to a given disturbance by numerically solving Eqs. (2.1) and (2.2) via a step-by-step integration method such as the modified Euler method. Specifically, the generators' "swing curves" (rotor angle trajectories along time) can be computed according to (2.3). Besides, other important parameters such as voltage response can be calculated.

Since a disturbance is generally defined by a starting and ending time, to assess the system's dynamic response, the TDS needs to simulate the during-fault and post-fault trajectories of the system. In general, the during-fault period is quite short (e.g. 100 ms). On the other hand, the post-fault period may be much longer: typically, a system that does not lose synchronism after several seconds, i.e. is able to withstand the disturbance. The maximum simulation period is dependent on the characteristics of the power system and the completion of the modeling. It generally does not exceed 15 seconds for full detailed modeling, while 3 seconds for simple modeling [1].

As introduced in Section 2.3, the stability criteria based on TDS relies on the preselected threshold of the maximum deviation of generator rotor angles. However, it has been widely shown that this threshold is usually system-dependent and not easy to define: when it is set to a small value, the operation tends to be conservative and less economic; while if it is too relaxed, the transient stability may not be ensured.

Since the TDS aims to provide full system dynamic trajectories, its advantages and disadvantages are very clear, as summarized below [1]:

- Advantages:
  - Comprehensive information about system dynamics over time can be obtained, such as generator rotor angles, speeds, and dynamic voltage response.
  - Any power system model and fault scenarios can be considered.
  - The accuracy of the results can be high, as long as the models and the solution algorithms are correctly used.
- Disadvantages:
  - Stability degree cannot be quantified, namely, how severe the system stability cannot be assessed.
  - Less information can be provided for stability control design.

Besides, given the high dimension of the system DAE set, the TDS process is always CPU time-consuming. However, due to continuous advancement in computing power and numerical solution methods, it can be observed that a single simulation with complicated and high models only requires several seconds. Today, the computational speed for TDS can be very fast using commercial software packages, but its weaknesses identified above still hold.

In summary, TDS is a fundamental approach to study power system dynamic behaviors. It can provide the transient electrical quantities such as voltages, frequency, and rotor angles versus time when subjected to a disturbance within several seconds. As explained in Section 2.3, the full model of the system should be used for high accuracy, and an effective numerical solution algorithm should be used for high speed. Some prevailing commercial TDS software tools are DSATools, PSS/E, and Digsilent PowerFactory, where the first two are used in this book for obtaining system dynamic trajectories.

2.5 Extended Equal-Area Criterion (EEAC)

The principle of EEAC is to drive a full TDS engine to obtain the multi-machine trajectories (wherein complex system models can be considered), and then separate them into two exclusive clusters: one composed of critical machines (CMs), which are responsible for the loss of synchronism, and the other composed of non-critical machines (NMs), which correspond to the remaining machines. The two clusters of CMs and NMs are represented as two equivalent machine trajectories [1, 7]

$$\delta_C(t) = M_C^{-1} \cdot \sum_{i \in C} M_i \delta_i(t); \quad \omega_C(t) = M_C^{-1} \sum_{i \in C} M_i \omega_i(t) \tag{2.5}$$

$$\delta_N(t) = M_N^{-1} \cdot \sum_{j \in N} M_j \delta_j(t); \quad \omega_N(t) = M_N^{-1} \sum_{j \in N} M_j \omega_j(t) \tag{2.6}$$

where C and N are the sets of CMs and NMs, M_C and M_N are the inertia of CMs and NMs, obtained as:

$$M_C = \sum_{i \in C} M_i; \quad M_N = \sum_{j \in N} M_j \tag{2.7}$$

Then, an OMIB trajectory is constructed as below:

$$\delta(t) = \delta_C(t) - \delta_N(t); \quad \omega(t) = \omega_C(t) - \omega_N(t) \tag{2.8}$$

$$P_m(t) = M \cdot \left[M_N^{-1} \cdot \sum_{i \in C} P_{mi}(t) - M_N^{-1} \cdot \sum_{j \in N} P_{mj}(t) \right] \tag{2.9}$$

$$P_e(t) = M \cdot \left[M_N^{-1} \cdot \sum_{i \in C} P_{ei}(t) - M_N^{-1} \cdot \sum_{j \in N} P_{ej}(t) \right] \tag{2.10}$$

$$M = (M_C \cdot M_N) \cdot (M_C + M_N)^{-1} \tag{2.11}$$

where δ and ω here are the rotor angle and angular speed of the OMIB; P_m and P_e are the mechanical power output and electrical power of OMIB, respectively.

Subsequently, after the OMIB is constructed, the EAC is applied to calculate stability characteristics and assess transient stability degree:

- The "*time to instability*," t_u, which is the time when the generator loses its synchronism. In this situation, the curve of P_m and P_e crosses (seeing Figure 2.1a):

$$P_a(T_u) = P_m(T_u) - P_e(T_u) = 0, \quad \dot{P}_a(T_u) > 0 \tag{2.12}$$

For the extremely unstable case, e.g. the fault clearing time is much longer than the critical clearing time (CCT), there is no intersection between P_e and P_m curves (seeing Figure 2.1b). In this case, t_u is represented by a backward extrapolated intersection between P_e and P_m as dash line illustrated in Figure 2.1b.

- The "*time to first-swing stability*," t_r, which is the time in which system can be determined as first-swing stable. At this time, P_e stops propagation and returns before crossing P_m (seeing Figure 2.1c):

$$P_a(T_r) = P_m(T_r) - P_e(T_r) < 0, \quad \omega(T_r) = 0 \tag{2.13}$$

- The "*transient stability margin*," η, which is the numerical measurement of the transient stability degree. It is determined by the decelerating area minus the accelerating area of the OMIB $P - \delta$ plane:

$$\eta = A_{dec} - A_{acc}$$
$$= \begin{cases} -M(\omega(T_u))^2/2, & \text{if (2.12) meets} \\ |P_a|T_r\|(\delta(T_u) - \delta(T_r))/2, & \text{otherwise} \end{cases} \tag{2.14}$$

$\eta \geq 0$ implies the system is stable; otherwise, unstable.

For an "extremely stable" case where (2.13) cannot be met, the stability margin can be set to 100. For an "extremely unstable" case where A_{dec} and A_{acc} do not exist, a substitute for instability margin can be the first nearest distance between P_e and P_e curves [1], which is shown in Figure 2.1b. The mathematical form can be expressed as:

$$\eta' = P_{amin} = -\min\{P_a(t) = P_m(t) - P_e(t), t > t_{cl}, P_a(t) > 0\} \tag{2.15}$$

In previous works [1, 8], it has been widely shown that the stability margin has a quasi-linear relationship with many key parameters such as fault clearing time and generation output. Based on this relationship, the CCT can be estimated through successive EEAC runs with different fault-clearing times [1, 6, 8]. Note that for different unstable cases, although the instability margin η and η' have different units, the quasi-linear relationship under each scenario persists [1]. Therefore, when estimating the CCT for an extremely unstable scenario, the process is to estimate a fault clearing time firstly to achieve zero η', then estimate the CCT to achieve zero η, this usually requires $2 \sim 4$ more computational iterations [1].

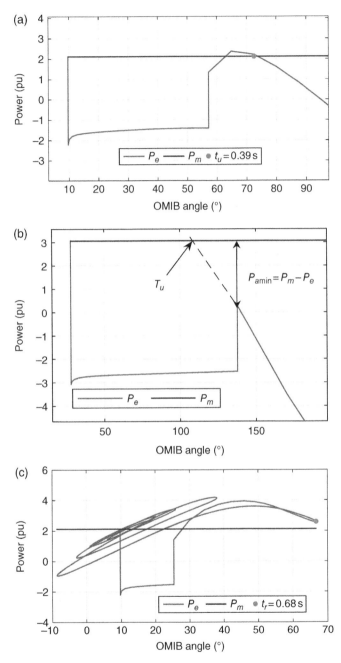

Figure 2.1 Illustration of OMIB $P - \delta$ plane: (a) unstable case, (b) unstable case (extremely), (c) stable case. *Source:* Xu et al. [9].

It is important to indicate that t_u and t_r can be used to early terminate the TDS to save computing time. During the TDS process, the machine angles are continuously checked against (2.12) and (2.13), if either condition is met, the TDS can be immediately terminated.

For grouping of CMs and NMs, several decomposition patterns of the two machine angle clusters are constructed based on the rotor angle deviations between adjacent machines' rotor trajectories [1, 8, 10] (e.g. top three largest angle deviations). Each candidate grouping of machines is replaced by the OMIB, and the candidate OMIB that has the lowest stability margin is declared as the final grouping. It should be noted that the transient stability concerns the synchronism among the synchronous machines, since wind power generators, such as DFIG, are asynchronous, they are not involved in the OMIB construction.

The inherent advantages of EEAC are the quantitative and exact TSA indicators, which enable sensitivity analysis and early termination of TDS. This method has been adopted by several industry-grade software packages, such as FASTEST [6] and DSATools [11], which have been practically adopted in many countries.

2.6 Trajectory Sensitivity Analysis

2.6.1 Basic Concept

Trajectory sensitivity analysis aims to calculate the sensitivities of the dynamic behavior of a hybrid nonlinear system with respect to parameter variations of small size [12]. It provides a way to quantify the variation of a trajectory resulting from (small) changes to parameters and/or initial conditions.

The dynamic behavior of a power system can be described by the following DAEs:

$$\dot{x} = f(x, y, \lambda) \tag{2.16}$$

$$0 = g(x, y, \lambda) \tag{2.17}$$

where x denotes dynamic state variables, e.g. generator angles and speeds; y denotes algebraic state variables, e.g. load bus voltage magnitudes and angles, and λ are parameter changes.

The *flow* of x and y can be defined as follows [12]:

$$x(t) = \phi(x_0, t, \lambda) \tag{2.18}$$

$$y(t) = \varphi(x_0, t, \lambda) \tag{2.19}$$

where $x(t)$ and $y(t)$ satisfy (2.16) and (2.17) with the initial condition:

$$\phi(x_0, t_0, \lambda) = x_0 \tag{2.20}$$

$$g(\phi(x_0, t_0), \varphi(x_0, t_0); \lambda) = 0 \tag{2.21}$$

To obtain the sensitivities of the *flows* ϕ and φ, the *Taylor* series expansions of (2.20) and (2.21) are formed [12]

$$\phi(x_0, t, \lambda + \Delta\lambda) = \phi(x_0, t, \lambda) + \frac{\partial\phi(x_0, t, \lambda)}{\partial\lambda}\Delta\lambda + \varepsilon^{\phi} \tag{2.22}$$

$$\varphi(x_0, t, \lambda + \Delta\lambda) = \varphi(x_0, t, \lambda) + \frac{\partial\varphi(x_0, t, \lambda)}{\partial\lambda}\Delta\lambda + \varepsilon^{\varphi} \tag{2.23}$$

where ε^{ϕ} and ε^{φ} are the higher order terms of the *Taylor* series expansion.

For small $\|\Delta\lambda\|$, the higher-order terms ε^{ϕ} and ε^{φ} can be neglected without much sacrifice of accuracy, giving

$$\Delta x(t) = \phi(x_0, t, \lambda + \Delta\lambda) - \phi(x_0, t, \lambda) \approx \frac{\partial\phi(x_0, t, \lambda)}{\partial\lambda}\Delta\lambda \equiv \Phi(x_0, t, \lambda)\Delta\lambda \tag{2.24}$$

$$\Delta y(t) = \varphi(x_0, t, \lambda + \Delta\lambda) - \varphi(x_0, t, \lambda) \approx \frac{\partial\varphi(x_0, t, \lambda)}{\partial\lambda}\Delta\lambda \equiv \Psi(x_0, t, \lambda)\Delta\lambda \tag{2.25}$$

where the time-varying partial derivatives Φ and Ψ are called *trajectory sensitivities* associated with the flows x and y [12].

For general nonlinear systems, the *flows* ϕ and φ given by (2.18) and (2.19) cannot be expressed in closed form. Hence, any change in initial conditions or parameters requires a complete re-simulation of the dynamic model. However, according to (2.24) and (2.25), when the changes are relatively small, the need for repeated simulations can be avoided by using the *approximate trajectories*:

$$\phi(x_0, t, \lambda + \Delta\lambda) \approx \phi(x_0, t, \lambda) + \Phi(x_0, t, \lambda)\Delta\lambda \tag{2.26}$$

$$\varphi(x_0, t, \lambda + \Delta\lambda) \approx \varphi(x_0, t, \lambda) + \Psi(x_0, t, \lambda)\Delta\lambda \tag{2.27}$$

For power system dynamic studies, the first-order trajectory sensitivities can be *analytically* calculated as a byproduct of TDS when using implicit integration techniques (such as the trapezoidal method) to minimize the computational burden [12–14]. Considering the independent nature of the trajectory sensitivity calculations for different parameters, parallel processing can be applied to save the total execution time [14].

In addition to the analytical approach, the trajectory sensitivities can also be *numerically* calculated via two successive TDS runs. The trajectory sensitivities for dynamic state variables are:

$$\Phi(x_0, t, \lambda) = \frac{\phi(x_0, t, \lambda + \Delta\lambda) - \phi(x_0, t, \lambda)}{\Delta\lambda} \tag{2.28}$$

where $\Delta\lambda$ is a small perturbation.

It has been reported in Ref. [15] that the two calculation approaches can yield almost identical results. The numerical approach is simpler and does not need additional programming efforts, but it requires multiple TDS runs so its computation efficiency is relatively lower.

Based on the trajectory sensitivities, the required control in the initial condition to stabilize an unstable case can be determined as:

$$\Delta\lambda' = \frac{\Delta\delta(t)}{\Phi(x_0, t, \lambda)} \tag{2.29}$$

where $\Delta\lambda'$ is the form of required control actions, which can be generation rescheduling; $\Delta\delta(t)$ is the required change for the unstable trajectories or stability margin to become stable.

By applying trajectory sensitivities, the transient stability constraints can be linearly constructed regarding the control variables, which will make the TSC-OPF be solved transparently and efficiently.

Figure 2.2 is an illustration of trajectory sensitivities versus generators' active power. In this figure, 10 OMIB trajectory sensitivities along the whole trajectory $(0 \sim 2$ seconds) are shown, where the values at $T_u = 0.5917$ second of the 10 generators are compared in an embedded bar chart. It can be found that only G38 has a positive sensitivity value, which means that reducing the output of the generator can decrease the OMIB angle. Meanwhile, the rest of the generators, which all

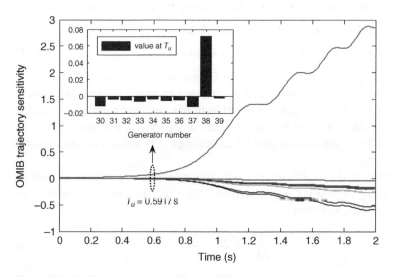

Figure 2.2 OMIB trajectory sensitivities with respect to generator active output – New England test system. *Source:* [16]/John Wiley & Sons.

have negative trajectory sensitivities, will have an adverse influence on stability when their output is reduced. It can be clearly seen that the trajectory sensitivities are quite transparent on the significance of the variables, namely, they can directly and quantitatively indicate how the variables affect system's transient stability.

The EEAC-based TSA indicators (η, t_u and t_r) provide inherent information, which enables exact sensitivity analysis for gradient information to stabilize the system. In the literature, Ref. [17] derives sensitivity coefficients of the stability margin with respect to the mechanical power of each generator using *Taylor* series expansion. In this book, in order to stabilize the power system while optimizing its operating state, we aim to linearize the stabilization constraints, which can be readily incorporated into an optimization model. Given the inherent advantages of the trajectory sensitivity approach, it will be applied in the subsequent chapters of this book.

Nomenclature

Indices

i, j indices for generators
t indices for time
n indices for operating point

Sets

S_G sets of all generators
C/N sets of critical machines and non-critical machines
$x/\mathbf{x}$ sets of dynamic state variables
$y/\mathbf{y}$ sets of algebraic state variables
$\mathbf{k}$ disturbance events

Variables

δ_i/ω_i angle and angular speed of i-th generator
P_{mi}/P_{ei} mechanical power and electrical power with i-th generator
δ_i^{COI} center of inertia angle
t_u' time when the generator loses its synchronism
t_r' time that system can be determined as first-swing stable
η stability margin
η' stability margin for extremely unstable case

ϕ	flows of $x(t)$
φ	flows of $y(t)$
$\varepsilon^{\phi}/\varepsilon^{\varphi}$	higher-order terms of the *Taylor* series expansion
Φ	trajectory sensitivities associated with the flows x
Ψ	trajectory sensitivities associated with the flows y
$\Delta\delta(t)$	required change for the unstable stability margin to become stable
A_{dec}	decelerating area
A_{acc}	accelerating area

Parameters

M_i	inertia coefficient of i-th generator
M_C/M_N	inertia of CMs and NMs
λ	parameter changes
$\Delta\lambda$	small perturbation
$\Delta\lambda'$	form of required control actions
t_0	pre-contingency state
ε	desired stability margin

References

1 Pavella, M., Ernst, D., and Ruiz-Vega, D. (2012). *Transient Stability of Power Systems: A Unified Approach to Assessment and Control*. Springer Science & Business Media.

2 Xu, Y., Dong, Z.Y., Xu, Z. et al. (2012). Power system transient stability-constrained optimal power flow: a comprehensive review. *IEEE Power and Energy Society General Meeting* 2012: 1–7.

3 Gan, D., Thomas, R.J., and Zimmerman, R.D. (2000). Stability-constrained optimal power flow. *IEEE Transactions on Power Systems* 15 (2): 535–540.

4 Papadopoulos, P.N. and Milanović, J.V. (2017). Probabilistic framework for transient stability assessment of power systems with high penetration of renewable generation. *IEEE Transactions on Power Systems* 32 (4): 3078–3088.

5 Pai, M. (2012). *Energy Function Analysis for Power System Stability*. Springer Science & Business Media.

6 Xue, Y. (1998). Fast analysis of stability using EEAC and simulation technologies. *POWERCON '98. 1998 International Conference on Power System Technology. Proceedings (Cat. No.98EX151)*, vol. 1, pp. 12–16, Beijing China. 18–21 August 1998: IEEE.

7 Xue, Y., Custem, T.V., and Ribbens-Pavella, M. (1989). Extended equal area criterion justifications, generalizations, applications. *IEEE Transactions on Power Systems* 4 (1): 44–52.

8 Ruiz-Vega, D. and Pavella, M. (2003). A comprehensive approach to transient stability control. I. near optimal preventive control. *IEEE Transactions on Power Systems* 18 (4): 1446–1453.

9 Xu, Y., Yin, M., Dong, Z.Y. et al. (2018). Robust dispatch of high wind power-penetrated power systems against transient instability. *IEEE Transactions on Power Systems* 33 (1): 174–186.

10 Pizano-Martianez, A., Fuerte-Esquivel, C.R., and Ruiz-Vega, D. (2010). Global transient stability-constrained optimal power flow using an OMIB reference trajectory. *IEEE Transactions on Power Systems* 25 (1): 392–403.

11 DSA. Dynamic security assessment software. http://www.dsatools.com. 29 January 2015.

12 Hiskens, I.A. and Pai, M.A. (2000). Trajectory sensitivity analysis of hybrid systems. *IEEE Transactions on Circuits and Systems I: Fundamental Theory and Applications* 47 (2): 204–220.

13 Hiskens, I.A. and Alseddiqui, J. (2006). Sensitivity, approximation, and uncertainty in power system dynamic simulation. *IEEE Transactions on Power Systems* 21 (4): 1808–1820.

14 Hou, G. and Vittal, V. (2012). Cluster computing-based trajectory sensitivity analysis application to the WECC system. *IEEE Transactions on Power Systems* 27 (1): 502–509.

15 Hou, G. and Vittal, V. (2013). Determination of transient stability constrained Interface real power flow limit using trajectory sensitivity approach. *IEEE Transactions on Power Systems* 28 (3): 2156–2163.

16 Xu, Y., Dong, Z.Y., Zhao, J. et al. (2015). Trajectory sensitivity analysis on the equivalent one-machine-infinite-bus of multi-machine systems for preventive transient stability control. *IET Generation Transmission and Distribution* 9 (3): 276–286.

17 Pizano-Martínez, A., Fuerte-Esquivel, C.R., Zamora-Cárdenas, E., and Ruiz-Vega, D. (2014). Selective transient stability-constrained optimal power flow using a SIME and trajectory sensitivity unified analysis. *Electric Power Systems Research* 109: 32–44.

3

Recent Large-Scale Blackouts in the World

3.1 Introduction

Power system blackouts do not often happen, but once happen, their impact is tremendous on the economy and society. In history, many large-scale blackouts have occurred around the world, and some of them are associated with stability problems. This chapter reviews several major blackouts including their process and root cause. Among them, the 2016 South Australia (SA) blackout is especially highlighted since it was caused by successive voltage disturbances triggering large-scale wind farms, which were supplying over 50% of the total power to the system. As a result, more dynamic reactive power resources may be needed to enhance the short-term voltage stability of the power system.

3.2 Major Blackouts in the World

3.2.1 Blackouts Triggered by Transmission Line Out-of-Service

3.2.1.1 U.S.–Canada Blackout (2003)

A widespread power blackout occurred on 14 August 2003, in most parts of the Midwest and northeast United States and Ontario, Canada [1]. This event affected about 50 million people 61 800 MW of electric load in the states of Ohio, Michigan, Pennsylvania, New York, Vermont, Massachusetts, Connecticut, New Jersey, and the Canadian province of Ontario, which sustained four days in some parts of United States and even more than one week in some parts of Ontario without full power restoration.

The event began at 12 : 15 Eastern Daylight Time (EDT) on 14 August 2003, when inaccurate input data rendered Midwest Independent System Operator (MISO) state estimator ineffective. At 13 : 31 EDT, FirstEnergy's (FE, which

Stability-Constrained Optimization for Modern Power System Operation and Planning,
First Edition. Yan Xu, Yuan Chi, and Heling Yuan.
© 2023 The Institute of Electrical and Electronics Engineers, Inc.
Published 2023 by John Wiley & Sons, Inc.

includes seven electric utility operating companies) Eastlake five generation unit tripped and shut down automatically. Shortly after 14:14 EDT, the alarm and logging system in FE's control room failed and was not restored until after the blackout. After 15:05 EDT, some 345 kV transmission lines in FE began to trip because they were touching the overgrown trees in the right-of-way area of the line. By about 15:46 EDT, when FE, MISO, and neighboring utilities began to realize that the FE system was in a dangerous situation, the only way to avoid power outages was to reduce at least 1500 MW around Cleveland and Akron Load. However, no such effort was made. By 15:46 EDT, the massive load shedding may be too late to have any impact. After 15:46 EDT, the loss of some key 345 kV lines of FE in northern Ohio led to the failure of its underlying 138 kV line network, which in turn caused FE's Sammis-Star 345 kV line loss at 16:06 EDT. With the loss of the Sammis-Star line, a cascade was triggered since it connected northern Ohio to eastern Ohio. Loss of the heavily overloaded Sammis-Star line instantly created major and unsustainable burdens on lines in adjacent areas, and the cascade spread rapidly as lines and generating units automatically tripped by protective relay action to avoid physical damage. The blackout rippled from the Cleveland–Akron area across much of the northeast United States and Canada as the facility trips.

3.2.1.2 Europe Blackout (2006)

On the night of 4 November 2006, at around 22:10, the union for the co-ordination of transmission of electricity (UCTE) interconnected grid was affected by a serious incident originating from the North German transmission grid that led to power supply disruptions for more than 15 million European households and a splitting of the UCTE synchronously interconnected network into three areas after tripping many high-voltage lines shown in Figure 3.1 [2]. The consumers were without power electricity about two hours on this date.

The mainland is in chaos, dozens of people are trapped in elevators, countless trains are stopped, and emergency calls to answer the phone are endless. The transmission system operator (TSO) took immediate action to prevent the unrest from turning into a power outage in Europe.

In total, more than 10 million people in northern Germany, France, Italy, Belgium, and Spain suffered power outages or were affected by power outages. In northern Germany, more than 100 trains were delayed for more than two hours due to power outages. Except for the southeastern part of the country, almost all of France is affected. In affected areas in France, firefighters were asked to respond to approximately 40 people trapped in elevators. In Belgium, only the surrounding area of Antwerp has been severely affected, as well as Ghent and Liège. The rest of the country still maintains electricity supply. Italy, which had experienced a similar blackout in 2003 that caused 95% of the country to lose power, was only affected in a few areas, mainly Piedmont, Liguria in northern Italy, and Puglia

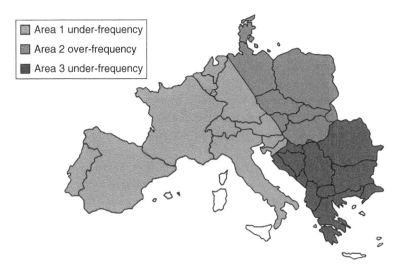

Figure 3.1 Schematic map of UCTE area splitting into three areas. *Source:* Ref. [2].

in southern Italy. In Spain, the news network Red Electrica was affected, as were the regions of Madrid, Barcelona, Zaragoza, and part of Andalusia.

There are mainly two reasons leading to the event. On 4 November, after closing the 380 kV double-circuit line Conneforde-Diele, the E.ON Netz power grid (including some of its tie lines) was not in the "$N-1$" safe state. Meanwhile, Inter-TSO coordination is insufficient. The preliminary plan to close the 380 kV Conneforde-Diele line from 01 : 00 to 5 : 00 on 5 November was properly prepared by the relevant TSO. However, E.ON Netz communicated this change in switching time to the relevant TSO very late, and did not make proper preparations and inspections during the operational planning stage to ensure the safe operation of the regional system. E.ON Netz did not prepare effective remedial measures with the neighboring TSO to maintain a sufficient safety margin on the Landesbergen-Wehrendorf line after closing the Conneforde-Diele line.

3.2.1.3 U.S. Blackout (2011)

This U.S. blackout was a widespread power outage that affected the San Diego–Tijuana area, southern Orange County, the Imperial Valley, Mexicali Valley, Coachella Valley, and parts of Arizona [3]. It occurred on Thursday, 8 September 2011, beginning at about 3 : 38 p.m. Pacific Daylight Time (PDT). It was the largest power failure in California's history.

On the afternoon of 8 September 2011, an 11-minute system failure occurred in the Pacific Southwest, which caused a chain of power outages, resulting in power outages of approximately 2.7 million customers. The power outage affected parts of Arizona, Southern California, and Baja California, Mexico. The San Diego area had all power outages, and nearly 1.5 million customers had power outages, some of which lasted as long as 12 hours. The breakdown occurred near the peak hours of the working day and caused traffic congestion for several hours. Schools and businesses are closed. Some flights and public transportation were interrupted. Water and sewage pumping stations were out of power, and beaches were closed due to sewage leaks. Millions of people do not have air conditioning in hot weather.

The loss of a 500 kV transmission line initiated the event. For the event of 8 September 2011, it indicated that the system was not operating in a safe $N - 1$ state. This failure is mainly due to the weaknesses of operational planning and real-time situational awareness in the two major areas. If done well, the system operator can actively operate the system in a safe $N - 1$ state under normal system conditions and restore the system to enter a safe $N - 1$ state as soon as possible, but no more than 30 minutes. Without adequate planning and situational awareness, the entity responsible for operating and supervising the transmission system cannot ensure reliable operation within the System Operational Limits (SOL) or prevent cascading outages in a single contingency.

The comparison of the two events on August 2003 and September 2011 can be found in the joint report issued by the US Federal Energy Regulatory Commission and the North American Electric Reliability Corporation.

Although the two blackouts were triggered by different initiating events tree touches in 2003 compared to a switching error in 2011. Both blackouts had common fundamental reasons:

1) The affected entities in both incidents did not conduct adequate long-term and operational planning studies to understand the vulnerabilities of their systems.
2) The affected entities in both incidents did not have sufficient situational awareness before and during the disturbance.
3) In addition to the above two fundamental reasons, both events were exacerbated by protection system relays that tripped facilities without allowing operators sufficient time to take mitigating measures.

3.2.1.4 South Australia Blackout (2016)

The SA region black system event on 28 September 2016 (Black System) led to 850 000 SA customers losing their electricity supply, affecting households, businesses, transport and community services, and major industries [4].

On Wednesday, 28 September 2016, a tornado with a wind speed of 190–260 km/h occurred in South Australia. The two tornadoes almost simultaneously damaged a single-circuit 275 kilovolts (kV) transmission line and a double-circuit 275 kV transmission line about 170 km apart.

The damage to these three transmission lines caused them to trip. A series of rapid and continuous faults caused the SA grid to experience six voltage dips in two minutes around 4.16 p.m.

With the increase in the number of transmission network faults, the power of the nine wind farms in north-central SA continued to decline because of the activation of protection functions. For eight of the wind farms, the protection settings of the wind turbines allowed them to withstand a preset number of voltage dips within two minutes. Activating this protection function can significantly reduce the sustained power of these wind farms. In less than seven seconds, it continuously reduced 456 megawatts (MW) of power generation.

The reduction of wind power output has resulted in a dramatic increase in imported power through the Heywood Interconnector. Approximately 700 milliseconds (ms) after the reduction of output from the last of the wind farms, the flow on the Victoria–SA Heywood Interconnector reached such a level that it initiated a special protection scheme to take the interconnector offline.

The SA power system then became separated ("islanded") from the rest of the national electricity market. After the system is separated, there is no substantial load shedding. The remaining power generation is much smaller than the connected load, and the island system frequency cannot be maintained. As a result, all supplies in SA were interrupted at 4.18 p.m. (Black System). AEMO's analysis shows that the subsequent system separation, frequency collapse, and subsequent black system were inevitable.

This is a typical example of blackout with high-level integration of wind farms. As shown in Figure 3.2, the penetration of wind generation in SA is occupying about 50% of the total generation. When the voltage drops, the protection

Figure 3.2   South Australia generation mix.

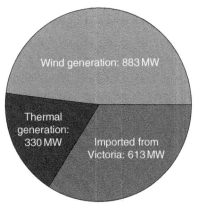

Wind generation: 883 MW

Thermal generation: 330 MW

Imported from Victoria: 613 MW

mechanism of wind turbines will be activated, namely, low voltage ride through (LVRT). In this case, most wind turbines successfully rode through the third voltage dip. However, the voltage disturbances were continuous. Then, the wind turbines reduced their output or were disconnected due to the activation of repetitive LVRT. As wind generation is the main source of the SA, the dramatic deficit of wind generation led to the imbalance of the generation and load and caused the Black System finally.

3.2.1.5 Venezuela Blackout (2019)

Beginning on the evening of 7 March 2019 (local time), power outages have taken place in most parts of Venezuela, including the capital, Caracas, exceeding 24 hours [5]. The power outage took place in Venezuela's 23 states, and 20 states of them had a full blackout, which left the Caracas subway out of operation, resulting in massive traffic congestion, and severely affecting the schools, hospitals, factories, airports, and so on, and the mobile phone and network could not be used normally.

In the early hours of the morning of the 8 March 2019, power began to be restored in parts of Caracas, followed by a gradual resumption of electricity supply in other parts of the region, but the power failure at noon on the 9th and 10th caused great panic. The long and extensive power failure caused serious losses to Venezuela, including the continuous suspension of schools for many days, the lack of access to some websites, and even the serious looting of shopping malls and supermarkets in some areas. The outage was Venezuela's longest and most widespread failure since 2012.

According to the media, the initial blackout was due to the failure of a major hydropower station in the southern state of Bolivar, which belongs to the Gurí hydropower station. But the reason for the station's failure is still ambiguous, as there are many claims from different sides of Venezuela.

3.2.2 Blackouts Triggered by Equipment Faults

3.2.2.1 Brazil Blackout (2018)

In Brazil, on the 21 March 2018, a severe fault has taken place in the Belo Monte Hydropower Station at 15:48 (local time), which caused power outages in at least 14 states in northern and northeastern Brazil, resulting in a loss of 18 000 MW load, or 22.5% of Brazil's national power system [6]. The most affected states are Rio Grande do Norte, Paraíba, and Maranhão, most of which have been in blackouts for hours. Public transport, including subways and trams, was halted and caused large-scale chaos. According to government authorities, the Xingu converter station's technical defects, which are the error of the over-current protection's set value of the section switch, lead to a chain reaction [7]. The blackout caused huge economic losses and serious social impacts.

3.2.2.2 Singapore Blackout (2018)

This power outage is the most severe failure in Singapore since 2004. In this case, one of the power generating units at Sembcorp Cogen Pte Ltd. (Sembcorp Cogen) was tripped due to an equipment fault [8]. When a power generating unit trips, the other units in operation will increase their electricity supply automatically. However, one of the power generating units owned by Senoko Energy Pte Ltd. (Senoko) was tripped while it was ramping up additional supply, due to the failure of a different equipment component. The tripping of the two power generating units resulted in insufficient electricity supply to meet demand. The protection devices in the power system kicked in as designed and automatically disconnected electricity to about 146 500 consumers to maintain the balance of the system.

3.2.2.3 U.S. Blackout (2021)

In February 2021, the state of Texas suffered a major power crisis due to three severe winter storms sweeping across the United States [9]. A massive electricity generation failure in the state of Texas, and resultant shortages of water, food, and heat. More than 4.5 million homes and businesses were left without power, some for several days. The process of the crisis is as below:

On 10 February, as low temperatures entered the ERCOT region, the total installed capacity of offline power plants increased from 14 400 MW to 25 850 MW, accounting for approximately 12–21% of the total installed capacity of ERCOT's 123 050 MW. Wind turbines suffered some of the earliest shutdowns and deratings due to ice accumulation on the blades, which were resulting from freezing precipitation and fog. As the temperature dropped further, other types of generators (mainly fueled by natural gas) were facing fuel supply problems before the power outage, and due to the reduction in fuel supply, they started to go offline or reduce their capacity as early as 10 February.

On Saturday, 13 February, ERCOT began deploying Response Reserve and issued emergency notifications for extreme cold weather events affecting the area. February 13 was also the first day when large generators began to accidentally shut down.

Late at night on 14 February, the electricity load or demand was close to the available power generation. As power generation could not be increased sufficiently to meet demand, the grid frequency began to drop. In this case, ERCOT initiates various contingency plans, such as calling reserves and shedding loads, and can automatically shed loads at a sufficiently low frequency.

On Monday, 15 February, at CST 00:15, ERCOT announced Energy Emergency Alert Level 1 (EEA 1), at CST 01:07 ERCOT moved to EEA 2, and at CST 01:20, ERCOT announced that it had entered the EEA 3 incident and began to "firm load shed" or "blackouts." ERCOT did not resume normal operations until 10:36 CST on Friday, 19 February.

3.2.3 Blackouts Triggered by Cyber Attack

3.2.3.1 Ukraine Blackout (2015)

A cyberattack took place in the Ukraine power grid on 23 December 2015 [10–12]. The incident is thought to be the first known successful hack aimed at utilities. The event highlights the vulnerability of highly automated, cyber information-based smart grid environments to coordinated cyberattacks. Hackers were able to successfully compromise information systems of three energy distribution companies in Ukraine and temporarily disrupt the electricity supply to the end consumers.

Attacks on energy sources are fairly commonplace but had never caused a blackout before this case. For several years, security experts have been publicly concerned that cyberattacks may weaken basic public systems, including electricity and water supply facilities, transportation systems, and communication networks. For hackers who want to affect a large number of people, the power grid is an obvious target, especially in a world that is increasingly dependent on a large number of electronic devices.

The attack was divided into several stages and initially involved hackers installing malicious software on the computer systems of Ukrainian power companies. This allowed attackers to remotely access these computers and allowed them to turn on the circuit breakers and shut down the power to 80 000 customers of the Prykarpattyaoblenergo utility company in western Ukraine. At the same time as the power outage, the attackers bombed the customer service telephone lines with fake calls to prevent customers from reporting the outage.

The incident happened during an ongoing conflict in Ukraine and is attributed to a Russian advanced persistent threat group known as "Sandworm". In total, up to 73 MWh of electricity was cut (or 0.015% of daily electricity consumption in Ukraine).

Table 3.1 shows the summary of the nine blackouts in recent 20 years. It contains the initial event, crucial event, and effect of each blackout, respectively.

Table 3.1 Summary of 9 blackouts in the world.

Blackouts	Initial event	Crucial event	Restoration time	Effects
U.S.–Canada (2003)	FE's computer failed and transmission lines tripped due to tree-to-line contacts	Sammis-Star 345 kV line tripped due to overloads	Most places were restored within 7 h some parts were restored within several days	Affected an area with an estimated 50 million people and 61 800 MW of electric load
Europe (2006)	A planned routine disconnection was advanced	Tripping of several high-voltage lines, which started in Northern Germany	Less than 2 h	Interruption of supply for more than 15 million households
U.S. (2011)	A 500 kV line was accidentally shut down	Instantaneous redistribution of power flows created sizeable voltage deviations	Up to 12 h to restore the load	Power outage affects approximately 2.7 million customers
South Australia (2016)	Three transmission lines tripped due to extreme weather	456 MW reduction of wind power due to successive voltage dips	80–90% of loads were restored within about 8–9 h; the remaining was gradually restored up to 13 d	All supply to the South Australia region was lost
Venezuela (2019)	Triggering of three 765 kV lines	Disconnection of the generators in the Guri dam	At least 10 d	The power outage took place in Venezuela's 23 states
Brazil (2018)	Section switch's relay protection acted due to an incorrect pre-set value	Failure of cutting six units of the Belo Monte Hydropower Station	About 7 h	Power outages in 14 states, a loss of 18 000 MW load

(Continued)

Table 3.1 (Continued)

Blackouts	Initial event	Crucial event	Restoration time	Effects
Singapore (2018)	One of the power generating units was tripped due to an equipment fault	Tripping of two power generating units	Within 38 min	Disconnection of electricity to about 146 500 consumers
U.S (2021)	Wind turbines suffered earliest outages as freezing precipitation and fog resulted in ice accumulation on blades	Other types of generations were starting to go offline in a single 24 h period due to the extreme weather	About 5 d	shortages of water, food, and heat. More than 4.5 million homes and businesses were left without power some for several days
Ukraine (2015)	A cyberattack was successfully aimed at utilities	Disruption of electricity supply to the end consumers.	About 6 h	225 000 customers lose power across various areas.

References

1 Muir, A. and Lopatto, J. (2004). Final Report on the August 14, 2003 Blackout in the United States and Canada: Causes and Recommendations.

2 UCTE (2007). Final Report System Disturbance on 4 November 2006.

3 FERC/NERC (2012). Arizona-Southern California outages on September 8, 2011: Causes and Recommendations.

4 A. E. M. Operator (2017). Black System South Australia 28 September 2016.

5 2019 Venezuelan blackouts. https://en.wikipedia.org/wiki/ 2019_Venezuelan_blackouts.

6 Costa, L. and Goy, L. Tens of millions in northern Brazil hit by massive power outage. https://www.reuters.com/article/us-brazil-power/tens-of-millions-in-northern-brazil-hit-by-massive-power-outage-idUSKBN1GX3CN (accessed 22 March 2018).

7 Liu, Y. (2019). Analysis of Brazilian blackout on March 21st, 2018 and revelations to security for Hunan Grid. *2019 4th International Conference on Intelligent Green Building and Smart Grid (IGBSG),* Hubei, China (06–09 September 2019). pp. 1–5: IEEE.

8 Energy Market Authority (EMA). EMA working with industry to establish cause behind electricity supply disruption on 18 September. www.ema.gov.sg/ media_release.aspx?news_sid=20180919FeaM3saiZSkV (accessed 19 September 2018).

9 King, C.W., Rhodes, J.D., Zarnikau, J., et al. (2021). The timeline and events of the February 2021 Texas electric grid blackouts. The University of Texas at Austin Energy Institute.

10 Department of Homeland Security (DHS). Hackers behind Ukraine power cuts, says US report. https://www.bbc.com/news/technology-35667989 (assessed 26 February 2016).

11 Liang, G., Weller, S.R., Zhao, J. et al. (2017). The 2015 Ukraine blackout: implications for false data injection attacks. *IEEE Transactions on Power Systems* 32 (4): 3317–3318.

12 Finkle, J. U.S. firm blames Russian 'Sandworm' hackers for Ukraine outage. https://www.reuters.com/article/us-ukraine-cybersecurity-sandworm-idUSKBN0UM00N20160108 (assessed 8 January 2016).

Part II

Transient Stability-Constrained Dispatch and Operational Control

List of Acronyms

ANN	artificial neural network
BD	benders decomposition
C&CG	column and constraint generation
CC	corrective control
CCT	critical clearing time
COI	center of inertia
CT	classification tree
DAEs	differential-algebraic equations
DE	differential evolution
DRIP	data rich information poor
DSA	dynamic stability assessment
DSC	dynamic security controls
DT	decision tree
EA	evolution algorithm
EC	emergency control
EEAC	extended equal area criterion
ELM	extreme learning machine
ELS	emergency load shedding
GA	generic algorithm
IP	interior point
IS	intelligent system
ISD	isolated stability domain
LP	linear programming
LR	*Lagrangian* relaxation
LS	learning set
MILP	mixed-integer linear programming
MIP	mixed-integer programming

Stability-Constrained Optimization for Modern Power System Operation and Planning,
First Edition. Yan Xu, Yuan Chi, and Heling Yuan.
© 2023 The Institute of Electrical and Electronics Engineers, Inc.
Published 2023 by John Wiley & Sons, Inc.

NN	neural network
NSE	network steady-state security evaluation
OMIB	one machine infinite bus
OP	operating point
OPF	optimal power flow
PC	preventive control
PD	pattern discovery
PDF	probability density function
PSO	particle swarm optimization
RT	regression tree
SCOPF	security constrained optimal power flow
SCUC	security-constrained unit commitment
SI	sensitivity index
SIME	single-machine equivalent
SVM	support vector machine
TDS	time-domain simulation
TEF	transient energy function
TOAT	*Taguchi*'s orthogonal array testing
TS	testing set
TSA	transient stability assessment
TSC	transient stability control
TSC-OPF	transient stability constrained optimal power flow
TSCUC	transient stability constrained-unit commitment
TSI	transient stability index
TSRO	two stage robust optimization
UC	unit commitment

4

Power System Operation and Optimization Models

4.1 Introduction

This chapter introduces the power system operation and its optimization models, as well as the engineering practice. Firstly, an overview and the framework of the power system operation are introduced. Then, two basic optimization models for power system operation decision-making are introduced, including unit commitment (UC) and optimal power flow (OPF), as well as their security-constrained formulations. These optimization models serve as the base for the stability-constrained power system operation and planning methods described in the subsequent chapters. Finally, the engineering practice for power system operation and management is introduced.

4.2 Overview and Framework of Power System Operation

Power system operations is a decision-making process for scheduling and dispatching the controllable resources in the system, including real and reactive power generations and network topology switching, to meet the power demand on the timescale from one day to minutes prior to the power delivery.

In practice, three operation objectives are usually considered in the power system: (i) maintaining continuous supply of power with an acceptable quality to all the consumers in the system, i.e. **security**; (ii) minimizing the total system operation cost such as total power generation cost or total power losses in the network, i.e. **economy**; and (iii) minimizing the environmental impacts such as the carbon emission via harnessing as much as possible clean energy sources such as hydro, wind, and solar power, i.e. **low-carbon**. However, the three objectives are usually

Stability-Constrained Optimization for Modern Power System Operation and Planning,
First Edition. Yan Xu, Yuan Chi, and Heling Yuan.
© 2023 The Institute of Electrical and Electronics Engineers, Inc.
Published 2023 by John Wiley & Sons, Inc.

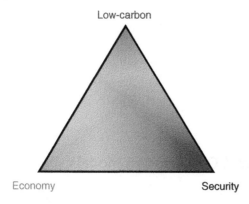

Low-carbon

Economy

Security

Figure 4.1 Conflicting triangle for power system operation.

conflicting with each other, so-called "conflicting triangle" (as shown in Figure 4.1).

Therefore, to practically balance among these requirements, the operation decisions should be made on an optimization basis with specific operational objectives and prevailing operational limits of the power system. In general, two basic optimization models have been widely used for power system operation decision-making, which are known as UC and OPF.

As illustrated in Figure 4.2, three typical operation stages are implemented in a real-world power system: (i) day ahead; (ii) hourly or 30- to 15-minutes ahead; and (iv) real-time.

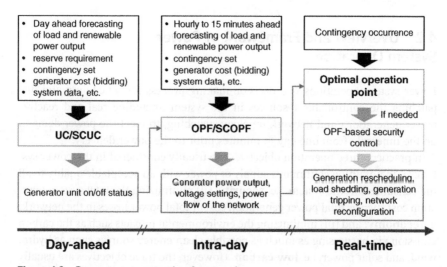

Figure 4.2 Power system operation framework.

- Since the start-up and shut-down of the generator units require several hours even one day, the generation unit states should be scheduled at least one day in advance. In the day-ahead operation time window, the UC program is run to determine the generator unit states (on or off) considering the day-ahead forecasted load demand, renewable power sources such as wind power and solar power output, as well as the generation reserve requirement for the next day.

- After the generator unit states are decided for the day ahead, the OPF program is run for intra-day operation. With the hourly or 15-minutes ahead load demand and renewable power output forecasting, the OPF aims to optimize the generator power output and voltage settings considering the power flow of the power network. Meanwhile, to meet the security requirement, credible contingencies such as N-1 faults can be incorporated into the UC and OPF models, yielding security-constrained UC (SCUC) and security-constrained OPF (SCOPF). As a preventive control (PC) action, running these security-constrained models can deliver an operating point (OP) that is able to withstand the considered contingencies should they really occur. Note that these actions are "preventive" since they are implemented for the pre-contingency OP.

- During the real-time operation, it may still happen that a contingency could lead to security violation, e.g. branch overflow and bus under/over-voltage, if the contingency is not considered preventively or the actual OP deviates largely from the SCUC/SCOPF solution. In such case, corrective control (CC) actions should be taken for the post-contingency OP, such as load shedding, generation tripping, generation rescheduling, and network reconfiguration, to restore the security of the power system.

For different electricity markets in the world, the specific UC, SCUC, OPF, and SCOPF models, dispatch intervals, forecasting lead-time, and market clearance time horizon could be very different. Without losing generality, the next chapter will introduce the typical mathematical models.

4.3 Mathematical Models for Power System Optimal Operation

4.3.1 Optimal Power Flow (OPF)

The OPF is a basic optimization model for power system operational dispatch and static security control. Its compact mathematical model is as follows:

$$\min_{\mathbf{u}} F(\mathbf{x}, \mathbf{u}, \mathbf{y}) \tag{4.1}$$

$$\text{s.t. } \mathbf{g}(\mathbf{x}, \mathbf{u}, \mathbf{y}) = 0 \tag{4.2}$$

$$\mathbf{h}(\mathbf{x}, \mathbf{u}, \mathbf{y}) \leq 0 \tag{4.3}$$

where $\mathbf{u}$ denotes the control variable (also called decision variable), such as active power output and voltage settings of the generator units, and $\mathbf{x}$ is the state variable (also called dependent variable), such as bus voltage magnitudes and angles as well as branch power flow. $\mathbf{y}$ is the parameter such as load demand, network topology, and network parameters. Note that the parameters can be (or assumed to be) deterministic if they can be accurately predicted such as the load demand; or stochastic if they are less predictable, such as the wind and solar power output that is naturally uncertain.

For normal operation, the objective function (4.1) is usually the total cost of active power generation or total power loss in the network. For security control, different objectives can be considered such as the total security control cost, e.g. generation rescheduling cost or load shedding cost. The equality constraint set (4.2) $\mathbf{g}$ is power balance at each node based on power flow equations. The inequality constraint set (4.3) $\mathbf{h}$ denotes the network operating limits (such as branch flow limits and voltage limits) and limits on control variables (such as generator capacity).

The full formulation of a conventional OPF model is as follows:

$$\min \sum_{i}^{N_G} \left(a_i P_{Gi}^2 + b_i P_{Gi} + c_i\right) \tag{4.4}$$

$$\text{s.t.} \begin{cases} P_{Gi} - P_{Di} - V_{Bi} \sum_{j=1}^{S_B} V_{Bj}\left(G_{ij} \cos\theta_{ij} + B_{ij} \sin\theta_{ij}\right) = 0 \\ Q_{Gi} - Q_{Di} - V_{Bi} \sum_{j=1}^{S_B} V_{Bj}\left(G_{ij} \sin\theta_{ij} - B_{ij} \cos\theta_{ij}\right) = 0 \end{cases} \tag{4.5}$$

$$\underline{P_{Gi}} \leq P_{Gi} \leq \overline{P}_{Gi} \quad \{Gi \in S_G\} \tag{4.6}$$

$$\underline{Q_{Gi}} \leq Q_{Gi} \leq \overline{Q}_{Gi} \quad \{Gi \in S_G\} \tag{4.7}$$

$$\underline{V_{Bi}} \leq V_{Bi} \leq \overline{V}_{Bi} \quad \{Bi \in S_B\} \tag{4.8}$$

$$\underline{S_{Ll}} \leq S_{Ll} \leq \overline{S}_{Ll} \quad \{Ll \in S_L\} \tag{4.9}$$

where a_i, b_i and c_i are the generation cost coefficients of the i-th generator S_{Ll} is the apparent power for l-th branch. V_{Bi} and V_{Bj} are the voltage magnitudes of the i-th bus and the j-th bus, respectively. P_{Gi} and Q_{Gi} are the active and reactive power of the i-th generator. P_{Di} and Q_{Di} are the active and reactive power loads of the i-th load bus. G_{ij} and B_{ij} are the conductance and susceptance from the i-th bus to j-th bus. θ_{ij} is the voltage angle difference between i-th bus and j-th bus.

The objective function (4.4) is to minimize the generation cost of synchronous generators. The equality and inequality constraints of (4.5) and (4.6)–(4.9) correspond to **g** and **h**, respectively.

It should be noted that the optimal power flow problem is a nonlinear and non-convex problem. The most effective solution algorithm is the interior point (IP) method that has been widely implemented in commercial and open-source software tools such as MATPOWER [1].

4.3.2 Security-Constrained Optimal Power Flow (SCOPF)

To satisfy the security requirement of the power system against credible contingencies (such as $N - 1$ faults), the security constraints can be incorporated into the OPF problem, yielding SCOPF models.

4.3.2.1 Preventive SCOPF

In its conventional form, the SCOPF aims to optimize an OP that is secure in both pre-contingency and post-contingency states without using the post-contingency control measures (such as generation rescheduling and load shedding), except for the system's automatic response, e.g. relay protection, automatic generation control, automatic tap changers, etc. Its compact mathematical model is as follows:

$$\min_{\mathbf{u}_0} f_0(\mathbf{x}_0, \mathbf{u}_0, \mathbf{y}_0) \tag{4.10}$$

$$\text{s.t.} \quad \mathbf{g}_k(\mathbf{x}_k, \mathbf{u}_0, \mathbf{y}_0) = 0 \ (k = 0, ..., K) \tag{4.11}$$

$$\mathbf{h}_k(\mathbf{x}_k, \mathbf{u}_0, \mathbf{y}_0) \leq 0 \ (k = 0, ..., K) \tag{4.12}$$

where subscript k denotes the kth system configuration ($k = 0$ corresponds to pre-contingency configuration, and $k = 1$, ..., K corresponds to the kth post-contingency configuration). The objective function (4.10) corresponds to (4.4), which is the total generation cost of the pre-contingency state. The constraints (4.11) correspond to (4.5), and (4.12) correspond to (4.6)–(4.9).

It is important to note that P-SCOPF restricts no operation limit violations that exist in both pre-contingency and post-contingency conditions. Since all the security limits are satisfied by only pre-contingency control variables $\mathbf{u}_0$, this kind of SCOPF is also called "preventive" SCOPF (P-SCOPF). It should also be noted that since the contingencies may not really happen, incorporating the security constraints into the OPF model would heavily tighten the feasible region, which results in a conservative operation state with higher operating cost than the conventional OPF.

4.3.2.2 Corrective SCOPF

Different from P-SCOPF, "corrective" SCOPF (C-SCOPF) allows the use of post-contingency CC actions, such as generation rescheduling or load shedding, to regain the security of the post-contingency power system. Its compact model is as follows:

$$\min_{\mathbf{u}_0} f_0(\mathbf{x}_0, \mathbf{u}_0, \mathbf{y}_0) \tag{4.13}$$

$$\text{s.t.} \quad \mathbf{g}_k(\mathbf{x}_k, \mathbf{u}_k, \mathbf{y}_0) = 0 \ (k = 0, ..., K) \tag{4.14}$$

$$\mathbf{h}_k(\mathbf{x}_k, \mathbf{u}_k, \mathbf{y}_0) \leq 0 \ (k = 0, ..., K) \tag{4.15}$$

$$|\mathbf{u}_k - \mathbf{u}_0| \leq \Delta \mathbf{u}^{\max} \tag{4.16}$$

where $\mathbf{u}_k$ is the control variable after the kth contingency. (4.16) is called coupling constraints, which denotes the limitation of the control variable changes, i.e. the maximum allowed adjustment of the control variables between the pre-contingency and the kth post-contingency state. For corrective generation rescheduling, $\Delta \mathbf{u}^{\max}$ is the generation ramping rate in response to contingency k.

It should be noted that, when $\Delta \mathbf{u}^{\max}$ is set zero, the above model becomes a P-SCOPF because this means $\mathbf{u}_k = \mathbf{u}_0$; when $\Delta \mathbf{u}^{\max}$ is $+\infty$, it becomes a conventional OPF since the post-contingency control variables $\mathbf{u}_k$ are independent of the pre-contingency state $\mathbf{u}_0$.

4.3.2.3 Preventive-Corrective SCOPF

It is important to note that although post-contingency CC actions are included in the C-SCOPF model, the objective function (4.13) only considers the pre-contingency operating cost f_0. This is based on the assumption that the probability of the use of CC is relatively small, and its cost can be negligible in long term. However, this may not be valid since the expected CC cost is non-negligible considering the probabilistic nature of the contingencies. Besides, the C-SCOPF model lacks sufficient coordination between preventive and corrective actions and tends to rely on CC to cover as many as possible contingencies.

To address the above problems, the preventive-corrective SCOPF (PC-SCOPF) model has been proposed as follows [2]:

$$\min_{\mathbf{u}} \underbrace{\left[f(\mathbf{x}, \mathbf{u}) - f(\widehat{\mathbf{x}}, \widehat{\mathbf{u}}) \right]}_{PC \ cost} + \underbrace{\sum_{k \in S_n} p_k \times C_k(\mathbf{x}, \mathbf{u})}_{CC \ cost} \tag{4.17}$$

$$\text{s.t.} \quad \mathbf{g}_k(\mathbf{x}_k, \mathbf{u}_k, \mathbf{y}_0) = 0 \ (k = 0, ..., K) \tag{4.18}$$

$$\mathbf{g}_k^0(\mathbf{x}_k^0, \mathbf{u}_0, \mathbf{y}_0) = 0 \ (k = 0, ..., K) \tag{4.19}$$

$$\mathbf{h}_k(\mathbf{x}_k, \mathbf{u}_k, \mathbf{y}_0) \leq 0 \ (k = 0, ..., K) \tag{4.20}$$

$$\mathbf{h}_k^0\left(\mathbf{x}_k^0, \mathbf{u}_0, \mathbf{y}_0\right) \le 0 \ (k = 0, ..., K) \tag{4.21}$$

$$\left|\mathbf{u}_k - \mathbf{u}_0\right| < \Delta\mathbf{u}^{\max} \tag{4.22}$$

where the objective function (4.17) is summation of the PC cost and expected CC cost, also called expected security control cost (ESCC). Since the PC actions are mainly implemented during the generation dispatching stage, its control cost can be expressed as the increased cost of the optimized operating state $(\mathbf{x}, \mathbf{u})$ over the original state $(\hat{\mathbf{x}}, \hat{\mathbf{u}})$, where $(\hat{\mathbf{x}}, \hat{\mathbf{u}})$ is the optimal operating state determined by a conventional OPF solution. For the CC, the expected control cost is defined as the probability-weighted sum of the control costs of the contingency set $\mathbf{S}_C$ that call for CC actions.

$C_k(\mathbf{x}, \mathbf{u})$ is the CC cost function of the contingency k for the OP $(\mathbf{x}, \mathbf{u})$, which can be modeled as a quadratic function of generation changes:

$$C_k(\mathbf{x}, \mathbf{u}) = \sum_{i \in S_G} r_{Gi} \times (P_{Gki} - P_{G0i})^2 \tag{4.23}$$

where subscript 0 represents pre-contingency state and k denotes the state after contingency k. p_k is the occurrence probability of contingency k. In practice, it is a time-varying variable with respect to circuit (line length, voltage level), environment (weather), and geographic (location) features.

Equations (4.19) and (4.21) correspond to the short-term (immediate) state reached just after the contingency happens and before the CC acts, and they impose the existence and viability of the intermediate post-contingency state. ε is the parameter defining how much the short-term post-contingency security constraints can be temporarily relaxed from the permanent limits before the CC acts, e.g. 120% of the line rating in the normal state. The inclusion of (4.19) and (4.21) aims to ensure the existence of a viable short-term post-contingency equilibrium, which can effectively prevent the risk of voltage collapse.

4.3.3 Unit Commitment (UC)

UC is a basic optimization model for day-ahead power system operation and electricity market clearance. It aims to determine the generators operating states, including their turn-on/off statuses and power outputs, with the objective of minimizing the total production cost while meeting prevailing operational limits. Conventionally, the prevailing constraints include power balance, unit minimum on/off time limits, ramping up/down limits, and system spinning reserve requirements.

Its compact mathematical model can be formulated as follows:

$$\min_{\mathbf{x}, \mathbf{y}} c^T\mathbf{x} + d^T\mathbf{y} \tag{4.24}$$

$$\text{s.t. } \mathbf{Ax} + \mathbf{Ey} \leq b \tag{4.25}$$

$$\mathbf{x} \in \{0,1\}, \mathbf{y} \geq 0 \tag{4.26}$$

where the binary variable $\mathbf{x}$ is the decision vector of commitment such as on/off and start-up/shut-down status of the generator unit under every time interval. $\mathbf{y}$ is the continuous decision vector of dispatches such as generation output, resource reserve levels, or load consumption levels under every time interval. c and d are the cost vectors, while $\mathbf{A}$, $\mathbf{E}$, and b are the coefficient matrices.

The objective function (4.24) aims to minimize the sum of the commitment cost $c^T\mathbf{x}$ and the dispatch cost $d^T\mathbf{y}$ within the operation time horizon. Constraint (4.25) limits the commitment and dispatch decisions, such as minimum up and down, start-up/shutdown constraints, power balance constraints, reserve requirements and reserve capacity. Constraint (4.26) denotes the binary value of the unit on/off status.

The full formulation of a conventional UC model is as follows:

$$\min_{P,I} \sum_{t=1}^{NT}\sum_{i=1}^{NG}[C_i(P_{it}) \cdot I_{it} + CU_{it} + CD_{it}] \tag{4.27}$$

where P_{it} is the active power output of unit i in period t. I_{it} is the binary variable for on/off states of unit i in period t. CU_{it} and CD_{it} are start-up and shut-down costs of unit i in the period of t. NG and NT are the number of units and dispatching periods.

$C_i(P_{it})$ is the generation cost function, which is usually modeled as:

$$C_i(P_{it}) = a_i \cdot I_{it} + b_i \cdot P_{it} + c_i \cdot (P_{it})^2 \tag{4.28}$$

The operational constraints are formulated as:

a) Power balance

$$\sum_{i=1}^{NG} P_{it} \cdot I_{it} = D_t, \ \forall t \tag{4.29}$$

where D_t is the system load demand in period t.

b) Generation limits

$$P_i^{\min} \cdot I_{it} \leq P_{it} \leq P_i^{\max} \cdot I_{it}, \ \forall t \tag{4.30}$$

c) Spinning reserve limits

$$\sum_{i=1}^{NG} R_{it} \cdot I_{it} \geq R_t^S, \ \forall t \tag{4.31}$$

where R_{it} is the spinning reserve of unit i in period t and R_t^s is the system spinning reserve requirement in period t.

d) Ramping limits

$$\begin{cases} RD_i \le P_{i(t+1)} - P_{it} \le RU_i \\ SD_i \le P_{i(t+1)} - P_{it} \le SU_i \end{cases}, \quad \forall t, \forall i \tag{4.32}$$

where RD_i and RU_i are the ramping down and up limit of unit i, respectively. SD_i and SU_i are the shut-down and start-up ramping limit of unit i.

e) Minimum up and down time limits

$$\begin{cases} \left(X^{\mathrm{on}}_{i(t-1)} - T^{\mathrm{on}}_i\right) \cdot \left(I_{i(t-1)} - I_{it}\right) \ge 0 \\ \left(X^{\mathrm{off}}_{i(t-1)} - T^{\mathrm{off}}_i\right) \cdot \left(I_{i(t-1)} - I_{it}\right) \ge 0 \end{cases}, \quad \forall t, \forall i \tag{4.33}$$

where X^{on}_{it} and X^{off}_{it} are the ON and OFF time of unit i in period t.

Essentially, the UC is a large-scale, mixed-integer, combinational, and nonlinear programming problem that is highly difficult to solve. Nowadays, two major solution methods have been widely applied: one is based on the Lagrange Relaxation (LR), and the other is to reformulate the problem into a mixed-integer linear programming (MILP) model and employ a MILP solver.

The basic idea of the LR method is to relax system-wide demand (4.29) and reserve requirements (4.31) using Lagrange multipliers and to form a hierarchical optimization structure. However, due to the nonconvexities of the UC problem, heuristic procedures are needed to find feasible solutions, which may be suboptimal. By contrast, the MILP approach is becoming more popular since a variety of compact and tight MILP formulations have been reported, e.g. [3–5], and the high-performance commercial solvers (such as CPLEX and GUROBI) can be directly employed for higher-quality solutions.

4.3.4 Security-Constrained Unit Commitment (SCUC)

Similar to the SCOPF, the SCUC incorporates the security constraints into the UC model in order to satisfy the network and contingency security requirements. Its mathematical form is as follows:

$$\min_{\mathbf{x}_0, \mathbf{y}_0} c^T \mathbf{x}_0 + d^T \mathbf{y}_0 \tag{4.34}$$

$$\text{s.t. } \mathbf{A}\mathbf{x}_0 + \mathbf{E}\mathbf{y}_0 \le b \tag{4.35}$$

$$\mathbf{F}\mathbf{x}_0 + \mathbf{H}\mathbf{y}_0 + \mathbf{G}\mathbf{z}_k \le h \; (k = 0, ..., K) \tag{4.36}$$

$$\mathbf{x}_0 \in \{0, 1\}, \; \mathbf{y}_0 \ge 0 \tag{4.37}$$

where subscript 0 denotes the value of variables in the base case and $k = 1, ..., K$ stands for the value of variables for contingency k. $\mathbf{z}$ denotes the state variables of the power networks. Constraint set (4.35) is the same as (4.25). Constraint set

(4.36) denotes the network steady-state security constraints under base case ($k = 0$) and k-contingency ($k > 0$).

4.3.5 Solution Strategy for Security-Constrained UC/OPF

For security-constrained UC/OPF models, given its even larger problem dimension, the solution is highly challenging. The most effective solution strategy is to decompose the original problem into a master problem and a range of slave sub-problems, known as Bender's decomposition [6]. The master problem corresponds to a conventional UC/OPF model without the security constraints. Given the UC/OPF results solved from the master problem, the slave subproblems check the security constraints for each contingency. If security violation exists, Benders cuts (or similarities) are generated and added to the master problem for the next-round solution. The whole problem is iteratively solved between the master and slave problems until no violation exists. The major advantage of this decomposition method is that it can decompose the large problem into a series of smaller and tractable problems, diminishing significantly the complexity of the whole problem and enabling parallel computing to reduce the total execution time. In the subsequent chapters, such decomposition idea will be used to handle the stability constraints.

4.3.6 Renewable Energy Sources in UC and OPF

It is important to point out that the above models are all deterministic, i.e. all the parameters are exactly known. This is valid based on the assumption that the forecasting of the load demand and the intermittent renewable energy sources (RES) such as wind and solar power are accurate and/or the penetration level of the RES is low. However, given the fast-growing and larger-scale integration of RES into the modern power system, the stochastic nature and the uncertainties of RES should be well modeled and addressed in the UC and OPF (as well as the security-constrained formulations). In general, there are two types of optimization methods for uncertainty handling, i.e. stochastic programming and robust optimization. In general, the former relies on a probability distribution function (PDF) such as the Weibull distribution to sample a large number of scenarios to approximate the uncertain variables and optimizes the expected value of the objective function; the latter employs an interval (the lower and upper bound values) to model the uncertain variable and optimizes for the worst case of the objective function. Based on them, the corresponding power operation methods have been widely studied in recent years [7, 8]. In the subsequent chapters, several stability-constrained power system operation and planning methods considering uncertainties (which mainly focus on dynamic load model uncertainty and wind power output uncertainty) will be systematically introduced. The presented methods are based on the principles of stochastic programming and robust optimization.

4.3.7 Smart Grid Elements

Moreover, with the development of smart grid technologies and their deployments, demand response, flexible load (such as controllable electric vehicle charging, and smart appliance load.), and energy storage system (ESS) have added more controllable resources to support the power system operation, which have also been incorporated into the above steady-state optimal operation models, as some early research works are [9–11]. However, very few (if none) works have considered these new controllable resources in the stability-constrained optimization problems. In this book, these "smart grid elements" are not considered.

4.4 Power System Operation Practices

In addition to forecasting of load demand and renewable power generation and running the UC and OPF models, the practical power system operation needs to take into consideration of many other realistic factors and requirements. While different nations and utilities may have different requirements and concerns, the general summary can be listed and shown in Figure 4.3 [12]:

- Document system-operating diagrams such as single-line drawings, panel directories, equipment layouts, and site plans. Moreover, vendor drawings and maintenance histories, instruction manuals, and plant drawings may also need to be documented.
- Proper administrative procedures to reduce operating errors, such as comprehensive documentation, planning switching operations in detail, securing

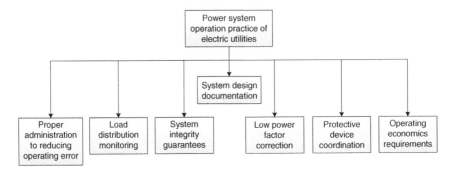

Figure 4.3 Power system operation practice classification.

equipment from unintentional operation, and defining operating responsibility clearly. In addition, key interlocking of circuit breakers should be applied where switching sequences have to be done in a specific order, or a system failure can be caused by closing a circuit breaker when the system is in the wrong configuration.

- The criticality of the load should be classified to the facility operation. In addition, load distribution monitoring is required for system operators with respect to nominal electrical load measurements outputs such as watts, vars, and amperes to ensure there are no overloaded circuits.
- The whole operational electrical system should be kept in service as much of the time as possible, which means the system integrity should be considered and protected from some conditions, such as considering external or environmental outside forces, isolating equipment location, addressing protection action during congested construction/maintenance activity and proposing the best maintenance plan.
- Low power factor should be aware by system operators. Evaluation and correction of the low power factor such, as adding power-factor correction capacities, should be conducted properly.
- System protective devices coordination study should be conducted. In addition, the system overload or short circuit study should be also performed, and the protective devices should be sized and set based on the study results. Then, the system operators should gain a copy of this study and validate that all the protective devices are set correctly. The coordination of the system should be reviewed periodically before any changes in the electrical systems.
- Operating an electrical system economically is required. Based on the energy conversation, economic operation, i.e. minimizing any utility demand charge and power losses can be achieved by various methods of monitoring and controlling the energy flow. The operator should be aware of the demand profile to operate at peak effectiveness and to avoid any unnecessary demand charges.

It should be noted that switching procedures, which are a general part of any power system, are also important for power system operation in practice. As every switching operation should be reached for the sake of safely changing electrical power or isolating the power system, all switching should only be performed with written orders by qualified personnel. In addition, all switching orders should be comprehensively reviewed and approved by all personnel that will be involved in the operation. Moreover, the created switching procedure should be tested at least once prior to its practical execution to validate the correction of the procedure.

In power system operations, it is not just the system operators or managers who are responsible for system operation and control. Anyone who communicates with the power supply and distribution system of an industrial or commercial facility, even household consumers, has some responsibility for the control and operation of that system, especially for modern power systems, which integrate renewables and advanced intelligent technologies and facilities.

Since renewables are penetrated in power systems remarkably with the development of power electronic technologies, the system operation becomes more uncertain and intermittent, and the security issue may emerge. Household consumers can take over this responsibility of minimizing the demand charging under the uncertainties through demand response technology, which is credit to the implementation of smart metering. Therefore, the consumers, as well as the system operators, can benefit from it.

Nomenclature

Indices

i, j indices for buses and generators and units
l index for lines
t index for the time period
0 index for basecase
k index for contingency

Sets

S_B set of all buses
S_C set of credible contingencies
S_G set of all the generators
S_L set of lines
NG/NT number of units and dispatching time periods
$\mathbf{u}$ set of control variable
$\mathbf{x}$ set of state variable
$\mathbf{y}$ set of parameter

Variables

S_{Ll} apparent power for l-th branch
V_{Bi}/V_{Bj} voltage magnitudes of the i-th bus and the j-th bus
P_{Gki} active power output of generator i (MW) after contingency k
P_{G0i} active power output of generator i (MW) at pre-contingency state
P_{Gi}/Q_{Gi} active and reactive power of the i-th generator
P_{it} active power output of unit i in period t
θ_{ij} voltage angle differences between i-th bus and j-th bus
I_{it} binary variable for on/off states of unit i in period t

R_{it} spinning reserve of unit i in period t

X_{it}^{on}/X_{it}^{off} ON and OFF time of unit i in period t

Parameters

$a_i, b_i c_i$	generation cost efficient
G_{ij}/B_{ij}	conductance and susceptance from the i-th bus to j-th bus
C_i	generation cost function of unit i
CU_{it}, CD_{it}	start-up and shut-down cost of unit i in period of t
D_t	system load demand in period t
p_k	occurrence probability of contingency k
P_i^{min}/P_i^{max}	active generation limits of unit i in period of t
r_{Gi}	cost coefficient for corrective generation rescheduling of generator i (\$/MW/hr)
R_t^s	system spinning reserve requirement in period t
RD_i/RU_i	ramping down and up limit of unit i
SD_i/SU_i	shut-down and start-up ramping limit of unit i
T_i^{on}/T_i^{off}	minimum up and down time of unit i
ε	vector defining maximal allowed adjustment of the control variables between the pre- and post-contingency state for contingency k (MW).

References

1 Zimmerman, R.D., Murillo-Sanchez, C.E., and Thomas, R.J. (2011). MATPOWER: steady-state operations, planning, and analysis tools for power systems research and education. *IEEE Transactions on Power Systems* 26 (1): 12–19.

2 Xu, Y., Dong, Z.Y., Zhang, R. et al. (2014). Solving preventive-corrective SCOPF by a hybrid computational strategy. *IEEE Transactions on Power Systems* 29 (3): 1345–1355.

3 Carrión, M. and Arroyo, J.M. (2006). A computationally efficient mixed-integer linear formulation for the thermal unit commitment problem. *IEEE Transactions on Power Systems* 21 (3): 1371–1378.

4 Ostrowski, J., Anjos, M.F., and Vannelli, A. (2012). Tight mixed integer linear programming formulations for the unit commitment problem. *IEEE Transactions on Power Systems* 27 (1): 39–46.

5 Yang, L., Zhang, C., Jian, J. et al. (2017). A novel projected two-binary-variable formulation for unit commitment in power systems. *Applied Energy* 187: 732–745.

6 Shahidehopour, M. and Yong, F. (2005). Benders decomposition: applying Benders decomposition to power systems. *IEEE Power and Energy Magazine* 3 (2): 20–21.

7 Zheng, Q.P., Wang, J., and Liu, A.L. (2015). Stochastic optimization for unit commitment – a review. *IEEE Transactions on Power Systems* 30 (4): 1913–1924.

8 Sun, A. and Lorca, A. (2014). Adaptive robust optimization for daily power system operation. *Power Systems Computation Conference* 2014: 1–9.

9 Khodayar, M.E., Wu, L., and Shahidehpour, M. (2012). Hourly coordination of electric vehicle operation and volatile wind power generation in SCUC. *IEEE Transactions on Smart Grid* 3 (3): 1271–1279.

10 Zhao, C., Wang, J., Watson, J.P., and Guan, Y. (2013). Multi-stage robust unit commitment considering wind and demand response uncertainties. *IEEE Transactions on Power Systems* 28 (3): 2708–2717.

11 Jabr, R.A., Karaki, S., and Korbane, J.A. (2015). Robust multi-period OPF with storage and renewables. *IEEE Transactions on Power Systems* 30 (5): 2790–2799.

12 IEEE (2011). *Recommended Practice for the Operation and Management of Industrial and Commercial Power Systems*, IEEE Std 3007.1-2010, 1–31. IEEE.

5

Transient Stability-Constrained Optimal Power Flow (TSC-OPF): Modeling and Classic Solution Methods

5.1 Mathematical Model

In the TSC-OPF model, the transient stability requirement is considered as an additional constraint in the classical OPF model. Conventionally, the generalized model is formed as follows:

$$\min F(\mathbf{x}, \mathbf{u}, \mathbf{y}) \tag{5.1}$$

$$\text{s.t. } \mathbf{g}(\mathbf{x}, \mathbf{u}, \mathbf{y}) = 0 \tag{5.2}$$

$$\mathbf{h}(\mathbf{x}, \mathbf{u}, \mathbf{y}) \leq 0 \tag{5.3}$$

$$\text{TSI}_k(\mathbf{x}, \mathbf{u}) \geq \varepsilon, \{k \in C\} \tag{5.4}$$

where Eqs. (5.1)–(5.3) represent the conventional OPF model. The conventional OPF is introduced in Chapter 4, shown in Eqs. (4.4)–(4.9).

The constraint set (5.4) represents the transient stability constraints regarding a transient stability index (TSI) and the associated threshold ε against a pre-defined contingency list C. The stability requirement is defined by the TSI and ε. Generally, they are considered as the maximum acceptable rotor angle deviation [1–8]. However, different approaches to TSC-OPF can have distinct TSIs. Three different TSIs are introduced in Chapter 2. In addition, the different TSIs applied in TSC-OPF approaches will be reviewed in detail subsequently.

The computation of TSC-OPF is very difficult because of the highly nonconvex and nonlinear characteristics of the power system's transient stability. Generally, the difficulties appear in two subproblems:

1) how to add the transient stability constraints to the conventional OPF model, i.e. how to express Eq. (5.4); and
2) how to solve the optimization problem after the transient stability constraints are included, i.e. (5.1)–(5.4).

Stability-Constrained Optimization for Modern Power System Operation and Planning,
First Edition. Yan Xu, Yuan Chi, and Heling Yuan.
© 2023 The Institute of Electrical and Electronics Engineers, Inc.
Published 2023 by John Wiley & Sons, Inc.

In this chapter, three types of approaches for TSC-OPF are introduced: discretization-based method, direct method, and evolution algorithm (EA)-based method. For the first two methods, classic programming algorithms, such as linear programming (LP) and interior point (IP) methods, are applied for subproblem (2). However, in terms of dealing with subproblem (1), their philosophy is substantially distinguishing, the former is dependent on the conventional TSA principle, namely, the system dynamics are represented by rotor angle swing equations. The latter one, on the other hand, depends on direct TSA to stabilize the system to meet the stability constraints. This is the underlying naming principle that this chapter adopts for the first two methods. For the last method, it is realized by modern heuristic programming techniques, known as EAs, to solve subproblem (2), and it's also different from the first two methods as it's free from TSA methods to tackle stability constraints in subproblem (1). All three methods are reviewed in the following sections.

5.2 Discretization-based Method

Based on the traditional TSA, the discretization-based method constructs the transient stability constraint as rotor angle swing equations. For classic generator model, it's formed as below:

$$
\begin{cases}
M_i \dfrac{d\omega_i}{dt} = P_{mi} - P_{ei} \\
\dfrac{d\delta_i}{dt} = \omega_i
\end{cases}, \quad \{i \in S_G\}
\tag{5.5}
$$

The model is formulated in Chapter 2 Eq. (2.3). By applying the discretization-based method, TSI is usually the rotor angle deviation. The common expression is in the form of center of inertia (COI), which is shown in Eq. (2.4).

The stability constraints expressed by (5.5) and COI are differential with respect to time t. By forming the TSC-OPF model in this way, a large set of differential-algebraic equations (DAEs) will be encountered. However, classic programming algorithms are only able to handle algebraic equations. In order to solve such a TSC-OPF model, the differential equations need to be converted into algebraic form. After that, the TSC-OPF becomes a standard nonlinear programming problem and can be solved by classic optimization algorithms, for example, IP method.

For the discretization-based method, the kernel is how to convert the differential swing equations into algebraic form and make the problem scale compatible without loss of computation requirements. In Ref. [1], a numerical discretizing scheme was first applied to convert classic two-order swing equations from the differential form into numerically equivalent difference form. And then, the classic LP algorithm was able to solve the problem.

The general principle of discretization conversion is presented [1]. Firstly, let:

$$\frac{d\omega_i}{dt} = D_i(P_{mi}, P_{ei}) \tag{5.6}$$

Then, the DAEs can be converted by trapezoidal rules:

$$\delta_i^{n+1} - \delta_i^n - \frac{h}{2}\left(\omega_i^{n+1} + \omega_i^n\right) = 0 \tag{5.7}$$

$$\omega_i^{n+1} - \omega_i^n - \frac{h}{2}\left(D_i^{n+1} + D_i^n\right) = 0 \tag{5.8}$$

where h is the integration step length, $n = 1, 2, \ldots, n_{end}$ is the integration step counter. The stability constraints, i.e. the expression of COI in Eq. (2.4) can be rewritten as below:

$$\delta_i^{COI} = \delta_i^n - \frac{\sum\limits_{i=1}^{S_G} M_i \cdot \delta_i^n}{\sum\limits_{i=1}^{S_G} M_i} \tag{5.9}$$

By bounding the COI within a pre-defined degree, the stability constraints of (5.5) and (2.4) can be converted into (5.6)–(5.9), which are all algebraic forms. Then, by integrating them into classical OPF, the TSC-OPF can be solved.

Although the discretizing scheme proposed in Ref. [1] is indeed a breakthrough for solving TSC-OPF, its computation burden is quite heavy as a huge number of variables and equations are introduced at each time step, resulting in oversized dimensions.

In order to improve the computation efficiency with the reduction of problem scale, the TSC-OPF problem was formed as a nonlinear optimization problem in functional space, rather than directly tackle it [2]. The authors proposed to use the function transcription techniques to convert the infinite-dimensional TSC-OPF into a finite-dimensional optimization problem in the Euclidean space, which can be viewed as an initial value problem for all disturbances and solved by any standard nonlinear programming technique adopted by OPF. The transformed TSC-OPF problem has the same variables as those of OPF in form and is tractable even for large-scale power systems with a large number of practical stability constraints [2].

Based on the well-demonstrated principles of the above strategies, there are modified approaches in the literature to improve computation efficiency and accuracy. In Ref. [3], the authors proposed a modified formulation for expressing the electric power output of machines in the swing equation, by which the equality constraints are considerably reduced in number, making the model computationally suitable for considering multiple contingencies. In order to improve the computation efficiency of the function transcription process in Ref. [2], the adjoint equation method is employed in Ref. [4]. Later on, in order to further alleviate the computation burden of TSC-OPF in functional transcription framework, modified calculation of Jacobian and Hessian matrices and direct nonlinear primal-dual IP method are applied in Ref. [5]. More recently, the authors in Ref. [6] proposed an enhanced numerical discretization method that can reduce nearly 50% of the problem dimension and thus greatly enhance the computation efficiency of IP-based solution of TSC-OPF.

In general, the traditional method directly incorporates the stability constraints into OPF and solves the TSC-OPF problem as a whole. It keeps the mathematical model of rotor angle dynamics, and thus the characteristics of power system transient stability are remained. However, despite best efforts, this method still suffers from a heavy computation burden for large-scale power systems with numerous contingencies (especially for multi-contingency conditions), and convergence difficulties could also be encountered with a detailed system model. Moreover, the traditional method lacks transparency about the salient parameters responsible for the instability and the underlying reasons for the advocated solution. Furthermore, the discretization of swing equations could bring considerable approximations and thus inaccuracies in calculating TSC-OPF. Last but not least, since the method relies on classic programming techniques, high-sensitivity problems of initial conditions may emerge, which may cause local optimum solutions.

5.3 Direct Method

Different from the traditional method, the direct method is based on the underlying stability mechanism and directly stabilizes the system on related variables under contingencies.

In Ref. [7], based on the trajectory sensitivity technique, the significance of generation output with respect to rotor angle deviation is calculated. The real power output is shifted from the least advanced to the most advanced generators by which the system's transient stability can be achieved. After generation rescheduling for real power output, the systems other variables are determined by conventional OPF.

In Ref. [8], SIME (for single-machine equivalent), also called EEAC [9], has been applied to solving TSC-OPF. Transient stability requirements are met by direct shifting active generation from CMs to NMs, and the power re-dispatch for NMs and other system variables is determined by conventional OPF.

Figure 5.1 depicts the computation process of the direct method. Firstly, without considering stability constraints, an initial operating point (OP) is determined by a

Figure 5.1 Direct method for
TSC-OPF computation process.

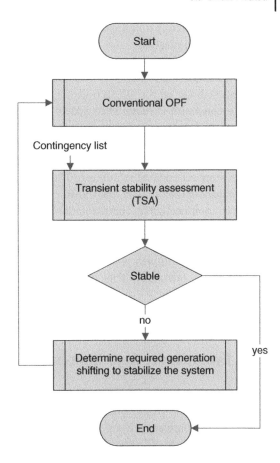

conventional OPF. If the initial OP is not stable with respect to a contingency list, direct TSA methods are then performed to calculate the required generation shifting to remain stable. Fixing the generation shifting, other system variables are determined by conventional OPF. To ensure stability, an iterative procedure is required until the current OP is determined as stable. As defined in Ref. [7], this computation process is a "sequential" approach.

In contrast with traditional methods, the direct method is more efficient as it does not need to solve a large number of DAEs, and its major computation burden is only from determining the required generation shifting.

Therefore, it can easily consider multi-contingency conditions and has no limitations to model complexity. More importantly, it can provide explicit control mechanisms, which can be considered as the reference for system operators and power market participants. However, the direct method can only provide sub-optimal solutions to TSC-OPF, the reason is that the generation shifting is determined by merely trajectory sensitivity values or calculations on OMIB, which are without optimization.

For the sake of achieving global optimum while still applying SIME/EEAC method, the authors in Ref. [10] investigated a strategy that models the system's transient stability on the OMIB trajectory. Based on this, only a single transient stability constraint is required to add to the OPF formulation representing the multi-machine system. According to transient stability margin η, the stability requirement is expressed in critical condition, namely, make η larger than zero, so as to avoid unnecessary over-stabilizing. It's shown in Ref. [10] that this can provide more economical solutions to TSC-OPF. However, due to the highly non-linear nature of the problem, the transforming of multi-machine into OMIB shall introduce non-negligible approximations and inaccuracies. Meanwhile, some recent work has shown that non-monotonicity between the variation of stability margin and critical parameters, especially under multi-swing unstable conditions, and some other complex phenomena, such as isolated stability domain (ISD), sometimes do exist [11]. This could baffle the solution of the OMIB-based TSC-OPF.

5.4 Evolutionary Algorithm-based Method

In addition to the analytical methods for TSC-OPF, advanced heuristic optimization strategies, known as evolution algorithms (EAs), are also promising for solving TSC-OPF problem [12–15], yielding EA-based methods.

EAs are the trade-off between stochastic search and optimization algorithms, which are derived from Darwin's natural evolution principles and have been developed for more than 40 years. In general, it contains three mainstream algorithms: genetic algorithm, evolutionary programming, and evolutionary strategy [16]. These approaches can be commonly characterized by population-based, steered random search, and iterative development. In terms of the procedure of search and optimization, the key principle is the selection of the fittest to generate successively better results over generations and to reach the optimal solution finally. Different from classical calculus-based optimization techniques, EAs do not have a limitation for specified formulations of the objective function and the constraints, namely, complicated, nonlinear, and nonconvex problems such as conventional OPF problems can be solved [17]. Comprehensive introduction and derivation about EAs can be found in Ref. [18].

For solving TSC-OPF based on EA method, stability requirements are met by conducting rigorous TSA on the population. After that, only stable individuals are selected in the evolution process. Figure 5.2 illustrates the common

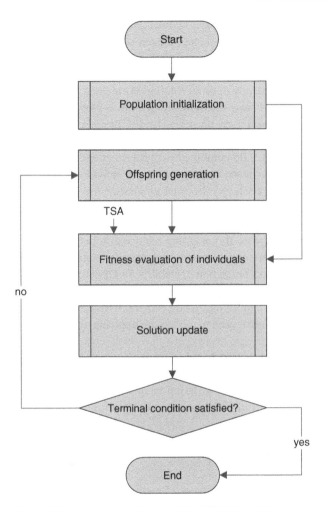

Figure 5.2 EA-based method for TSC-OPF computation process.

computation process of the EA-based method. However, there are various EAs adopted in literatures such as the particle swarm optimization (PSO) technique used in Ref. [12], the genetic algorithm (GA) employed in Ref. [13], and a relatively novel EA, differential evolution (DE) algorithm adopted in Ref. [14].

For the procedure of the EA-based method to TSC-OPF, firstly, an initial population (a set of OPs that are obtained by running power flow) is generated, which usually refers to select control variables. And then, the initial population is evaluated by fitness function, which refers to TSA, to check if the individuals meet the transient stability requirements. Various TSA tools can specify the stability requirements. Then the solution to TSC-OPF is updated, if the solution meets the termination criterion, the computation will stop; otherwise, the population will evolve to produce offspring followed by repeated fitness evaluation. The evolution process will continue until the termination criterion is met. The termination criterion is typically set when a relatively optimal solution is obtained.

EA-based method for TSC-OPF can theoretically achieve a globally optimal solution due to its strong global searching property [12–15]. Besides, the stability requirements can be strictly satisfied with any TSA method. For example, in Ref. [12, 13], TDS is used; in Ref. [14], TDS combined with transient energy function (TEF) method is adopted. However, since there may be many individuals, the computation burden on TSA can be very heavy. In order to alleviate the TSA computation time, during the fitness evaluation process, an artificial neural network (ANN) for TSA is proposed in Ref. [15]. The idea is to train ANN by using a pre-prepared database and to capture the mapping relationship between power system operating point parameters and corresponding stability status. After the ANN is trained well, fast and direct prediction of stability for OPs can be realized [19]. In Ref. [15], the results indicate that the overall computation speed for TSC-OPF can be significantly improved. Moreover, it becomes more tractable for multi-contingency conditions and detailed system models by applying the EA-based method.

Nevertheless, the limitations of the EA-based method are obvious: (1) a large number of iterations are needed, especially for modern complex power systems, as it is a stochastic searching method; (2) suffering inconsistent solutions, especially when a large number of control variables are to be optimized in practice.

5.5 Discussion and Summary

As introduced in the above sections, three types of methods for TSC-OPF have their own philosophy for solving stability constraints, respectively. Especially, the EA-based method has different optimization strategies compared with the other two. A summary of the three types of methods is described below:

Discretization-based method: it is to construct a classic swing equation for representing dynamic constraints of power systems, and then try to convert this differential equation into an algebraic form so as to apply traditional programming techniques to solve the optimization problem. Based on this method, mathematical strictness can be guaranteed. However, it will suffer from excessive computational burden and convergence problems for detailed models. Meanwhile, it should be noted that the accuracy of the method will be affected by the inevitable conversion of swing equations, even if the most advanced techniques are applied to it. Therefore, only a local optimum can be obtained.

Direct method: it is to solve TSC-OPF through direct TSA methods, such as SIME, which tries to directly stabilize the system by reducing active generation outputs from related generators and optimizing other variables by a conventional OPF. In terms of this method, the computational efficiency is high, and system dynamics are not approximated; thus, multi-contingency can be readily taken into account with only linear increments of computation time. However, only a near-optimal solution can be obtained since the generation shifting is not optimized. It's worth mentioning that the approach reported in Ref. [10] is a significant improvement of the direct method. A "global" model is built on the basis of the OMIB framework for representing stability constraints. However, it should be noted that the global optima could still be hindered, as when deriving the OMIB, the inevitable errors still exist.

EA-based method: this is substantially different from the above two deterministic approaches, it solves the TSC-OPF by taking a modern heuristic searching algorithm. The stability constraints are satisfied by rigorous TSA. Based on this method, system modeling and multi-contingency can be readily considered as the direct method does. Since EAs own strong global searching capability, they can theoretically find the global optima of TSC-OPF. However, its demerits include requirements of considerable iterations to obtain a good solution and inconsistent solutions in practice.

5.5.1 Numerical Simulation Results on Benchmark Systems

All of the reviewed approaches above have been successfully demonstrated on test and/or real power systems. In this section, the simulation results on some benchmark systems obtained by them are exhibited, for more direct comparisons of them.

To the best of our knowledge, two systems were commonly used as testing systems for the reviewed approaches: one is the WSCC 3-machine 9-bus system [20], and the other is the New England 10-machine 39-bus system [21]. In literature, most of the TSC-OPF calculations take the total generation cost ($/hr) as the optimization objective. In Tables 5.1 and 5.2, the reported TSC-OPF solution results in terms of total generation cost by different methods on the two testing systems are exhibited, respectively.

According to the reported results, it can be observed that, on average, the EA-based method performs the best among the three kinds of methods. This is because of its strong global searching capacity. The direct method by global SIME [10] also provides high-quality solutions among the methods, owing to the use of stability margin η to define the stability requirement, it can avoid over-stabilizing so as to save cost. For the direct method by trajectory sensitivity technique, the solutions are relatively inferior, as admitted in Ref. [7], and it can only provide

Table 5.1 Reported simulation results for WSCC 3-machine 9-bus system.

Fault	Method	Result ($/hr)
Fault at bus 7, tripping line 7–5 after 0.35 s	Direct method by trajectory sensitivity [7]	1191.56
	Direct method by global SIME [10]	1135.20
	EA-based method by DE [14]	1140.06
Fault at bus 9, tripping line 9–6 after 0.30 s	Direct method by trajectory sensitivity [7]	1179.95
	EA-based method by DE [14]	1148.58

Table 5.2 Reported simulation results for New England 39-bus system.

Fault	Method	Result ($/hr)
Fault at bus 4, tripping line 4–5 after 0.25 s	Direct method by trajectory sensitivity [7]	61 826.53
	EA-based method by DE [14]	61 021.04
Fault at bus 21, tripping line 21–22 after 0.16 s	Traditional method by discretizing scheme [22]	62 263.00
	EA-based method by DE [14]	60 988.25
Fault at bus 29, tripping line 26–29 after 0.10 s	Traditional method by discretizing scheme [22]	61 148.00
	Direct method by global SIME [10]	60 916.80

near-optimum due to the unoptimized generation shifting. The traditional method by discretizing scheme is also relatively worse than the EA-based method and global SIME method, as analyzed before, this can be attributed to the approximation in the discretization and use of classic local programming technique.

Nomenclature

Indices

i index for generators
k index for contingency
t index for the time period

Sets

S_G set of all the generators
g   set of equality constraints
h   set of inequality constraints
u   set of control variable
x   set of state variable
y   set of parameters

Variables

δ_i/ω_i angle and angular speed
P_{mi}/P_{ei} mechanical input and electrical output powers

Parameter

h integration step length
M_i machine's inertia constant
ε threshold against a pre-defined contingency

References

1 Gan, D., Thomas, R.J., and Zimmerman, R.D. (2000). Stability-constrained optimal power flow. *IEEE Transactions on Power Systems* 15 (2): 535–540.

2 Chen, L., Taka, Y., Okamoto, H. et al. (2001). Optimal operation solutions of power systems with transient stability constraints. *IEEE Transactions on Circuits and Systems I: Fundamental Theory and Applications* 48 (3): 327–339.

3 Yue, Y., Kubokawa, J., and Sasaki, H. (2003). A solution of optimal power flow with multicontingency transient stability constraints. *IEEE Transactions on Power Systems* 18 (3): 1094–1102.

4 Sun, Y., Xinlin, Y., and Wang, H.F. (2004). Approach for optimal power flow with transient stability constraints. *IEE Proceedings – Generation, Transmission and Distribution* 151 (1): 8–18. https://digital-library.theiet.org/content/journals/10.1049/ip-gtd_20040059.

5 Xia, Y., Chan, K.W., and Liu, M. (2005). Direct nonlinear primal–dual interior-point method for transient stability constrained optimal power flow. *IEE Proceedings – Generation, Transmission and Distribution* 152 (1): 11–16. https://digital-library.theiet.org/content/journals/10.1049/ip-gtd_20041204.

6 Jiang, Q. and Huang, Z. (2010). An enhanced numerical discretization method for transient stability constrained optimal power flow. *IEEE Transactions on Power Systems* 25 (4): 1790–1797.

7 Nguyen, T.B. and Pai, M.A. (2003). Dynamic security-constrained rescheduling of power systems using trajectory sensitivities. *IEEE Transactions on Power Systems* 18 (2): 848–854.

8 Ruiz-Vega, D. and Pavella, M. (2003). A comprehensive approach to transient stability control. I. Near optimal preventive control. *IEEE Transactions on Power Systems* 18 (4): 1446–1453.

9 Pavella, M., Ernst, D., and Ruiz-Vega, D. (2012). *Transient Stability of Power Systems: A Unified Approach to Assessment and Control*. Springer Science & Business Media.

10 Pizano-Martinez, A., Fuerte-Esquivel, C.R., and Ruiz-Vega, D. (2010). Global transient stability-constrained optimal power flow using an OMIB reference trajectory. *IEEE Transactions on Power Systems* 25 (1): 392–403.

11 Yin, M., Chung, C.Y., Wong, K.P. et al. (2011). An improved iterative method for assessment of multi-swing transient stability limit. *IEEE Transactions on Power Systems* 26 (4): 2023–2030.

12 Mo, N., Zou, Z.Y., Chan, K.W., and Pong, T.Y.G. (2007). Transient stability constrained optimal power flow using particle swarm optimisation. *IET Generation Transmission and Distribution* 1 (3): 476–483.

13 Chan, K.Y., Chan, K.W., Ling, S.H. et al. (2007). Solving multi-contingency transient stability constrained optimal power flow problems with an improved GA. *2007 IEEE Congress on Evolutionary Computation*, Singapore (25–28 September 2007). pp. 2901–2908: IEEE.

14 Cai, H.R., Chung, C.Y., and Wong, K.P. (2008). Application of differential evolution algorithm for transient stability constrained optimal power flow. *IEEE Transactions on Power Systems* 23 (2): 719–728.

15 Tangpatiphan, K. and Yokoyama, A. (2009). Adaptive evolutionary programming with neural network for transient stability constrained optimal power flow. *2009 15th International Conference on Intelligent System Applications to Power Systems*, Curitiba, Brazil (8–12 November 2009). pp. 1–6: IEEE.

16 Yang, G.Y., Dong, Z.Y., and Wong, K.P. (2008). A modified differential evolution algorithm with fitness sharing for power system planning. *IEEE Transactions on Power Systems* 23 (2): 514–522.

17 Yuryevich, J. and Kit Po, W. (1999). Evolutionary programming based optimal power flow algorithm. *IEEE Transactions on Power Systems* 14 (4): 1245–1250.

18 Lee, K.Y. and El-Sharkawi, M.A. (2008). *Modern Heuristic Optimization Techniques: Theory and Applications to Power Systems*. Wiley.

19 Xu, Y., Dong, Z.Y., Meng, K. et al. (2011). Real-time transient stability assessment model using extreme learning machine. *IET Generation, Transmission and*

Distribution 5 (3): 314–322. https://digital-library.theiet.org/content/journals/10.1049/iet-gtd.2010.0355.

20 Sauer, P.W. and Pai, M. (1998). *Power System Dynamics and Stability*. Prentice Hall.

21 Pai, M. (2012). *Energy Function Analysis for Power System Stability*. Springer Science & Business Media.

22 Layden, D. and Jeyasurya, B. (2004). Integrating security constraints in optimal power flow studies. *IEEE Power Engineering Society General Meeting* 1: 125–129.

6

Hybrid Method for Transient Stability-Constrained Optimal Power Flow

6.1 Introduction

Since transient stability depends on the generator's active power output, we can define generators' maximum outputs as a feasible solution region for transient stability-constrained (TSC) problem. Based on the feasible solution region, the conventional optimal power flow (OPF) model, including various control variables and steady-state limits, can be solved by classical programming techniques. It is obvious that global optimality can be realized by solving conventional OPF in the TSC solution region. However, the maximum TSC-feasible solution region is quite difficult to identify and define as the model is nonlinear and nonconvex. Fortunately, evolutionary algorithms (EAs), which own excellent searching capability and adaptability, can tackle this issue.

Therefore, this chapter applies a hybrid method to solve TSC-OPF via combining conventional OPF and EA, which divides the process into a two-layer inter-active optimization pattern. The outer layer is to search the maximum TSC-feasible solution region, namely, the upper active generation bound by EAs, while the inner layer is to find the optimal OP within the region by conventional OPF. For the obtained OPs from inner layer, strict transient stability assessment (TSA) is conducted on them, and only stable individuals are selected to meet transient stability requirements. By this way, the whole calculation of TSC-OPF can continuously generate better solutions in a global search yet deterministic computing mechanism.

The method is verified on the New England test system and a dynamic equivalent system of a real-world large power grid. Comparing with existing methods, it has provided more economical solutions and better adaptability to multi-swing instability and multi-contingency stabilizing. Besides, the method is free from the selection of the two optimization techniques and is also open to any TSA tools and transient stability indexes (TSIs). Moreover, complex system modeling is never a

Stability-Constrained Optimization for Modern Power System Operation and Planning,
First Edition. Yan Xu, Yuan Chi, and Heling Yuan.
© 2023 The Institute of Electrical and Electronics Engineers, Inc.
Published 2023 by John Wiley & Sons, Inc.

limitation and various state-of-the-art techniques can be adopted. Remarkably, this hybrid method can also be extended to other types of stability-constrained OPFs, as it is unified in terms of the optimization philosophy.

6.2 Proposed Hybrid Method

6.2.1 Mathematical Model

The mathematical formulation is obtained by recalling the generalized TSC-OPF model (5.1)–(5.4), where $\mathbf{x}$ and $\mathbf{u}$ represent state and control variables, respectively, $\mathbf{g}$ and $\mathbf{h}$ denote static constraints, and $\mathbf{TSI}_k$ represents transient stability constraint. F represents the objective function.

1) *Control variables*: the control variable $\mathbf{u}$ comprises two parts: $\mathbf{u}^O$ and $\mathbf{u}^I$, which are involved in the outer-layer and the inner-layer optimization process, respectively. Specifically, $\mathbf{u}^O$ are active generation output upper bounds, i.e. $\overline{P}_{Gi}$ in (4.6), and $\mathbf{u}^I$ can include all the control variables of conventional OPF presented in Section 4.3.
2) *Objective function*: if transient stability constraint is considered in the OPF, an additional cost will be spent since the feasible solution region is restricted and narrowed. Practically, system operators may refuse to implement the transient stability constraints and pay such cost as the contingencies may not occur. In this section, a mathematically equivalent but practically meaningful objective function for the hybrid method is applied, which is the increased cost, calculated by

$$f(NewOP) - f(IniOP) \equiv \Delta Cost \tag{6.1}$$

where $f(NewOP)$ and $f(IniOP)$ are the total generation costs of the new and initial OP, respectively, and both are calculated by Eq. (4.4). Here, the initial OP is solved by conventional OPF without considering stability constraints and it keeps constant during the whole computation, while the new OP denotes the OP generated in the computation process.

Then, the objective function can be written as:

$$\min \Delta Cost \tag{6.2}$$

It should be noted that although the objective function (6.2) is mathematically equivalent to Eq. (4.4) for the TSC-OPF model, it can be used as a terminal condition about the acceptable additional cost due to transient stability constraints.

3) *Steady-state constraints*: $\mathbf{g}$ and $\mathbf{h}$ denote those physical and operational limits of the power system, such as (4.5)–(4.9). It should be noted that in each inner-layer optimization, the constraint (4.6) is updated by the searching results of outer layer, $\mathbf{u}^O$.

4) *TSCs*: in the TSC-OPF, the transient stability constraints can be generally constructed by a TSI with respect to a threshold shown as (5.4).

5) *Terminal condition*: when the global optimal solution is achieved, the whole process can be terminated. There are two terminal conditions that assume a globally optimal solution is acquired: maximum iteration number is met, or objective function (6.1) reaches a pre-defined tolerance.

In practical operation, it should be indicated that the second terminal condition is significant: once the increased additional cost is acceptable, the whole computation can be terminated even without obtaining the globally real optima.

6.2.2 Computation Process

Figure 6.1 depicts the computation process for the hybrid TSC-OPF, where the solid lines indicate the outer-layer optimization process and the dashed line describes the detailed inner-layer optimization process.

The important steps are expressed as follows:

Step 1	Initialization: determine outer-layer control variables $\mathbf{u}^O$ and generate the initial population.
Step 2	Conduct inner-layer optimization: input each individual, $\mathbf{u}_i^O$, $i \in G$, where G represents the G-th generation, into the conventional OPF model and solve it. After that, output new OPs.
Step 3	Based on the new OPs, conduct TSA and calculate the objective function F.
Step 4	Update the current global best individuals (denoted as OP_{best}) according to the objective function calculated in Step 3, if a terminal condition is satisfied, stop and output results; otherwise, go to Step 5.
Step 5	Search max TSC-feasible solution region (defined by $\mathbf{u}^O$) by EA then go to Step 2.

6.2.3 Remarks

From a global perspective, the hybrid method makes full use of EA-enhanced stochastic search for maximum TSC-feasible solution region, on the other hand, without losing classical programming technique for deterministically optimizing system control variables for a conventional OPF. For a more visual understanding, the whole process can also be conceptually expressed in Figure 6.2.

The hybrid method provides the following advantages over the methods reviewed in Sections 5.2–5.4

1) In contrast with single classical programming-based approaches, the computation philosophy of the hybrid method is more straightforward, and the stability constraints can be rigorously performed via TSA procedure. Therefore, the

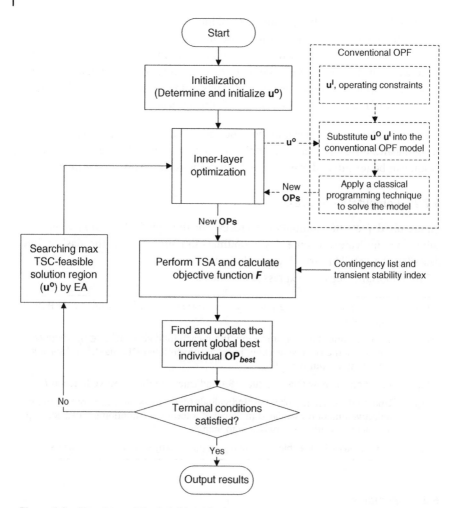

Figure 6.1 Flowchart of the hybrid method.

exactness of system dynamics will not be sacrificed, and complicated models or multi-contingency can be easily tackled. Moreover, the quality of the solution could be better as a global search process is applied for the hybrid method.

2) Compared with single EA-based approaches, only generator maximum active outputs are modeled as control variables for the hybrid method, thus the searching space is significantly reduced, that is, for an N_G-generator system, the hybrid method has N_G control variables at most, but prevailing EA-based methods have at least $2N_G$ control variables even without considering some practical control variables such as phase-shifter angles and transformer tap

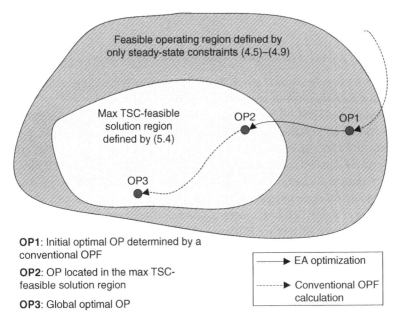

Figure 6.2 Conceptual expression of the optimization process by the hybrid method.

settings. In addition, the OP is practically obtained based on conventional OPF. Hence, all control variables of the system can be incorporated, and steady-state limits can be systematically handled without using a heuristic fitness function, which could remarkably improve the robustness of the method.

3) The hybrid method renders generality and flexibility, as it is suitable to any state-of-the-art TSA tools and TSIs for different transient stability requirements defined by different utilities, and it is also open to any incorporation of the deterministic programming and the EA.

6.3 Technical Specification

6.3.1 Application of TSI and TSA Tool

As introduced in Chapter 2, it is difficult to determine conventional TSI such as maximum rotor angle deviation due to its system-dependent nature. Besides, conventional TSI may result in conservative solutions for TSC-OPF. As shown in [1–3], an alternative index, transient stability margin η, is quite good since it can well reflect the inherent mechanism of transient stability. To achieve a more economical solution, over-stabilizing should be avoided by limiting a small positive η. Following this guideline, η is adopted to specify the TSI. Constraint (5.4) then becomes

$$0 \leq \eta_k \leq \varsigma_k, \quad (k= 1, 2, 3, ..., n) \tag{6.3}$$

where ς is a positive value, which can be flexibly determined according to different utilities, and k denotes the k-th contingency in consideration.

In this section, FASTEST [4], which is an industrial-grade power system stability analysis software, is employed for TSA with stability margin η as the output. For this software, full TDS and mature EEAC method for fast and quantitative TSA are integrated, where detailed modeling of generators, HVDC, SVC, etc. is not limited and stability margin η can be accurately calculated. It has been widely used in China and many other countries and has shown online compatible TSA speed [4]. Mathematical details of EEAC theory can be found in Chapter 2 and literature [5, 6]. It should be noted that other TSIs applied in practice such as critical clearing time (CCT) can also be used, and the CCT can also be directly calculated by FASTEST.

6.3.2 Solution Approach

For the hybrid method, the conventional OPF is solved by the IP method [7], and Differential Evolution (DE) algorithm is used as the EA for globally searching the maximum TSC-feasible solution region. As a relatively advanced EA, DE is quicker, more straightforward, and more robust compared with other EAs on various benchmark problems [8]. Its implementation in the hybrid method is introduced as follows.

1) *Initialization*: this step includes control variable selection and their initialization. To further reduce the searching space, namely, making the control variable less than N_G, prior knowledge such as the CMs determined by the EEAC method can be applied. These CMs referring to the initial OP can be selected as initial control variables and if new CMs appear during the later evolution process, these new generators can be added to the control variables. It should be illustrated that this is a remarkable enhancement as it can reduce the searching space without losing the capability in restoring stability. The initialization of control variables can be calculated by

$$\mathbf{y}_i^{G0} = \mathbf{y}_{i(L)} + \mathbf{rand}_i[0.5, 1] \cdot \left(\mathbf{y}_{i(H)} - \mathbf{y}_{i(L)}\right) \tag{6.4}$$

where $G0$ denotes the initial generation, $\mathbf{y}_{i(H)}$ and $\mathbf{y}_{i(L)}$ are lower and higher bounds of the control variables. Note that $\mathbf{y}_i$ corresponds to the element of $\mathbf{u}^O$. A relatively large randomness range $[0.5, 1]$ is set to ensure the power balance between generation and load.

2) *Mutation and recombination:* in this step, the diversity of the individuals are generated for seeking the global optima. The rule is

$$\mathbf{y}_i'^{G+1} = \mathbf{y}_i^G + f_2 \cdot \left(\mathbf{y}_{r3}^G - \mathbf{y}_i^G\right) + F \cdot \left(\mathbf{y}_{r1}^G - \mathbf{y}_{r2}^G\right) \tag{6.5}$$

where the randomly generated integers $r_1 \neq r_2 \neq r_3 \neq i$, F is the mutation constant, and f_2 controls the crossover constant. $f_2 \in [0,1]$ and remains constant throughout the evolution process.

3) *Individual selection and solution updating:* this step involves comparing individuals with each other to select and update current OP_{best}. The comparison rules are taken as below:

   Comparing two OPs:

   Scenario 1: if they both satisfy (6.3), select the one, which has a smaller $\Delta Cost$;
   Scenario 2: if they are unstable ($\eta < 0$), select the one, which has the larger η;
   Scenario 3: if they are stable ($\eta > 0$) but neither satisfy (6.3), select the one, which has the smaller η;
   Scenario 4: if only one of them satisfies (6.3), select this one.

   Note: The selected one is the OP_{best} in the current generation.

6.4 Case Studies

The hybrid method is firstly validated on the New England 10-machine 39-bus system [9], which is a benchmark system and we can compare results with existing methods. Besides, we also apply the hybrid method on a dynamic equivalent system of a real-world large power grid with 39 machines, 140 buses, 223 AC lines, and 4 HVDC lines [10] for demonstrating its practical application potential. For the first system, the classical generator model is considered in order to compare with the current literature; for the second system, detailed models of generators and HVDC lines are considered. In the case studies, TSA is performed on FASTEST 3.0 [4] and the conventional OPF is solved by MATPOWER 4.0 with IP solver [11].

6.4.1 New England 10-Machine 39-Bus System

Based on the system data [9] and fuel cost parameters [12], the initial optimal OP is obtained and shown in Table 6.1. Three contingencies are considered, which are indicated in Table 6.2. The stability margins with respect to the contingencies for the initial OP are also calculated and shown.

Five scenarios are conducted with TSC-OPF calculations. Three single contingency scenarios are considered: C1, C2, and C3, respectively; then two multi-contingency scenarios are considered: C1 and C2 concurrently, and C1, C2, and

Table 6.1 Initial optimal OP of New England system.

Generator	Generation cost function ($/h)	P_{max} (MW)	Generation (MW)	Total cost ($/h)
G1	$0.0193P^2 + 6.9P$	350	242.68	60920.39
G2	$0.0111P^2 + 3.7P$	650	566.43	
G3	$0.0104P^2 + 2.8P$	800	642.20	
G4	$0.0088P^2 + 4.7P$	750	629.80	
G5	$0.0128P^2 + 2.8P$	650	508.10	
G6	$0.0094P^2 + 3.7P$	750	650.55	
G7	$0.0099P^2 + 4.8P$	750	558.23	
G8	$0.0113P^2 + 3.6P$	700	535.25	
G9	$0.0071P^2 + 3.7P$	900	829.42	
G10	$0.0064P^2 + 3.9P$	1200	977.04	

Table 6.2 Contingency list for New England system.

Contingency	Fault location	Fault duration	Tripped line	Stability margin
C1	Bus 4	0.25 s	Line 4–5	−17.46
C2	Bus 21	0.16 s	Line 21–22	−32.44
C3	Bus 29	0.10 s	Line 29–26	−32.99

C3 simultaneously. The population size is set to 30, which is three times the maximum number of control variables. Meanwhile, the maximum generation number is set to 30 as the terminal condition. The stability margin, which is the TSI in the hybrid method, is restricted to [0, 10]. Since the stability margin has a quasi-linear relationship with CCT, a larger stability margin will lead to a larger CCT [5, 6]. But for redundancies here, the upper TSI limit is set to 10. To avoid multi-swing instability, the TDS time is set up to 10 seconds. The integration step is 0.01 second.

By application of the hybrid method, the generation output, $\Delta Cost$, and stability margin under the five scenarios are indicated in Table 6.3. For comparison, the reported $\Delta Cost$ in the current literature is given in the last row. It can be noticeably observed that the proposed hybrid method provides overall more economical solutions than reported results except when compared with Ref. [2] on C2, but the difference is subtle, which can be negligible. Besides, all the increased costs of the hybrid method are significantly small. This owes to the cooperation of the two optimization techniques as well as the stability margin-based constraints in searching

Table 6.3 TSC-OPF calculation results of New England System.

Contingency	C1	C2	C3	C1, C2	C1, C2, C3
G1 (MW)	244.89	246.32	245.03	250.65	252.59
G2 (MW)	570.20	572.73	570.40	576.19	576.20
G3 (MW)	646.21	648.91	646.41	656.79	660.33
G4 (MW)	634.96	638.00	634.70	647.97	652.12
G5 (MW)	494.50	513.70	511.45	494.50	494.50
G6 (MW)	655.13	621.50	655.31	623.50	623.50
G7 (MW)	562.53	531.50	562.70	532.00	532.00
G8 (MW)	539.46	541.01	540.21	548.92	552.57
G9 (MW)	807.50	838.21	788.40	807.50	788.00
G10 (MW)	983.38	987.56	983.76	1000.02	1005.59
Stability margin	8.0	9.6	9.8	9.4 (C1) 7.8 (C2)	4.5 (C1) 9.7 (C2) 8.9 (C3)
$\Delta Cost$ ($/h)	7.78	19.83	15.29	34.28	50.86
$\Delta Cost$ ($/h) in the literature	833.65 [12] 84.53 [13] 16.16 [2]	1190.0 [14] 51.74 [13] 19.19 [2]	176 [14]	1566.72 [12]	Never reported

the solutions. It is also important to note that all the contingencies have been controlled within the required stability margin range. Remarkably, the multi-contingency scenarios are tackled without encountering much more complexity and the solutions are quite economical, which are unmatched by reported results.

Figure 6.3 illustrates the generators' output change (MW) with respect to the initial OP under each scenario. It is generally found that the generations of CMs are reduced, which is consistent with previous findings [15]. However, different from analytically determining the generation shifting, the proposed hybrid is based on a standard optimization process to obtain optimal generation shifting; thus, the solutions are much more economical.

In terms of the computation efficiency, the total required computing time can be calculated by: maximum generation number × population size × (OPF solving time + TSA computing time), note that the computing time of DE is negligible. It should be indicated that the maximum generation number and population size setting are essential for computation efficiency since the computing times for OPF and TSA depend on specific tools and computation hardware as well as integration step and simulation time. In Figure 6.4, the convergence speed for five different scenarios is indicated. The OP_{best} during the evolution is also shown. It can be

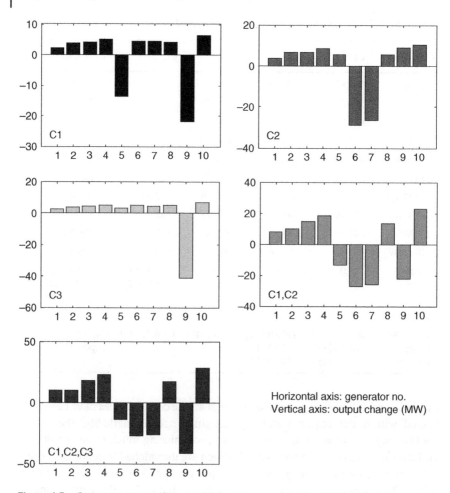

Figure 6.3 Generator output changes (MW) with respect to the initial OP.

found that the evolution converges very fast. Particularly, the single-contingency scenarios only need about 15~20 generations to achieve good enough solutions, which is much less than that of classical single EA-based methods [13, 16]. For the multi-contingency scenarios, 25~30 generations could be sufficient. This owes to the conventional OPF used in the hybrid method and the strategical selection of initial control variables. On a normal single PC with a 2.66 GHz CPU and 2 G RAM, for one OP, the OPF solution time is 0.12 second and the TSA for 10s' simulation time (integration step is 0.01 second) is 0.46~0.75 second.

In contrast with the literature [1–3], it should be noted that the number of iterations of the proposed hybrid method is relatively larger. However, it is more adaptive and flexible to multi-swing instability and multi-contingency conditions since

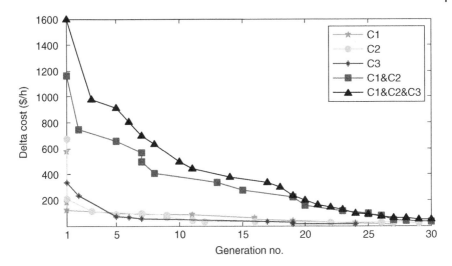

Figure 6.4 Convergence curves for the New England system test.

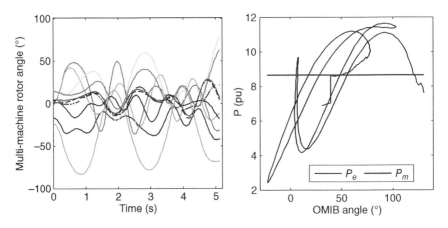

Figure 6.5 Unstable trajectories of initial OP under C1.

the complex dynamics of the system are not a concern during its computation. Particularly, a multi-swing unstable case, which occurs under C1, is illustrated here. Figure 6.5 shows the multi-machine's trajectory (left side) and corresponding P-angle plane (right side) of the initial OP under C1. It can be seen that the system is stable during its first several swings but becomes unstable after five seconds (note that all the reported methods only considered first-swing stability for this fault [2, 12, 13]). Figure 6.6 shows the trajectories of the optimal OP by our hybrid method, which remains multi-swing stable at 10 seconds.

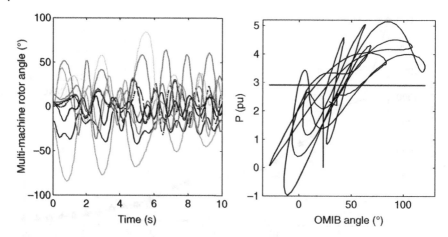

Figure 6.6 Stable trajectories of optimal OP under C1.

Another test is also conducted to compare the proposed hybrid method with Ref. [1]. To be consistent with Ref. [1], there are some modifications for the system data: the load at bus 30 is changed from 628 to 680 MW. FASTEST is used to scan all the bus three-phase faults with a fault clearing time of 0.1s on the initial OP. As consistent with [1], the fault occurs at bus 29, cleared by tripping lines 29–28, is the most severe, whose stability margin is −52.12. Table 6.4 indicates the results by applying the proposed hybrid TSC-OPF method.

It can be found that, with the same contingency and CCT, the hybrid method still provides a more economical solution.

6.4.2 39-Machine 120-Bus System

For the sake of demonstrating the practical application of the proposed hybrid method, a dynamic equivalent system of a real-world large power grid is investigated. The system has 39 machines, 120 load buses, 223 AC lines, and 4 HVDC lines, whose data are from Ref. [10] and generation cost coefficients are arbitrarily given. In the OPF model, the power transmitted through HVDC lines is kept constant. A severe three-phase fault occurring at a power exchange corridor is set and the initial CCT is 0.16 second. We require that the CCT should be increased to 0.2 second. The unstable trajectories of the initial OP under 0.2 second fault clearing time are given in Figure 6.7.

There are a total of 39 generators in the system and 23 CMs are selected as initial control variables. The population size is set to 90, and two termination criteria are adopted: when ΔCost is smaller than 1.5% of the initial OP's cost; and when the

Table 6.4 Result comparison on New England system.

Method	Proposed		Ref. [1]	
	Initial OP	Solved OP	Initial OP	Solved OP
G1 (MW)	245.22	248.95	244.03	247.89
G2 (MW)	570.82	577.11	569.82	576.89
G3 (MW)	646.88	653.62	645.97	653.41
G4 (MW)	636.81	645.07	635.24	643.31
G5 (MW)	513.35	518.97	512.12	517.81
G6 (MW)	656.08	664.08	654.87	662.66
G7 (MW)	563.42	571.15	652.54	569.79
G8 (MW)	539.26	547.65	538.12	545.22
G9 (MW)	835.54	767.50	835.07	765.65
G10 (MW)	984.38	994.80	981.84	994.87
CCT (ms)	66	101	62	101
Generation cost ($/h)	61 740.47	61 783.06	61 558.0	61 602.1
$\Delta Cost$ ($/h)	42.59		44.1	

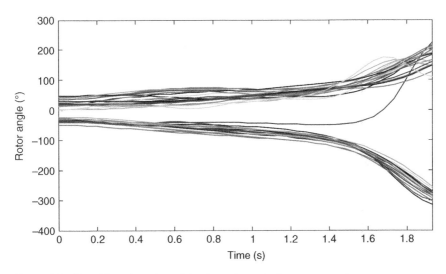

Figure 6.7 Unstable trajectories of the dynamic equivalent power system.

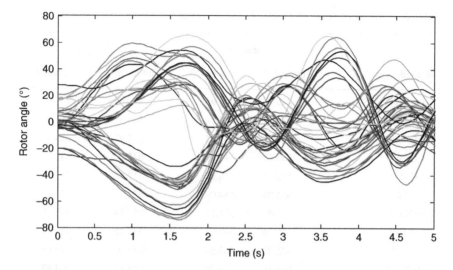

Figure 6.8 Stable trajectories of the dynamic equivalent power system.

maximum generation number 50 is reached. TSI, which is the stability margin here, is restricted to [0, 20]. Meanwhile, TDS time is set to five seconds and the integration step is 0.05 second for the TSA.

After the 41st generation, a solution whose $\Delta cost$ is 1.43% of the initial OP's cost is obtained. The CCT of this new OP is 0.214 second, which is slightly larger than the required 0.2 second due to a redundant stability margin (say, [0, 20]) restricted in the TSC-OPF. Figure 6.8 indicates the stable rotor angle trajectories of the new OP under 0.2 second fault clearing time.

For this practical large system, assuming the computation environment is the same, the OPF solution time for one OP is 0.28 second and TSA for five seconds' simulation time is 0.75~1.02 seconds.

6.4.3 Discussion

In the case studies shown above, it is demonstrated that the proposed hybrid method has a strong capability in searching global optima. On the benchmark system, the hybrid method provides more economical and efficient solutions than existing methods. In contrast with the classical single EA-based method, the required generation number is significantly reduced. Compared to analytical methods, its adaptability to multi-swing instability and multi-contingency conditions is more prominent as the complexity of system dynamics is not a concern.

Considering the dynamic equivalent system of a real-world grid, its application potential is also shown.

6.4.4 Computation Efficiency Improvement

It should be indicated that the proposed hybrid method needs a relatively larger number of iterations. Hence, subject to the OPF and TSA speed, the total computational time could be larger compared with some analytical methods such as Refs. [1–3]. However, there are a couple of strategies to speed up the computation efficiency.

Firstly, according to the parallel characteristic of EAs, distributed computation design can be employed to parallelize the TSC-OPF computation. This has already been illustrated in Refs. [13, 17]. In doing this, the whole computing time can be linearly reduced.

Secondly, it is possible to reduce the elapsed time during TSA. For example, TDS-based TSA can apply an early termination function or set a larger integration step to save simulation time. On the other hand, alternative TSA strategies such as intelligent system (IS) techniques, whose computation speed is much faster than TDS, can be employed [18].

Last but not the least, in the proposed hybrid method, all the individuals have been conducted by TSA, but it is feasible and reasonable to strategically select only a part of them for TSA so that amount of computing time can be saved [13].

Nomenclature

Indices

i	index for generators
k	index for contingency
G_0	index for initial generation

Sets

$\mathbf{g}$	set of equality constraints
$\mathbf{h}$	set of inequality constraints
$\mathbf{u}$	set of control variable
$\mathbf{u}^O$	set of outer-layer of the control variable
$\mathbf{u}^I$	set of inner-layer of the control variable
$\mathbf{x}$	set of state variable
$\mathbf{y}$	set of parameters

Variables

P_{Gi} active generation output of i-th generator
η_k transient stability margin of k-th contingency
y_i element of $\mathbf{u}^o$

Parameters

ς threshold for transient stability margin
$y_{i(H)}$, $y_{i(L)}$ lower and higher bounds of the control variables
F mutation constant
f_2 crossover constant

References

1 Pizano-Martianez, A., Fuerte-Esquivel, C.R., and Ruiz-Vega, D. (2010). Global transient stability-constrained optimal power flow using an OMIB reference trajectory. *IEEE Transactions on Power Systems* 25 (1): 392–403.

2 Zarate-Minano, R., Cutsem, T.V., Milano, F., and Conejo, A.J. (2010). Securing transient stability using time-domain simulations within an optimal power flow. *IEEE Transactions on Power Systems* 25 (1): 243–253.

3 Pizano-Martinez, A., Fuerte-Esquivel, C.R., and Ruiz-Vega, D. (2011). A new practical approach to transient stability-constrained optimal power Flow. *IEEE Transactions on Power Systems* 26 (3): 1686–1696.

4 Xue, Y. (1998). Fast analysis of stability using EEAC and simulation technologies. *POWERCON '98. 1998 International Conference on Power System Technology. Proceedings (Cat. No.98EX151)* 1: 12–16.

5 Xue, Y., Custem, T.V., and Ribbens-Pavella, M. (1989). Extended equal area criterion justifications, generalizations, applications. *IEEE Transactions on Power Systems* 4 (1): 44–52.

6 Pavella, M., Ernst, D., and Ruiz-Vega, D. (2012). *Transient Stability of Power Systems: A Unified Approach to Assessment and Control.* Springer Science & Business Media.

7 Zhu, J. (2015). *Optimization of Power System Operation.* Wiley.

8 Vesterstrom, J. and Thomsen, R. (2004). A comparative study of differential evolution, particle swarm optimization, and evolutionary algorithms on numerical benchmark problems. *Proceedings of the 2004 Congress on Evolutionary Computation (IEEE Cat. No.04TH8753)* 2: 1980–1987.

9 Pai, M. (2012). *Energy Function Analysis for Power System Stability.* Springer Science & Business Media.

10 Yan, X., Lin, G., Zhao Yang, D., and Kit Po, W. (2010). Transient stability assessment on China Southern Power Grid system with an improved pattern discovery-based method. *2010 International Conference on Power System Technology*, Hangzhou (24–28 October 2010), pp. 1–6: IEEE.

11 Zimmerman, R.D., Murillo-Sanchez, C.E., and Thomas, R.J. (2011). MATPOWER: steady-state operations, planning, and analysis tools for power systems research and education. *IEEE Transactions on Power Systems* 26 (1): 12–19.

12 Nguyen, T.B. and Pai, M.A. (2003). Dynamic security-constrained rescheduling of power systems using trajectory sensitivities. *IEEE Transactions on Power Systems* 18 (2): 848–854.

13 Cai, H.R., Chung, C.Y., and Wong, K.P. (2008). Application of differential evolution algorithm for transient stability constrained optimal power flow. *IEEE Transactions on Power Systems* 23 (2): 719–728.

14 Layden, D. and Jeyasurya, B. (2004). Integrating security constraints in optimal power flow studies. *Proc. IEEE PES. General Meeting*.

15 Ruiz-Vega, D. and Pavella, M. (2003). A comprehensive approach to transient stability control. I. Near optimal preventive control. *IEEE Transactions on Power Systems* 18 (4): 1446–1453.

16 Mo, N., Zou, Z.Y., Chan, K.W., and Pong, T.Y.G. (2007). Transient stability constrained optimal power flow using particle swarm optimisation. *IET Generation Transmission and Distribution* 1 (3): 476–483.

17 Meng, K., Dong, Z.Y., Wong, K.P. et al. (2010). Speed-up the computing efficiency of power system simulator for engineering-based power system transient stability simulations. *IET Generation Transmission and Distribution* 4 (5): 652–661. https://digital-library.theiet.org/content/journals/10.1049/iet-gtd.2009.0701.

18 Dong, Z., Xu, Y., Zhang, P., and Wong, K. (2013). Using IS to assess an electric power system's real-time stability. *IEEE Intelligent Systems* 28 (4): 60–66.

7

Data-Driven Method for Transient Stability-Constrained Optimal Power Flow

7.1 Introduction

In recent years, a promising alternative strategy, intelligent system (IS), has been recognized to surmount some deficiencies of the basic approaches and provide additional value [1, 2]. In the area of dynamic stability assessments (DSA), neural network (NN) [3], support vector machine (SVM) [4], decision tree (DT) [5], and extreme learning machine (ELM) [6] techniques have been employed with satisfactory performance. By learning from a dynamic security database, the nonlinear relationship between the power system operating parameters (input) and the corresponding security index (output) can be extracted and reformulated into an IS. In the online application phase, the system's security can be determined in real-time as soon as the input is fed. The IS-based DSA not only provides much faster online assessment speed, but also outperforms conventional approaches in terms of data requirement, generalization capacity, and extensibility [2].

Along with the well-established DSA applications, the IS technique has also shown encouraging potential in designing dynamic security controls (DSCs) [7–11]. In the literature, most prevailing methods depend on DT technology, which is to exploit its classification rules and then derive corresponding control schemes against dynamic insecurities. In Ref. [7], a DT-based online preventive control of isolated power systems is presented. In Ref. [8], DT is utilized for predicting one-shot stabilizing controls. In Ref. [9], load-shedding strategies are inferred from the inverse reading of the DT. In Ref. [10], a hybrid method using NN and DT is designed for preventive generation rescheduling. In Ref. [11], the dynamic secure boundaries are approximated by a linear combination of DT rules, which are then applied for generation rescheduling and load shedding. In contrast with conventional methods, the DT-based DSC strategies are faster, more flexible, and interpretable.

Stability-Constrained Optimization for Modern Power System Operation and Planning, First Edition. Yan Xu, Yuan Chi, and Heling Yuan.
© 2023 The Institute of Electrical and Electronics Engineers, Inc.
Published 2023 by John Wiley & Sons, Inc.

Inspired by the promising direction of IS-based DSCs, this chapter proposes two data-driven solution methods for TSC-OPF. (i) a new DT-based online preventive control strategy for transient instability prevention of power systems: Firstly, a distance-based feature estimation algorithm called *RELIEF* is applied to identify the most critical generators that are decisive for transient stability restoration. Then the generator's active outputs are used as features to construct a DT. By interpreting the DT, its splitting rules regarding transient stability restoration are formulated as inequalities and are incorporated into a conventional optimal power flow (OPF) model. By solving the OPF, the preventive control can be realized quite efficiently suitable for online applications and the cost can be minimized; (ii) a PD-based method for preventive dynamic security control of power systems: Firstly, critical operating variables regarding system's dynamic security are selected via a distance-based feature estimation process. Then, PD is performed in the space of the critical variables to extract the subtle structure knowledge called patterns. The patterns are geometrically non-overlapped hyper-rectangles, representing the system's dynamic secure/insecure regions, and can be explicitly presented to provide decision support for real-time security monitoring and situational awareness. By formulating the secure patterns into a standard OPF model, the preventive control against dynamic insecurities can be efficiently and transparently attained. Both methods are validated on New England 10-machine 39-bus test system with single- and multi-contingency cases.

7.2 Decision Tree-based Method

7.2.1 DT for Classification

DT [12] as a popular supervised automatic learning technique, is a tree-structured predictive model for classification or regression on unknown *targets* given their *features* (or *predictors*). In this section, the classification tree (CT) rather than the regression tree (RT) is employed since the transient stability is to be represented as categorical class labels, i.e. "stable" and "unstable" with respect to a contingency.

A CT contains tree nodes and branches. Branches are the connectors between nodes, which can be divided into two types: "internal node" representing a feature that characterizes an attribute of the problem, and "terminal node" denoting a class label of the problem. In structure, each "internal node" is connected with two successors and an "if-then" question is asked for reaching a successor, while the "terminal node" has no successors.

In Figure 7.1, a typical CT for power system stability assessment is illustrated. The tree is trained by a stability database (DB), which consists of 1000 instances.

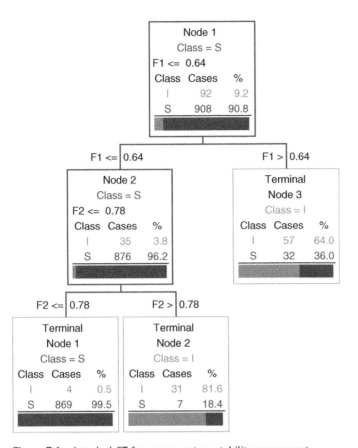

Figure 7.1 A typical CT for power system stability assessment.

For each instance, it corresponds to an OP, which is defined by power system operational features and its stability labels with respect to a contingency ("S" class for stable or "I" class for unstable). The stability assessment of an unknown OP consists of dropping its features (F1 and F2) down the tree from the root node, i.e. Node 1 at the topmost of the tree until a terminal node is reached along a path. Such a classification procedure can be done very quickly, and thus it can be used for online applications to enable earlier detection of the risk of instability.

With the development of a CT, the training data are divided into a learning set (LS) and a testing set (TS). Feature identification is the beginning for the growth of a CT, which is to recognize a proper feature as the root node that can slip the LS into distinct classes optimally. Then all the children nodes are searched again for the best split to separate the remaining subsets. The process is continued recursively until no diversity is increased by splitting the children nodes, yielding a *maximal tree*. In order to mitigate over-fitting, the maximal tree is then pruned into a series of smaller trees and evaluated on the TS with the consideration of classification accuracy. The *optimal tree* is the one giving the lowest misclassification cost on the TS. For the tree in Figure 7.1, there are two internal nodes (where Node 1 is the root node) and three terminal nodes.

During the tree induction, the selection of nodes is according to the measurement of the information content (entropy-based) that a feature can provide. A feature with high information content tends to better partition the instances into homogenous regions. Generally, *information gain, gain ratio*, and *Gini* are popular algorithms to evaluate the features in DT growing [13]. More technical details regarding DTs are introduced in Refs. [12, 13].

Compared with other IS approaches, DT can provide explicit classification rules, namely, nodes and their thresholds and associated paths. Based on these rules, the stability assessment can be transparent and interpretable. Moreover, the rules provide insight into critical system operating variables relating to stability, i.e. the variables shown at tree nodes. In the subsequent section, control schemes are designed for instability prevention via appropriately deriving and interpreting the tree rules.

7.2.2 Preventive Control Strategy Based on DT

On the basis of DT, an online preventive control strategy for transient instability prevention of power systems is designed and introduced in this section.

7.2.2.1 Problem Description

As Eq. (5.5) indicates, transient stability strongly depends on active power outputs of generators. In practice, generation rescheduling is usually an effective method for transient stability controls. The procedure is to modify the OP by shifting active power output from some generators to others so that it can ensure the transient stability subjected to potential contingencies. The kernel problems are the direction and the amount that generation rescheduling shall undergo.

As a machine learning technique, DT is promising to provide the generation rescheduling strategy for transient stability enhancement. In Ref. [11], an oblique DT is developed and the stability region is defined as the linear combination of the features provided by the tree. Stability control is achieved by driving an unstable OP to the stable region. However, due to the high nonlinear characteristics of the problem, the sensitivity of the inferred stability region with respect to the variation of a tree is quite high. Besides, the transparency is lost to a large extent when the tree is oblique. In this section, we designed a more effective and transparent strategy: given a well-developed tree, by interpreting the tree rules, active power generation can be shifted from "unstable" terminals to "stable" terminals, along a branch.

Taking the DT in Figure 7.1 for example, assume an unknown OP is followed along the path "Node 1→Node 2→Terminal Node 2," yielding an "unstable" conclusion, if the value of Node 2 of this OP can be manipulated toward "Terminal Node 1," an alternative stability condition, say "stable" can be obtained. By this way, stability control can be achieved. Since this way is very straightforward and efficient, online application for providing transparent control is possible and feasible.

7.2.2.2 Feature Space Selection-Critical Generators Identification

As mentioned previously, the DT technique is based on information evaluation criteria to select tree nodes. The features with higher information contents are selected as tree nodes. However, the information-based algorithms poorly illustrate the distinguishing capability of features in distance space, which is important for the effectiveness and economy of the generation rescheduling in the context of transient stability control. Besides, in order to avoid complicated structure of the tree, which could be difficult for people to interpret, fewer features to construct the tree are strongly required.

In this section, a distance-based feature estimation algorithm, called *RELIEF*, is adopted to identify critical generators. The critical generators are then used to grow a tree. *RELIEF* statistically evaluates the quality of features based on how well their values distinguish among instances near each other. It not only considers the difference in features' values and classes but also the distance between the instances. Hence, the similar instances will be close while dissimilar ones will be separated based on the good attributes. Tree growing can benefit from this property as homogenous instances tend to be gathered while dissimilar instances tend to be separated by the good features.

The principle of the *RELIEF* is to iteratively update the weight of each feature as below [14]:

$$W[A] = W[A] - diff(A, R_i, H)/m + diff(A, R_i, M)/m, i \in m \tag{7.1}$$

where A represents a feature, R_i is the instance sampled in this iteration, H denotes the nearest instance from the same class as R_i (nearest hit), while M is the nearest instance from the different class as R_i (nearest miss), and m is the amount of sampled instances guaranteeing all weights are in the interval $[-1, 1]$. Function $diff(A, I1, I2)$ is to calculate the difference between the values of feature A for two instances $I1$ and $I2$:

- For discrete feature

$$diff(A, I1, I2) = \begin{cases} 0; & value(A, I1) = value(A, I2) \\ 1; & \text{otherwise} \end{cases} \tag{7.2}$$

- For continuous feature

$$diff(A, I1, I2) = \frac{|value(A, I1) - value(A, I2)|}{\max(A) - \min(A)} \tag{7.3}$$

To handle noises, incomplete data, and multi-class problems, *RELIEF*-F [15], which is an extension of *RELIEF* with weight updating algorithm, is shown below:

$$W[A] = W[A] - \sum_{j=1}^{k} diff\left(A, R_i, H_j\right)/(m \cdot k)$$

$$+ \sum_{C \neq class(R_i)} \left[\frac{P(C)}{1 - P(class(R_i))} \cdot \sum_{j=1}^{k} diff\left(A, R_i, M_j(C)\right) \right]/(m \cdot k) \tag{7.4}$$

where C denotes a class label, $P(\cdot)$ is the prior probability of a class, and k is a user-defined parameter.

Different from basic *RELIEF* that finds the nearest one hit and miss, *RELIEF*-F finds k nearest hits and misses to average their contribution for updating the weight, and consequently, the risk of miss estimation is reduced, and the introduction of $P(\cdot)$ leads to estimating the ability to separate each pair of classes.

For regression problems, i.e. the class of the instance is continuous, rather than judging whether two instances belong to the same class or not (absolute difference), the probability of difference is used [16]. This probability is modeled with the relative distance between the predicted class values of the two instances. Function $diff(A, I1, I2)$ is reformulated using the probability that predicts values of two instances:

$$P_{diff\,A} = P(\text{diff.value of } A \mid \text{nearest instance}) \tag{7.5}$$

$$P_{diff\,C} = P(\text{diff.prediction} \mid \text{nearest instance}) \tag{7.6}$$

$$P_{diff\,C \mid diff\,A} = P(\text{diff.prediction} \mid \text{diff.value of } A \\ \text{and nearest instance}) \tag{7.7}$$

Then $W[A]$ in (7.3) can be calculated based on *Bayes* rule:

$$W[A] = \frac{P_{diffC \mid diffA} P_{diffA}}{P_{diffC}} - \frac{\left(1 - P_{diffC \mid diffA}\right) P_{diffA}}{1 - P_{diffC}} \tag{7.8}$$

In this section, the generators with higher *RELIEF*-F evaluation weights can be selected as features for DT construction.

7.2.2.3 Generation Rescheduling Within OPF

To minimize the overall cost of preventive control, the generation rescheduling can be realized by conventional OPF. The detailed OPF model is introduced in Section 4.3.

To improve the transient stability, active power outputs of critical generators can be restricted in (4.6). According to the DT splitting rules, the new OP can be obtained through conventional OPF. By this way, the preventive control can be effectively achieved. In the meantime, the cost can be minimized as (4.4) shows.

7.2.2.4 Computation Process

The DT splitting rules applied for preventive controls can be constructed as follows:

$$R_C = \{T \in S : N_i \leq \eta_i, i \in \vartheta\} \tag{7.9}$$

where R_C represents the tree growing for contingency C, T denotes the terminal nodes of a tree, S means "stable" class, N_i and η_i are the node i and the corresponding threshold, respectively, ϑ denotes the critical generator set.

The rules should be prepared off-line via automatic tree growing and the online computation process of the preventive control is as follows:

1) Conduct TSA on current OP subject to contingency C, if it is "stable," stop the computation; otherwise, go to Step 2;
2) Modify tree splitting rules (7.9) as $R_C = \{T \in S : N_i \leq d_i \times \eta_i, i \in \vartheta\}$, and set $d_i = 1$;
3) Substitute the modified splitting rules into OPF model, i.e. substitute (7.9) into (4.6);
4) Solve the OPF, if it is convergent, obtain a new OP and go to Step 5; otherwise, go to Step 6;
5) Conduct TSA on the new OP subject to contingency C, if it is "stable," stop the computation; otherwise, go to the next step;
6) Update $d_i = d_i + \varepsilon$ and return to Step 3.

It should be emphasized that ε in Step 6 is a user-defined parameter, it is used to increase the generation shifting in case the transient stability can't be ensured from the original tree threshold η.

7.3 Pattern Discovery-based Method

As introduced in Section 7.2, the DT-based DSC strategies are faster, more flexible, and interpretable in contrast with conventional approaches. However, DT induction is a *supervised* learning process that fully depends on the prior knowledge of the data for tree growing. Under some special conditions, for example, class imbalance, the tree may become sensitive to subtle changes in the data. Besides, when a tree is growing complexly or deeply in structure, it is necessary to prune the tree in some subjective sense so that explicit rules can be obtained for use.

Under a similar knowledge extraction and utilization framework, a more robust statistical learning technique, called PD [17] is employed to find a preventive control method for transient instability prevention. Different from DT, PD belongs to *unsupervised* learning. It statistically discovers the hidden structure in a database and provides objective, transparent, and interpretable knowledge called *patterns* for specific use [17–19]. The *patterns* are geometrically non-overlapped hyperrectangles in *Euclidean* space, which are easy to present and interpret. When discovered in a power system critical feature space, the stable/unstable regions can be represented by them. Meanwhile, The *patterns* can provide decision support for real-time stability monitoring and situational awareness. By explicitly formulating the patterns into a standard OPF model, the preventive control can be realized transparently and efficiently.

7.3.1 Pattern Discovery

7.3.1.1 Overview
In Refs. [17, 18], PD was first studied as an efficient knowledge extraction technique. It can discover nonlinear and multimodal patterns of high order at a fast speed and rank them based on their statistical significance for comparison, interpretation, and assessment. Therefore, a better decision can be made according to a better understanding of the data. In academic and engineering fields, PD has been successfully applied to solve plentiful statistical learning problems and alleviate DRIP (data rich information poor) embarrassment effectively [17–19]. In previous work, PD has been applied for rule-based DSA [20, 21].

Based on the argument that the knowledge for a certain class (or group) is the significant event associations inherent in the data of that class or group, PD aims to find the statistically significant subset in *Euclidean* space in an unbiased and exhaustive manner [17]. In general, residual analysis and optimization are the two elements in the discovery process. The former recognizes significant organized information by statistically scaling the degree of the difference between actual occurrence and expected occurrence of instances, and the latter is to search out all the subtle information in the space. The fundamentals and computation procedure of PD are introduced as follows, and more theoretical details and proofs could be found in Refs. [17, 18].

7.3.1.2 Key Definition
Assume a continuous data set Ω in the N-dimensional *Euclidean* space $\mathfrak{R}^N$, let $\mathbf{X} = \{X_1, X_2, ..., X_N\}$ represent its feature set, and each feature X_i, $1 \leq i \leq N$, takes

on values from its domain d_i, $d_i \subset \mathfrak{R}$, the following definitions are made for PD [17, 18].

Event: an event, E, is a *Borel* subset [22] of $\mathfrak{R}^N$, while a *Borel* subset geometrically forms a N-dimensional hyper-rectangle in $\mathfrak{R}^N$, defined by

$$E = I_1 \times I_2 \times \cdots \times I_N = \{\mathbf{X} : X_i \in I_i, 1 \leq i \leq N\} \tag{7.10}$$

where $I_i = (a_i, b_i]$ is a one-dimensional semi-closed interval along the i-th feature, $-\infty < a_i < b_i < \infty$.

Volume: the volume of an event, v, is the hyper-volume occupied by the *Borel* subset. Let L_i represent the length of the i-th interval I_i of event E, $L_i = |b_i - a_i|$, the volume of E is calculated by

$$v(E) = \prod_{i=1}^{N} L_i \tag{7.11}$$

Observed frequency: the observed frequency of an event E, o_E, is the actual number of instances that fall inside the volume occupied by E.

Pattern: a pattern means a statistically significant event. Let $\vartheta(\cdot)$ be a test statistic relating to a specified discovery criterion c and θ_c^{α} be the critical value of the statistical test at a significant level of α. An event E is regarded to be significant, i.e. a pattern, if it satisfies the following condition:

$$\vartheta(E) \geq \theta_c^{\alpha} \tag{7.12}$$

Residual: since the statistic $\vartheta(\cdot)$ is to test the significance of the pattern candidates [17, 18], the residual of an event E is the difference between its actual occurrence, i.e. observed frequency, and its expected occurrence:

$$\delta_E = o_E - e_E \tag{7.13}$$

where e_E is the expected occurrence, or expected frequency, under the pre-assumed model estimated by the given data set.

7.3.1.3 PD by Residual Analysis and Recursive Partitioning

It is typical that the estimation of the expected frequency e_E is under uniform random distribution [18] (other criteria for this are also available [17]). It implies that if instances are randomly distributed in the space, there is no significant structure information in the space. Equivalently, if the residual of an event is large, which means the difference between observed frequency and uniform random distribution is large, there will be more structure information in it. According to the above definitions, PD can be considered as an optimization problem since it searches all the significant events in the instance space [17].

Here, patterns are discovered by adopting a residual analysis combined with a recursive partitioning procedure [18]. The procedure contains recursively partitioning the instance space with residual evaluation of each hyper-rectangle, until all the significant events, i.e. patterns, are recognized. The following describe its main computation steps:

1) The instance space is divided into Q^N events, where N is the number of feature dimensions and Q is the number of partitions for each feature.
2) Boundaries of each event are refined by adjusting the event boundaries to coincide with the minimum and maximum coordinates of the contained events. By this way, the events can be non-overlapped on boundaries except ones that are really overlapped.
3) The residual value of each event is calculated by the following Eq. [18]:

$$\hat{r}_j = \frac{\left(n_{1j} - n_{2j}\right)}{2 \times c_j^{1/2}} \tag{7.14}$$

where $\hat{r}_j$ is the residual of the event E_j, n_{1j} denotes the count of the observed contained samples of event j, n_{2j} is the expected frequency of event j, and c_j is the estimated asymptotic variance of the numerator calculated by [18]

$$c_j = \frac{1}{4}\left(n_{1j} + n_{2j}\right)\left(1 - \frac{\left(n_{1j} + n_{2j}\right)}{2 \times n_{1+}}\right) \tag{7.15}$$

where n_{1+} is the total count of the instances in the whole space.

4) The significance of each event is evaluated based on the following criterion: at the $\alpha \times 100$ percent significant level, an event is *significant* if $\hat{r}_j \geq z_{1-\alpha/2}$; is *negative significant* if $\hat{r}_j \leq z_{\alpha/2}$; is insignificant if $|\hat{r}_j| < z_{1-\alpha/2}$, where Z_α is the value of the standard normal deviation, $Z \sim N(0,1)$, such that $P(Z \leq Z_\alpha) = \alpha$.
5) For every significant event, Steps 1–4 are repeated to obtain a more accurate description of the pattern, until termination conditions [18] are met.

In Figure 7.2, an example of PD in a two-dimensional *sine* distribution instance space is indicated. The upper window shows the first partition, and the lower window shows the last partition. Obviously, the structure information in the instance space has been well identified as the rectangles.

7.3.2 Preventive Control Strategy Based on PD

7.3.2.1 General Description
In this section, a preventive control method for dynamic security is developed by PD technique. In the offline stage, a critical generator feature space is firstly selected from a dynamic security database by means of a distance-based feature estimation procedure. Then, PD is conducted in the feature space to extract

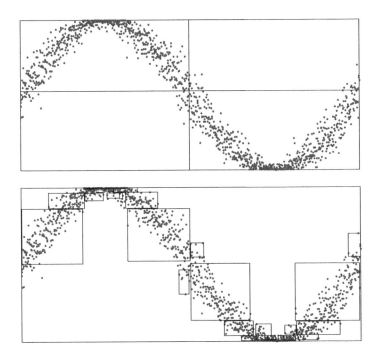

Figure 7.2 An example of PD by residual analysis and recursive partitioning.

patterns comprehensively and unbiasedly. The unions of these patterns represent the dynamic secure/insecure regions of power systems. Therefore, based on the secure/insecure regions obtained during the offline stage, in the online stage, decision support for real-time dynamic security monitoring and situational awareness can be provided. Once an insecure OP is detected through DSA, the preventive control can be activated by driving the insecure OP into the secure region. To be systematic and economical, the patterns are formulated as generator output constraints into a classical OPF. Then, the new OPF problem can be solved to obtain a transparent and efficient preventive control scheme.

7.3.2.2 Feature Space Selection-Critical Generator Identification

For transient stability control, generator active power outputs are also selected as the features. It is necessary to select critical generators as feature space for knowledge discovery. Because (i) only a subset of generators is responsible for the loss of synchronism [23, 24], and (ii) the irrelevant generators have very little knowledge and may generate noise that impedes the knowledge extraction.

Here, a distance-based feature estimation algorithm, *RELIEF* [14, 15] is also applied to identify the critical generator features. Typically, the features having the largest weights should be selected as critical features. The details are introduced in Section 7.2.2.2.

7.3.2.3 Pattern-based Dynamic Secure/Insecure Regions

Based on the critical generator feature space, PD is conducted to extract patterns. For each pattern, a security class label is determined based on the possibility of different instances occurring in the pattern. For pattern E, its class label can be determined by the following rule:

$$
\begin{cases}
\dfrac{M_S}{M_S + M_I} > \lambda \rightarrow \text{"secure"} \\[2mm]
\dfrac{M_I}{M_S + M_I} \geq \lambda \rightarrow \text{"insecure"}
\end{cases}
\tag{7.16}
$$

where M_S and M_I represent the number of "secure" and "insecure" instances in E, respectively, and λ denotes the threshold to measure the occurrence possibility of a class in a pattern (50% is set here).

After assigned security labels, the patterns are the structure knowledge of the power system transient stability characteristics in the critical generator feature space. The secure and insecure regions under the contingency, $\mathbf{R}^S$ and $\mathbf{R}^I$, can be represented by the unions of corresponding patterns:

$$
\mathbf{R}^S = \bigcup_{i=1}^{K} E_i^S \quad \mathbf{R}^I = \bigcup_{j=1}^{J} E_j^I
\tag{7.17}
$$

where E_i^S and E_j^I represent the i-th secure and j-th insecure patterns, respectively.

Since the forms of patterns are explicitly mathematical and geometrical, rule-based decision support for dynamic security monitoring and situational awareness can be provided. Typically, an OP located in an insecure region would indicate an insecure status. When the dimension of the feature space is low, for example, two or three, the monitoring can be visualized.

7.3.2.4 Computation Process

For preventive control of an insecure OP, generation rescheduling is activated to drive it to the pattern-based secure region. By constructing a set of inequality constraints based on patterns and incorporating the constraints into a classical OPF, the generation rescheduling can be calculated economically. The detailed OPF model is introduced in Section 4.3.

Specifically, preventive control is to formulate active generation output limits defined by a secure pattern into (4.6) and solve (4.4)–(4.9). It should be noted that the embedded pattern to the OPF does not provide additional complexity for OPF solving; thus any classical programming algorithm can be applied. Since there may be many separate patterns in the secure region, it is necessary to decide which pattern to use in the OPF. Here, the nearest secure pattern in the distance space could be firstly selected to avoid over-control. However, the optimality cannot be guaranteed by the term of "nearest" after OPF computations since the problem is nonlinear. Namely, the obtained OP may not be completely secured, because the nearest secure pattern cannot provide sufficient security margin. In this situation, a DSA must be conducted to verify the obtained OP. If the new OP is still insecure, another secure OP pattern should be used.

Figure 7.3 indicates the flowchart of the complete computation procedure. By applying a contingency on the initial OP, once it is determined to be insecure

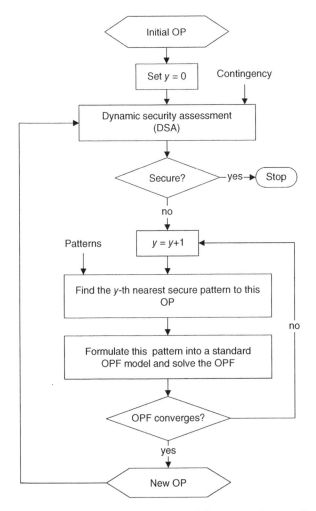

Figure 7.3 Computation flowchart of the proposed preventive control.

by DSA, its first-nearest secure pattern is constructed into the OPF, and a new OP is obtained via the OPF solution. If the new OP is secure, the computation is terminated; otherwise, the next-nearest secure pattern is used. The whole process will repeat until a secure OP is obtained. The term "*y*-th nearest" is determined by ranking the *Euclidean* distance between the OP and the centroids of the patterns. The *Euclidean* distance between two points R and R' in an N-dimensional space is calculated by

$$D(R, R') = D(R', R) = \sqrt{\sum_{i=1}^{N} (X_i - X_i')^2} \tag{7.18}$$

where X_i and X_i' denote the values of the i-th feature of instance R and R', respectively.

The centroid of a pattern is calculated by

$$\tau_i = \frac{X_{i1} + X_{i2} + \cdots + X_{in}}{n} \tag{7.19}$$

where τ_i denotes the value of the i-th feature of the centroid, and X_{in} denotes the value of the i-th feature of the n-th instance in the pattern. It should be noted that the features in distance calculation need to be normalized into the same range, e.g. [0, 1].

7.4 Case Studies

7.4.1 Test System and Simulation Software

The studied data-driven preventive control methods are validated on New England 10-machine 39-bus test system. In Figure 7.4, the one-line diagram of the test system is indicated. The system data are obtained from Ref. [25], and the fuel cost parameters are taken from Ref. [23].

In the simulation, power flow and OPF are solved by MATPOWER 4.0 package [26], TSA is conducted by TDS applying PSS/E software under distributed computation architecture [27], DT is developed using CART 6.0 package [28], and the PD is realized in JAVA programming platform.

In Table 7.1, the initial OP on the test system is illustrated, which is solved by OPF without preventive controls. The total active power load is 6097.1 MW, and the total generation cost is 60 920.39 $/h.

In this chapter, a continuous transient stability index (TSI) is applied to represent the transient stability level of an OP subjected to a contingency [11]:

$$TSI = \frac{360 - |\delta_{max}|}{360 + |\delta_{max}|} \times 100 \tag{7.20}$$

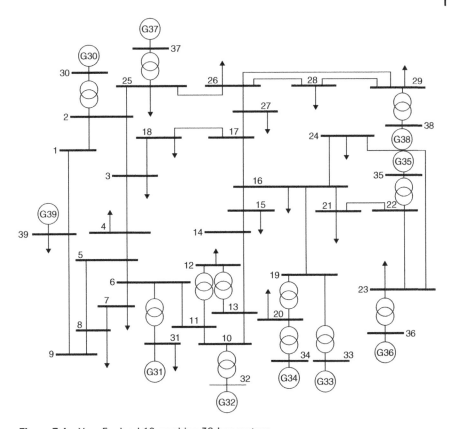

Figure 7.4 New England 10-machine 39-bus system.

Table 7.1 Initial operating point of 10-machine 39-bus system.

Generator	Active power output (MW)	Total cost ($/h)
G30	242.68	
G31	566.43	
G32	642.20	
G33	629.80	
G34	508.10	
G35	650.55	60 920.39
G36	558.23	
G37	535.25	
G38	829.42	
G39	977.04	

where δ_{max} is the maximum angle deviation between any two generators during the transient period (this chapter sets 4 seconds). If $TSI > 0$, the OP is defined as "stable," otherwise, the OP is "unstable."

7.4.2 Database Preparation

Dynamic security database will have a strong impact on the efficiency of the proposed data-driven methods. It contains a large number of instances, each associating a pre-contingency OP and the corresponding TSI [6]. The OP is characterized by features, such as steady-state operating parameters; the TSI can be a discrete class label, e.g. "stable" or "unstable," or a continuous value, e.g. stability margin, with respect to a contingency.

To be effective, the representative operating region should be illustrated by the database. Practically, the database can be obtained from historical DSA archives and/or be generated through offline simulations. To generate a database, provided the forecasted load levels are known, various OPs can be calculated according to the knowledge of generation scheduling information [6]. The number of OPs required for satisfactory performance on a specific power system could be experimentally determined. Under the pre-defined contingencies, the transient stability indices of the produced OPs can be calculated through TDS.

7.4.3 DT-based Method and Numerical Simulation Results

7.4.3.1 Database Generation

A severe three-phase fault (called Fault 1) occurs at bus 4 and is cleared by tripping line 4–5 after 0.3 s. For the initial OP, the TSI under this contingency is -72.88, and as indicated in Figure 7.5, the rotor angle curve is showing that one of the generators loses synchronism with others under this contingency. It should be noted that in Figure 7.5 the relative rotor angles are portrayed in the center of inertia (COI) framework, this will be also followed in later simulations.

In order to generate a stability database to train a DT, a large number of OPs should be acquired and simulated for the test system. Firstly, various OPs are acquired by solving power flow under each load/generation pattern, which is obtained by varying the load and generator outputs for various combinations in a neighboring range to the initial OP. Then, the obtained OPs are simulated vis TDS subjected to the same contingency to check their stability status under the same criterion (7.20). Finally, 1000 instances are obtained in total, where 231 are "stable" and 769 are "unstable."

It should be noted that the database generation should be finished offline prior to the online implementation of the method. Systematical schemes for database generation could also be found in Ref. [6].

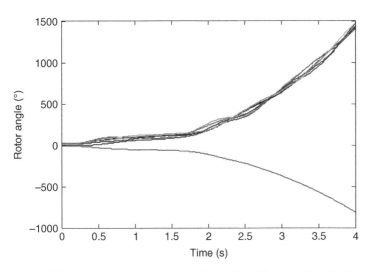

Figure 7.5 Rotor angle swing curves of the initial OP under Fault 1 without preventive control.

7.4.3.2 Critical Generators

RELIEF-F algorithm is used to evaluate the 10 generator active power features, G30 to G39, of the test system, and the weight of each generator is shown in Figure 7.6.

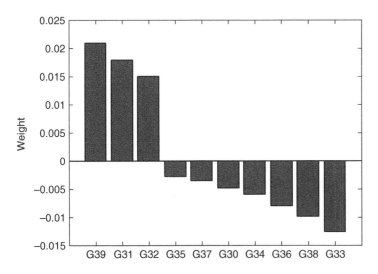

Figure 7.6 Weight of each generator feature under Fault 1.

In Figure 7.6, it can be observed that G39, G31, and G32 have positive weight, indicating they have positive instance distinguishing capability, which means they have the most impact on the transient stability characteristic under Fault 1. Hence, the three generators are selected as critical features to train the DT.

On the contrary, the other features have negative weight, which means that they cannot discriminate the instances in the distance space, and thus have a subtle impact on the stability.

To better understand the concept and show the effectiveness of the *RELIEF* algorithm, 2 two-dimensional distance spaces are depicted in Figure 7.7, where the

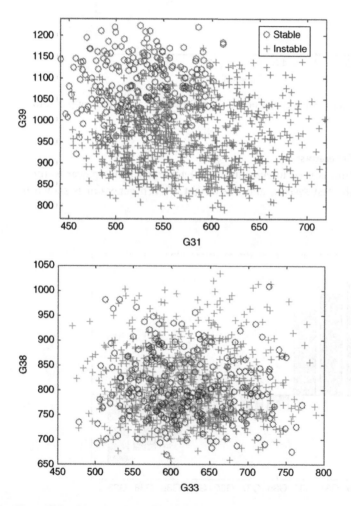

Figure 7.7 Critical and noncritical generator distance spaces.

upper window consists of the first two critical features while the lower window consists of the last two noncritical features. It is obvious that in the upper window, the OPs are marginally separated, but in the lower window, they are significantly overlapped. Therefore, from Figure 7.7, it is concluded that tuning the critical features can directly change the system stability status.

7.4.3.3 DT and Rules

Based on the tree critical generator features, a CT is developed by applying algorithms integrated in CART package, *Gini* is selected as the splitting method, and 10-folder cross-validation criterion is selected for testing and pruning the tree.

Figure 7.8 shows the optimal tree, where PG39 and PG31 denote the active power output of G39 and G31.

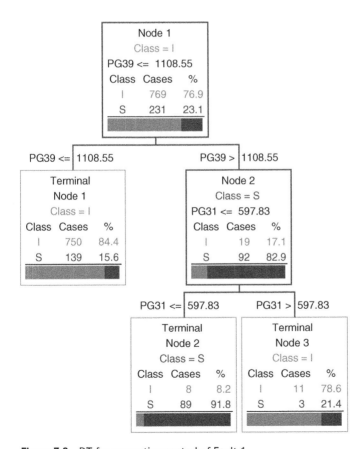

Figure 7.8 DT for preventive control of Fault 1.

Based on the tree, the preventive control rules are constructed as below:

$$R_{Fault1} = \{P_{G39} > 1108.55 \text{ and } P_{G31} \leq 597.83\} \tag{7.21}$$

7.4.3.4 Single-contingency Controls
By applying the proposed DT-based preventive control method on the initial OP, a new, stable OP is obtained after one iteration. The new OP is shown in Table 7.2 and its rotor angle swing curves are shown in Figure 7.9.

Table 7.2 New stable OP (under Fault 1).

Generator	Active power output (MW)	Total cost ($/h)
G30	233.04	
G31	550.09	
G32	625.20	
G33	612.91	
G34	496.55	
G35	634.22	61 057.51
G36	542.96	
G37	522.27	
G38	810.20	
G39	1108.55	

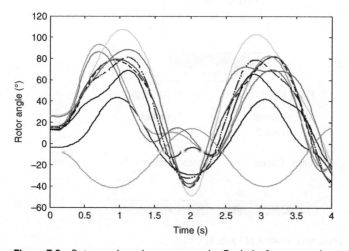

Figure 7.9 Rotor angle swing curves under Fault 1 after preventive control.

From Table 7.2, the critical generators' active power outputs are restricted within the DT rules; correspondingly, Figure 7.9 shows that the new OP can maintain transient stability under Fault 1, and its *TSI* is 41.8. An additional generation cost 137.12 $/h is required due to the stability constraints for Fault 1.

7.4.3.5 Multi-contingency Controls

Most prevailing research only considers one single contingency for transient stability preventive control [9, 11]. However, in practice, it is necessary to include more than one contingency. Typically, the most possible and harmful contingencies that lead to instability should be controlled. In order to achieve multi-contingency control for the proposed DT-based method, the rules for each contingency are separately derived and then incorporate these rules simultaneously in the OPF. To better indicate this, another contingency is defined (named Fault 2), which is a three-phase short-circuit fault at bus 28 and is cleared by tripping line 28–29 after 0.12 second. Under this contingency, the initial OP is also unstable as the TSI is −91.6 and the rotor angle swing curves are shown in Figure 7.10.

The feature estimation result is given in Figure 7.11 and accordingly, G38 is identified as a critical generator. Meanwhile, Figure 7.12 shows the corresponding developed DT.

The rule under Fault 2 can be determined as simply as below:

$$R_{Fault2} = \{P_{G38} \leq 706.44\} \tag{7.22}$$

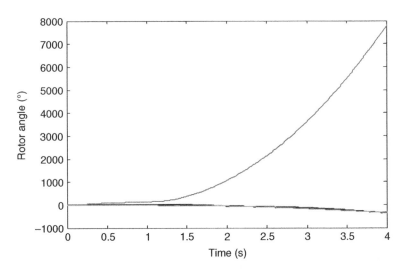

Figure 7.10 Rotor angle swing curves of the initial OP under Fault 2 without preventive control.

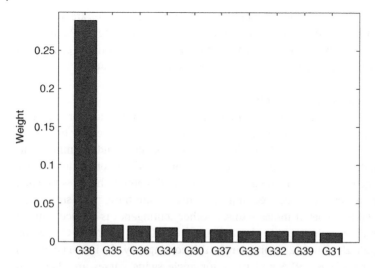

Figure 7.11 Weight of each generator feature under Fault 2.

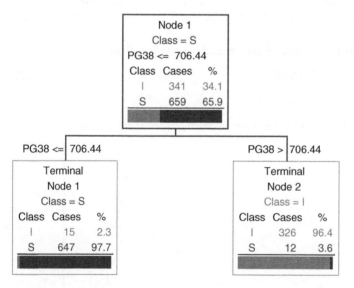

Figure 7.12 DT for preventive control of Fault 2.

Therefore, it is convenient to handle the multi-contingency by incorporating the rules (7.21) and (7.22) simultaneously in the OPF model. During the computation, TSA is conducted on the two contingencies and both contingencies should be stable. After one iteration, the new, stable OP is obtained for two contingencies. The

Table 7.3 New stable OP (under multi-contingency).

Generator	Active power output (MW)	Total cost ($/h)
G30	240.48	
G31	562.69	
G32	638.59	
G33	627.98	
G34	506.85	
G35	648.81	61 160.53
G36	556.61	
G37	536.78	
G38	706.44	
G39	1108.55	

new OP results, including generation output and costs, are shown in Table 7.3. Meanwhile, the rotor angle swing curves under the two contingencies are indicated in Figure 7.13.

It is obvious that the new OP can withstand both contingencies and maintain transient stability. The *TSIs* under the two contingencies are 38.7 and 62.4, respectively. The generation cost increases by 240.14 $/h due to the stability requirement for the two contingencies.

7.4.3.6 Conclusive Remarks

The simulation results clearly verify the effectiveness of the proposed DT-based method. In contrast with the conventional analytical methods, the DT-based stability control is explicit, transparent, and interpretable.

In the online implementation of the control strategy, the computational burden is acceptable as OPF is solved during the process, which normally only spends several seconds. The OPF computation time is about 0.17 second here. Meanwhile, a TSA is conducted to check if the new OP is stable, which only costs 1.3 second for one TSA using TDS. It should be noted that in this case study, only one iteration is required to obtain the stable OP. Therefore, the online implementation is quite efficient. The major computational burden of the method is offline database generation, where a large number of OPs are simulated, and then TSAs are performed on them. It should be mentioned that feature evaluation and DT development are quite computationally efficient. Nevertheless, all of this can be performed at the offline stage, which is not time-critical. Moreover, distributed computing techniques [27] can be employed to increase the computation speed.

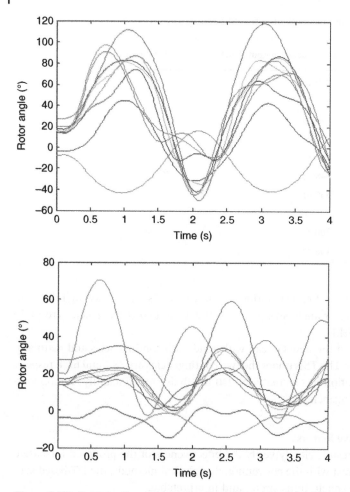

Figure 7.13 Rotor angle swing curves after multi-contingency preventive control (upper window-Fault 1, lower window-Fault2).

7.4.4 PD-based Method and Numerical Simulation Results

7.4.4.1 Database Generation

Fault 3 is defined as a three-phase fault that occurs at bus 22 and is cleared by tripping line 22–21 after 0.14 second. As shown in Figure 7.14, the system becomes unstable at the initial OP under this contingency. The TSI is −90.7.

In order to generate a database, the P and Q demands at each load bus are varied randomly within the ±3% range of the initial OP. Then, under each load distribution, a large number of OPs with various generation scenarios are generated in two ways: (i) the generation outputs for each generator are randomly varied and OPs

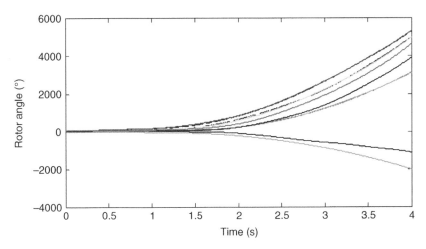

Figure 7.14 Rotor angle swing curves of the initial OP under Fault 3.

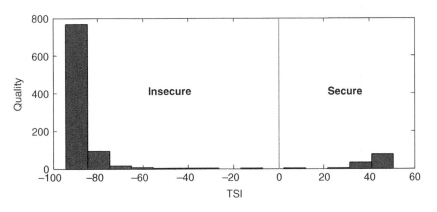

Figure 7.15 Histogram of TSI distribution of the generated database (Fault 3).

are obtained by solving the power flow model; (ii) the generation cost coefficient for each generator are randomly altered and OPs are obtained by solving OPF model. Note that only converged cases are saved for use. By this way, 1000 OPs are generated and their TSIs under Fault 3 are then obtained vis DSA. In Figure 7.15, the TSI distribution of the OPs is indicated, where 113 are "secure" and 887 are "insecure."

7.4.4.2 Critical Generators
The 10 generator features are evaluated by the regressive *RELIEF* algorithm. The weight of each feature is indicated in Figure 7.16, where it is obvious that generators G36 and G35 are much larger than others in weight, showing they have a

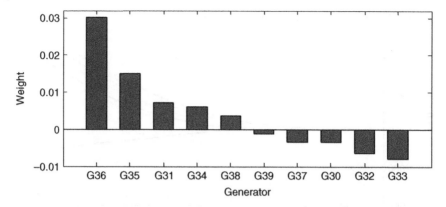

Figure 7.16 *RELIEF* feature estimation result (Fault 3).

strong impact on transient stability under Fault 3. Hence, G36 and G35 are selected to formulate the critical feature space for PD. Meanwhile, it can be also found that some generators have negative weights, which means that they have negative distinguishing capabilities, namely, they are mixing rather than separating the instances in the space.

In order to better understand and illustrate the concept and effectiveness of the *RELIEF* algorithm, the critical feature space formed by G36 and G35 and a noncritical feature space constructed by G32 and G33 are visually compared. Figure 7.17 portrays the noncritical feature space, and Figure 7.18 (note that the patterns are also indicated as the black rectangles) depicts the critical feature space. It can be clearly observed that in Figure 7.17 the OPs of different classes are heavily overlapped, which means no (or very little) structural information exists in the space. On the contrary, the OPs in Figure 7.18 are significantly separated, which provides adequate structural knowledge about the security characteristics.

7.4.4.3 Pattern Discovery Results

According to Figure 7.18, there are a total of 28 events discovered in the critical feature space. Among them, 27 are assessed as significant, i.e. patterns. These 27 patterns encompass 996 instances in total and occupy 75% volume of the whole instance space. As shown in Figure 7.18, non-overlapped rectangles are constructed by the patterns in the critical generator space.

As shown in Figure 7.19, The patterns are assigned security labels and their centroids are calculated, where "+" and "o" mean the centroid of insecure and secure patterns, respectively. From the figure, it can be seen that 4 patterns are secure and 23 are insecure.

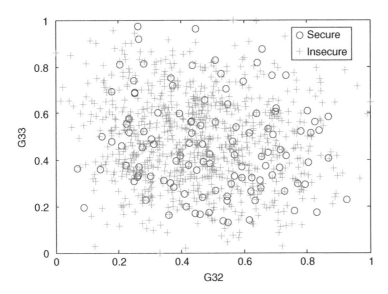

Figure 7.17 Noncritical feature space constructed by G32 and G33.

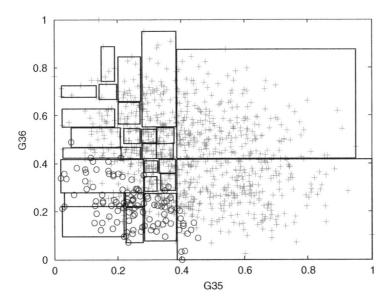

Figure 7.18 Different classes of instances and discovered patterns.

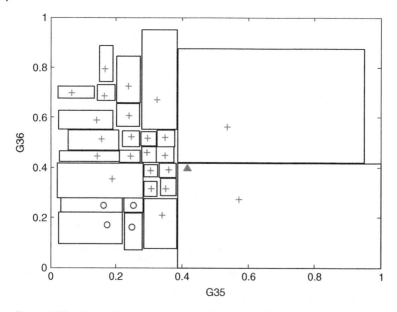

Figure 7.19 Secure/insecure regions and corresponding centroids.

7.4.4.4 Preventive Control Results

Under Fault 3, the obtained patterns can represent the corresponding secure/insecure regions. Meanwhile, the feature space can illustrate a visual dynamic security monitoring as it is two-dimensional. For example, indicated in Figure 7.19, the initial OP (marked as the red triangle) is in the insecure region, which is consistent with its actual security status. To ensure the security of the system subjected to contingencies, preventive control should be applied to drive it to the secure region. According to the computation process shown in Figure 7.3, the nearest pattern of the initial OP is applied first, whose region is $(0.2245, 0.2826] \times (0.2223, 0.2787]$ in the space of G35 and G36 (normalized value). By formulating them into the OPF model and solving it, the 1st new OP is obtained. After DSA, the 1st new OP is still insecure, as indicated in Figure 7.20. However, the TSI of the 1st new OP becomes -67.4, which has been significantly improved compared with the initial OP. Then, the next nearest secure pattern is applied, yielding the 2nd new OP. After DSA, it is shown that this OP becomes secure with 40.5 of the TSI. In Figure 7.21, the rotor angles are indicated. For this new secure OP, its generation cost is 61 003.14 $/h. An additional 82.75 $/h is spent on the preventive control strategy.

In order to further validate the proposed PD-based control method, 4 secure patterns are used to control all the 887 generated insecure OPs. All of them become secured by the control method, which further verifies the effectiveness of the secure patterns. Results show that 14 OPs are secured by their 1st secure pattern,

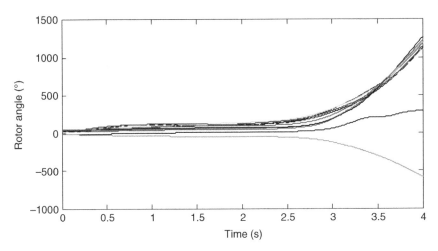

Figure 7.20 Rotor angle swing curves of the 1st new OP under Fault 3.

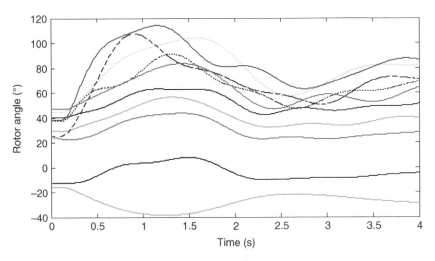

Figure 7.21 Rotor angle swing curves of the 2nd new OP under Fault 3.

737 OPs by their 2nd nearest secure pattern, 108 OPs by their 3rd nearest secure pattern, and 28 OPs by their 4th nearest secure pattern.

7.4.4.5 Multi-contingency Controls

Multi-contingency is also studied in the PD-based method. The kernel of the method is to find a general critical feature set for all the faults, and the patterns extracted in the general critical feature space can be the collective secure/insecure regions under the faults. Then the multiple faults can be simultaneously controlled in a single computation run.

To do this, another contingency, Fault 4, is considered, which is a three-phase fault at bus 28 cleared by tripping line 28–29 after 0.10 second. Under Fault 4, the TSI of the initial OP is −93.1. Meanwhile, there are 236 secure cases and 764 insecure cases in the database. By taking the average value of the TSIs for Fault 3 and Fault 4, a general TSI is defined as the composite security measure under the two faults, which means the two faults can be simultaneously considered for estimating features. Regressive *RELIEF* is used to evaluate the generator features under the general TSI. From Figure 7.22, it can be seen that G38, G36, and G35 should be selected as critical generators for the two faults. While G36 and G35 are critical features of Fault 3, as indicated previously, which indicates that G38 should be the critical feature of Fault 4. Noticeably, it is consistent with the feature estimation result considering only Fault 4 shown in Figure 7.23.

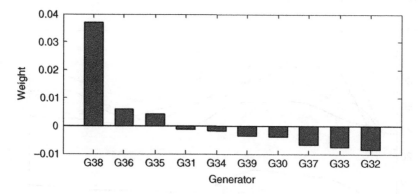

Figure 7.22 *RELIEF* feature estimation result (Fault 3 and Fault 4).

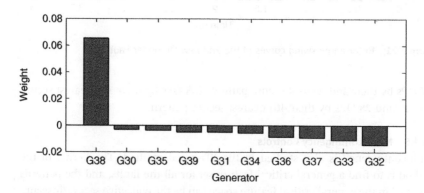

Figure 7.23 *RELIEF* feature estimation result (Fault 4).

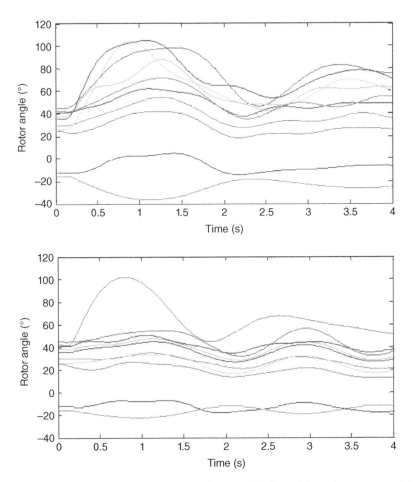

Figure 7.24 Rotor angle swing curves of the new OP for multi-contingency control (upper window-Fault 3, lower window-Fault 4).

In the space of G38, G36, and G35, there are 22 patterns, where 3 are secure and 19 are insecure. The union of the secure patterns represents the collective secure region under the two faults. After the 1st nearest secure pattern is applied, a new secure OP could be obtained and their TSIs with respect to Fault 3 and Fault 4 are 45.6 and 48.2, respectively. With this new OP, the generation cost is 61 184.2 $/h. In Figure 7.24, the rotor angle swing curves are indicated.

7.4.4.6 Discussion
Based on the discovered patterns, the initial OP can be stabilized for both single and multi-contingency conditions. The control mechanism of the method is transparent and interpretable. Besides, its computational efficiency is high.

In addition, it can be observed that the acquired secure patterns have a significant impact on the success of the control. From the case studies, most generated OPs are insecure due to the two quite severe contingencies, which cause there are only several secure patterns could be discovered: 4 for Fault 3 and 3 for Fault 4. In real applications, there are many ways that can enrich the secure patterns, such as generating more OPs or using historical DSA archives. By doing this, the quality of the PD-based method can be further improved. Since the PD process is quite efficient, the database enrichment/updating can be performed online using up-to-date operating information.

Nomenclature

Indices

i index for features
j index for events
n index for instances

Sets

ϑ sets of critical generators
$\mathbf{X}$ feature set

Variables

A feature
R_i instance sampled in this iteration
H nearest instance from the same class as R_i (nearest hit)
M nearest instance from the different class as R_i (nearest miss)
m amount of sampled instance
R_C tree growing for contingency C
T terminal nodes of a tree
S "stable" class
E a event, *Borel* subset of $\mathfrak{R}^N$
I_i one-dimensional semi-closed interval along the i-th feature
v volume of an event
L_i length of the i-th interval I_i of event E
O_E observed frequency
$\hat{r}_j$ residual of the event E_j
n_{1j} count of the observed contained samples of event j

n_{2j} expected frequency of event j
c_j estimated asymptotic variance of the numerator
$E_i^S E_j^I$ the i-th secure and j-th insecure patterns
τ_i value of the i-th feature of the centroid
X_{in} value of the i-th feature of the n-th instance in the pattern
δ_{max} maximum angle deviation between any two generators during the transient period
TSI   transient stability index

Parameters

$P(\cdot)$ prior probability of a class
C class label
k user-defined parameter.
N_i, η_i node i and the corresponding threshold
ε user-defined parameter
$\vartheta(\cdot)$ test statistic
c specified discovery criterion
θ_c^α critical value of the statistical test at a significant level of α
e_E expected occurrence, or expected frequency
N number of feature dimension
Q number of partitions for each feature
n_{1+} total count of the instances in the whole space
Z_α value of the standard normal deviation
M_S/M_I number of "secure" and "insecure" instances in E
λ threshold to measure the occurrence possibility of a class in a pattern

References

1 Morison, K. (2006). On-line dynamic security assessment using intelligent systems. In: *2006 IEEE Power Engineering Society General Meeting*. IEEE 5 pp, https://doi.org/10.1109/PES.2006.1709501.
2 Dong, Z., Xu, Y., Zhang, P., and Wong, K. (2013). Using IS to assess an electric power system's real-time stability. *IEEE Intelligent Systems* 28 (4): 60–66.
3 Mansour, Y., Vaahedi, E., and El-Sharkawi, M.A. (1997). Dynamic security contingency screening and ranking using neural networks. *IEEE Transactions on Neural Networks* 8 (4): 942–950.

4 Moulin, L.S., Silva, A.P.A. d., El-Sharkawi, M.A., and Marks, R.J. (2004). Support vector machines for transient stability analysis of large-scale power systems. *IEEE Transactions on Power Systems* 19 (2): 818–825.

5 Sun, K., Likhate, S., Vittal, V. et al. (2007). An online dynamic security assessment scheme using phasor measurements and decision trees. *IEEE Transactions on Power Systems* 22 (4): 1935–1943.

6 Xu, Y., Dong, Z.Y., Meng, K. et al. (2011). Real-time transient stability assessment model using extreme learning machine. *IET Generation, Transmission & Distribution* 5 (3): 314–322. https://digital-library.theiet.org/content/journals/10.1049/iet-gtd.2010.0355.

7 Karapidakis, E.S. and Hatziargyriou, N.D. (2002). Online preventive dynamic security of isolated power systems using decision trees. *IEEE Transactions on Power Systems* 17 (2): 297–304.

8 Mei, K. and Rovnyak, S.M. (2004). Response-based decision trees to trigger one-shot stabilizing control. *IEEE Transactions on Power Systems* 19 (1): 531–537.

9 Voumvoulakis, E.M. and Hatziargyriou, N.D. (2008). Decision trees-aided self-organized maps for corrective dynamic security. *IEEE Transactions on Power Systems* 23 (2): 622–630.

10 Niazi, K.R., Arora, C.M., and Surana, S.L. (2004). A hybrid approach for security evaluation and preventive control of power systems. In: *IEEE Power Engineering Society General Meeting, 2004*, vol. 1, 1061–1067. IEEE https://doi.org/10.1109/PECON.2003.1437442.

11 Genc, I., Diao, R., Vittal, V. et al. (2010). Decision tree-based preventive and corrective control applications for dynamic security enhancement in power systems. *IEEE Transactions on Power Systems* 25 (3): 1611–1619.

12 Breiman, L., Friedman, J.H., Olshen, R.A., and Stone, C.J. (2017). *Classification and Regression Trees*. Routledge.

13 Han, J., Pei, J., and Kamber, M. (2011). *Data Mining: Concepts and Techniques*. Elsevier.

14 Kira, K. and Rendell, L.A. (1992). A practical approach to feature selection. In: *Machine Learning Proceedings 1992* (ed. D. Sleeman and P. Edwards), 249–256. San Francisco, CA: Morgan Kaufmann.

15 Kononenko, I. (1994). Estimating attributes: analysis and extensions of RELIEF. In: *Machine Learning: ECML-94*, Berlin, Heidelberg (ed. F. Bergadano and L. De Raedt), 171–182. Springer Berlin Heidelberg.

16 Robnik-Šikonja, M. and Kononenko, I. (2003). Theoretical and empirical analysis of ReliefF and RReliefF. *Machine Learning* 53 (1): 23–69.

17 Wong, A.K.C. and Yang, W. (2003). Pattern discovery: a data driven approach to decision support. *IEEE Transactions on Systems, Man, and Cybernetics, Part C (Applications and Reviews)* 33 (1): 114–124.

18 Chau, T. and Wong, A.K.C. (1999). Pattern discovery by residual analysis and recursive partitioning. *IEEE Transactions on Knowledge and Data Engineering* 11 (6): 833–852.

19 *Pattern Discovery Technologies Inc., Discover*e*. https://www.patterndiscovery.com/technology/discovere/ (accessed 27 February 2011)

20 Wang, T. w., Guan, L., and Zhang, Y. (2008). A modified pattern recognition algorithm and its application in power system transient stability assessment. In: *2008 IEEE Power and Energy Society General Meeting - Conversion and Delivery of Electrical Energy in the 21st Century*, 1–7. IEEE https://doi.org/10.1109/PES.2008.4596483.

21 Yan, X., Lin, G., Zhao Yang, D., and Kit Po, W. (2010). Transient stability assessment on China Southern Power Grid system with an improved pattern discovery-based method. In: *2010 International Conference on Power System Technology*, 1–6. IEEE https://doi.org/10.1109/POWERCON.2010.5666045.

22 Port, S.C. (1994). *Theoretical Probability for Applications*. Wiley.

23 Nguyen, T.B. and Pai, M.A. (2003). Dynamic security-constrained rescheduling of power systems using trajectory sensitivities. *IEEE Transactions on Power Systems* 18 (2): 848–854.

24 Ruiz-Vega, D. and Pavella, M. (2003). A comprehensive approach to transient stability control. I. Near optimal preventive control. *IEEE Transactions on Power Systems* 18 (4): 1446–1453.

25 Pai, M. (2012). *Energy Function Analysis for Power System Stability*. Springer Science & Business Media.

26 Zimmerman, R.D., Murillo-Sanchez, C.E., and Thomas, R.J. (2011). MATPOWER: steady-state operations, planning, and analysis tools for power Systems research and education. *IEEE Transactions on Power Systems* 26 (1): 12–19.

27 Meng, K., Dong, Z.Y., Wong, K.P. et al. (2010). Speed-up the computing efficiency of power system simulator for engineering-based power system transient stability simulations. *IET Generation, Transmission & Distribution* 4 (5): 652–661. https://digital-library.theiet.org/content/journals/10.1049/iet-gtd.2009.0701.

28 Salford Systems, *CART*. http://www.salfordsystems.com/cart.php (accessed 21 October 2010).

8

Transient Stability-Constrained Unit Commitment (TSCUC)

8.1 Introduction

As an effective tool to clear day-ahead electricity markets, a unit commitment (UC) program aims to determine the generators operating states, including their turn on/off statuses and power outputs, with the objective of minimizing the total production cost while meeting prevailing operational limits [1]. Conventionally, the prevailing constraints include power balance, unit minimum on/off time limits, ramping up/down limits, and system spinning reserve requirements. State-of-the-art methods for solving a UC are *Lagrangian* relaxation (LR) [1] and mixed-integer programming (MIP) [2].

In recent years, due to rapid load growth and unmatched infrastructure investments, power systems are being pushed to operate near their security limits. Therefore, it is quite necessary to consider the security constraints in the UC, that is, security-constrained UC (SCUC) problems [3]. Conventionally, SCUC aims to satisfy the *static security* criteria including network power flow limits and voltage limits at steady states (before and/or after a contingency). Presently, SCUC can be effectively solved by decomposition-based methods, such as Benders decomposition (BD) [3]. As a necessary extension of UC, SCUC has become an essential tool to balance the economy and security requirements in day-head electricity market operations.

While the static security requirements can be soundly satisfied in SCUC, today's power systems are also facing significant risk of *dynamic insecurity*. Previous investigations [4, 5], illustrated that lacking adequate dynamic performance is one key driving force for cascading failures and widespread blackouts around the world. As one essential dynamic performance criterion, *transient stability* is necessary to be considered. However, conventional SCUC only considers static security, it may not

Stability-Constrained Optimization for Modern Power System Operation and Planning,
First Edition. Yan Xu, Yuan Chi, and Heling Yuan.
© 2023 The Institute of Electrical and Electronics Engineers, Inc.
Published 2023 by John Wiley & Sons, Inc.

meet the transient stability requirement. To further enhance the system security and reduce the risk of blackouts, it is sensible and imperative to include transient stability in the day-ahead generation dispatch stage, which yields the transient stability-constrained UC (TSCUC) problem studied in this chapter.

Although very important, very limited work on TSCUC has been reported due to the tremendous complexity of the problem. On one hand, SCUC is a large-scale, multi-stage, mixed nonlinear integer programming problem. On the other hand, transient stability study usually calls for a vast number of differential-algebraic equations (DAEs) that are intractable for current programming algorithms.

To the best of our knowledge, Ref. [6] is the first and sole work on this topic to date (as of year 2014). In Ref. [6], the authors propose an augmented LR method combined with variable duplication technique to solve the TSCUC model. The whole problem is divided into a basic UC subproblem and a transient stability-constrained optimal power flow (TSC-OPF) subproblem. The first subproblem is solved by dynamic programming and the second by a reduced-space interior point method.

Although effective, the method reported in Ref. [6] can suffer from several practical limitations. First, the transient stability constraint is formulated as DAEs, and time-domain numerical discretization is applied to convert the differential equations into the algebraic form. It is clear that this would result in a dramatic explosion of the problem size proportional to the number of integration time steps times the number of generators. The length of the whole integration period is usually arbitrarily selected. Although a reduced-space interior method is used to solve the TSC-OPF model, the overall problem dimension and computational burden remain extensive, making parallel computing a necessity for implementing the method. Second, the transient stability is constrained by a rotor angle limit index, i.e. the maximum rotor angle deviation of each generator against the center of inertia (COI) during the transient period bounded by a predefined threshold. However, it has been widely shown that this threshold is usually system dependent and not easy to define: when it is set to a small value, the operation tends to be conservative and less economic; while if it is too relaxed, the transient stability may not be ensured [7, 8]. Meanwhile, it provides very little information about the system stability degree, which is an important metric for system operators.

8.2 TSC-UC model

The traditional UC model is introduced in Section 4.3, which is (4.27)–(4.33). In addition to the UC model, the TSC-UC model also consists of (i) steady-state security constraints and (ii) transient stability constraints.

8.2.1 Steady-State Security Constraints

Since this chapter mainly concerns the transient stability criterion, for the steady-state security, a DC network is applied and the transmission flow limits are considered as shown below:

$$
\begin{aligned}
-F_l^{\max} \leq F_{lt,k} = \sum_{i=1}^{NG} PTDF_{lt,i}^k \cdot P_{it} \\
- \sum_{j=1}^{ND} PTDF_{lt,j}^k \cdot D_{jt} \leq F_l^{\max}, \quad \forall t, \forall l, \forall k
\end{aligned}
\tag{8.1}
$$

where $k = 0$ denotes the base case, and $k = 1, 2, ..., n$ means k-th contingency case, $PTDF_{lt,i}^k$ is the power transfer distribution factor of bus i to line l for contingency k at period t, and D_{it} is the load demand at bus i.

Note that AC network constraints, including bus voltage limits [9], can also be used for the introduced method.

8.2.2 Transient Stability Constraints

Similar to the TSC-OPF, the transient stability constraints can also be modeled as a stability margin that is larger than zero for each contingency:

$$
0 < \eta_{t,k} \leq \varepsilon, \quad \forall t, \forall k
\tag{8.2}
$$

$\eta_{t,k}$ is calculated through EEAC method. It should be noted that a larger stability margin will have a more conservative OP and thus will lead to a higher operating cost. Hence, the stability margin should be limited within a small threshold ε according to practical needs.

By applying (8.2) to the stability constraints in the TSCUC model, the inherent instability mechanism and accuracy can be well maintained. In the meantime, it can provide the opportunity for decomposing the TSCUC model in a BD manner with high efficiency.

8.3 Transient Stability Control

Transient stability control (TSC) can be realized based on the EEAC and trajectory sensitivities. In the literature, it has been used for efficiently solving preventive TSC and TSC-OPF problems [8, 10–16]. In this section, an example of preventive TSC is derived and constructed in terms of EEAC and sensitivities.

For preventive actions, stabilizing an unstable system consists of modifying the pre-contingency conditions until the stability margin η becomes zero (or positive).

This can be achieved by increasing the decelerating area A_{dec} and/or decreasing the accelerating area A_{acc} of the OMIB $P - \delta$ representation. In practice, this can be realized by decreasing the OMIB mechanical power $P_m(t_0)$, that is:

$$\Delta P_m(t_0) = \frac{M}{M_C} \cdot \Delta P_C - \frac{M}{M_N} \cdot \Delta P_N \tag{8.3}$$

where t_0 denotes the pre-contingency state, ΔP_C and ΔP_N are the changes in the total power of CMs and NMs, respectively:

$$\Delta P_C = \sum_{i \in C} \Delta P_{mi}(t_0); \quad \Delta P_N = \sum_{j \in N} \Delta P_{mj}(t_0) \tag{8.4}$$

To maintain the power balance, the following condition should be satisfied while ignoring power loss:

$$\Delta P_N = -\Delta P_C \tag{8.5}$$

Substituting (8.5) into (8.3), we have:

$$\Delta P_m(t_0) = \left[\frac{M}{M_C} + \frac{M}{M_N} \right] \cdot \Delta P_C = -\left[\frac{M}{M_C} + \frac{M}{M_N} \right] \cdot \Delta P_N \tag{8.6}$$

Equations (8.3)–(8.6) illustrate that by shifting the active power output of CMs to NMs, the transient stability can be restored [7, 11, 17, 18].

Meanwhile, numerous examples have reported a quasi-linear relationship between changes in stability margin and OMIB mechanical power at pre-contingency state [7, 8, 11, 17, 18], that is:

$$\Delta \eta = \tau \cdot \Delta P_m(t_0) \tag{8.7}$$

where τ is the approximate linear sensitivity of the stability margin with respect to generation change.

In practice, the sensitivity value around the operating point n is numerically estimated via two successive EEAC runs:

$$\tau_n = \frac{\Delta \eta_{(n-2)} - \Delta \eta_{(n-1)}}{\Delta P_m(t_0)_{(n-2)} - \Delta P_m(t_0)_{(n-1)}} \tag{8.8}$$

With τ_n, the required generation shifting for TSC can be analytically calculated. Specifically, to control an unstable case, whose stability margin is $\eta_{us}(\eta_{us} < 0)$, if the desired stability margin is $\varepsilon(\varepsilon \geq 0)$, the required increment in stability margin is $\Delta \eta \geq -\eta_{us} + \varepsilon$. Combining (8.6) and (8.8), the required generation shifting between CMs and NMs can be calculated as:

$$\Delta P_C \geq \left(\frac{-\eta_{us} + \varepsilon}{\tau_n} \right) \cdot \left(\frac{M}{M_C} + \frac{M}{M_N} \right)^{-1} \tag{8.9}$$

Meanwhile, EEAC reveals that multiple contingencies having common instability modes, i.e. common CMs, can be simultaneously stabilized. This is to apply the most constraining power shifting, imposed by the severest contingency, to these common CMs only [7, 11, 17, 18]. Hence, the computational efforts for multi-contingency can be remarkably reduced.

8.4 Decomposition-based Solution Approach

8.4.1 Decomposition Strategy

To solve SCUC problems, there are many mature approaches. Specifically, decomposition-based methods such as BD have been widely applied. The principle of the method is to decompose the whole problem into a master problem and a set of slave subproblems. As for the master problem, it is a UC problem, which is to determine the commitment and dispatch of the units without considering the network security. After the UC results are obtained from the master problem, the slave subproblems check the security constraints. If there is a violation, Benders cuts (or similarities) will be generated and added to the master problem. The whole problem is iteratively solved between the master and slave problems until there is no violation. By using the BD method, the large-scale problem can be decomposed into a set of small and tractable problems, which remarkably reduces the complexity of the whole problem and can enable parallel computation to decrease the total calculation time.

In this chapter, the TSCUC problem is solved by a similar decomposition strategy. The description of the proposed decomposition strategy is depicted in Figure 8.1.

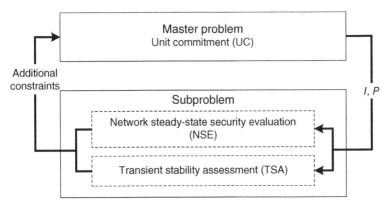

Figure 8.1 Decomposition strategy for TSCUC.

Let **x** represent the UC status I and generation dispatch P, and **y** represent the system state variables. The TSCUC problem can be rewritten as the following standard BD form:

$$\min \quad \mathbf{c}^T \mathbf{x} \tag{8.10}$$

$$\text{s.t.} \quad \mathbf{A}\mathbf{x} \geq \mathbf{b} \tag{8.11}$$

$$\mathbf{E}\mathbf{x} + \mathbf{F}\mathbf{y} \geq \mathbf{h} \tag{8.12}$$

$$0 \leq \eta \leq \varepsilon \tag{8.13}$$

where (8.10) corresponds to the cost function (4.27), (8.11) corresponds to the operational constraints (4.29)–(4.33) and the additional constraints generated from the subproblem, (8.12) corresponds to the network steady-state security constraints (8.1), and (8.13) corresponds to the transient stability constraints (8.2).

8.4.2 Master Problem

The master problem is to solve the UC model, i.e. (8.10)–(8.11). Then, the commitment I and generation dispatch P can be determined.

In the initial calculation, constraint (8.11) only includes the operational constraints (4.29)–(4.33), and the network steady-state security constraints and transient stability constraints are not considered. After the subproblems are solved, additional constraints are generated (if applicable) and added to (8.11) in subsequent iterative calculations to alleviate the security and transient stability violations.

It has been known that LR and MIP are two mature methods to solve a UC. A comprehensive comparison and discussion of the two methods for UC can be found in Ref. [3]. In this chapter, the MIP method is applied due to the availability of high-performance commercial MIP packages today, such as CPLEX [19] and GUROBI [20]. Most ISOs in the US are switching to the MIP method for UC calculation [21]. However, it should be indicated that the LR method can be used for the proposed approach here as well.

In order to apply the MIP, the nonlinear UC model should be converted into a tractable form, i.e. mixed-integer linear programming (MILP) for standard MIP solvers. In this chapter, a computationally efficient MILP formulation introduced in Ref. [2] is employed. This MILP formulation requires fewer binary variables and constraints than other reported models; thus, it can significantly save computational time.

8.4.3 Subproblem

The subproblem assesses the hourly network steady-state security and the transient stability of the solution $\hat{x}$ from the master UC problem. If the solution is violated, it will generate additional constraints: Benders cut and stabilization cut.

8.4.3.1 Network Steady-State Security Evaluation (NSE)

The NSE refers to both the base case and contingency cases. For each case, a linear programming (LP) model is built [3]:

$$\min \quad \nu(\hat{\mathbf{x}}) = \mathbf{1}^T \mathbf{s} \tag{8.14}$$

$$\text{s.t.} \quad \mathbf{Fy} + \mathbf{s} \geq \mathbf{h} - \mathbf{E}\hat{\mathbf{x}}, \quad \boldsymbol{\pi} \tag{8.15}$$

where $\mathbf{1}$ is the vector of ones, $\mathbf{s}$ is the slack vector used to check the violation of line flow constraints, and $\boldsymbol{\pi}$ is the *Lagrangian* multiplier vector of inequality constraints in (8.15). $\nu(\hat{\mathbf{x}}) > 0$ denotes the violation occurs, and the Benders cut is generated as:

$$\nu(\mathbf{x}) = \nu(\hat{\mathbf{x}}) - \boldsymbol{\pi}^T \mathbf{E}(\mathbf{x} - \hat{\mathbf{x}}) \leq 0 \tag{8.16}$$

$\boldsymbol{\pi}^T \mathbf{E}$ mathematically denotes the marginal decrement or increment of the objective function (8.14) when $\mathbf{x}$ is adjusted. In the subsequent iteration, (8.16) will be added to (8.11) of the master problem to mitigate the steady-state security violation.

8.4.3.2 Transient Stability Assessment (TSA)

For each contingency case, TSA is conducted through EEAC. The left side of the constraint (8.13) is evaluated: if the stability margin is negative, the stabilization cut, i.e. required generation shifting between critical machines and noncritical machines, is generated.

Moreover, to simultaneously stabilize multiple contingencies with the least number of additional constraints, the contingencies can be grouped based on their resulting instability modes. The contingencies having a common instability mode are divided into one group. Each contingency group is then represented by the severest contingency, which is the one with the smallest stability margin in that group. For each representative contingency, the stabilization cut is generated as:

$$\sum_{i \in C} P_{mi}(t_0) - \sum_{i \in C} \hat{P}_{mi}(t_0) \geq \frac{-\eta_{us} + \varepsilon}{\varsigma_n} \cdot \left[M \cdot (M_C)^{-1} + M \cdot (M_N)^{-1} \right]^{-1}$$

$$\tag{8.17}$$

where $\hat{P}_{mi}(t_0)$ denotes the generation output of unit i, which is obtained from the master problem. The comprehensive stabilization cut derivation can be found in Section 8.3.

The stabilization cut (8.17) takes information about how the generation dispatch should be shifted to preventively ensure transient stability.

Note that the derived stabilization cut is in the same mathematical form as the Benders cut, which means both of them are linear constraints. Therefore, they can be integrated together and added to the master problem. By doing this, the programming size of the problem can be reduced to as small as a traditional SCUC.

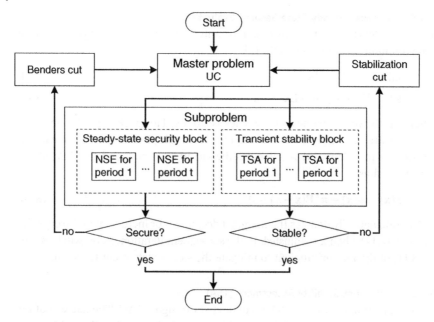

Figure 8.2 Computation flowchart.

8.4.3.3 Solution Procedure

In Figure 8.2, the computation flowchart of the TSCUC is depicted. The detailed steps are as follows.

Step 1	Solve the master problem (8.10)–(8.11) using MILP.
Step 2	When the UC solution is obtained from the master problem, perform the hourly NSE and hourly TSA for each contingency.
Step 3	If all the contingencies are both steady-state secure and transient stable, stop; otherwise, go to Step 4.
Step 4	For the steady-state security block, each insecure contingency will generate the Benders cut (8.16) by solving (8.14)–(8.15); for the transient stability block, each representative unstable contingency will generate the stabilization cut (8.17) based on the procedure introduced in Section 8.3.
Step 5	Add the generated Benders cut and stabilization cut to the master problem and return to Step 1.

8.5 Case Studies

8.5.1 Implementation of the Approach

The simulation is conducted on an ordinary 64-bit PC with 3.10 GHz CPU and 4.0 GB RAM. TDS is performed on the industrial-grade commercial power system

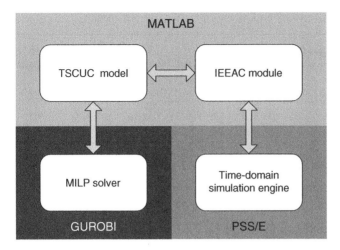

Figure 8.3 Implementation of the introduced approach.

simulation package PSS/E [22], and the EEAC algorithm is realized in the MATLAB platform. The PSS/E and MATLAB are connected by an interface in our previous work [23]. The optimization model is solved in MATLAB by invoking the commercial solver GUROBI [20], which has a built-in interface with MATLAB. In Figure 8.3, the implementation of the proposed approach is schematically presented.

8.5.2 New England 10-Machine 39-Bus System

The introduced approach is first conducted on the New England 10-machine system. The system network data and machine dynamic parameters are obtained from Ref. [24], and the 10-unit UC data are obtained from Ref. [25]. In order to accommodate the base loading level of the system, UC data have been modified with the increment of three times for the total load demand, generation limits, and ramping limits. Then, the total load is proportionally distributed to the load buses according to the base loading level.

8.5.2.1 SCUC Results

For comparison purposes, the SCUC model is first solved without considering the transient stability constraints. The total generation cost is US$ 1 590 910.6, and the detailed unit status and generation dispatch are given in Table 8.1 and Figure 8.4, respectively.

8.5.2.2 TSCUC Results-Single-contingency Case

A three-phase short-circuit at bus 21 is considered as C1. After 0.12 second, it is cleared. Under this contingency, it can be found that 17 out of 24 hours of the

Table 8.1 SCUC/TSCUC results (New England 10-machine 39-bus system).

Hour	Unit on/off status (G30–G39)										η	CMs
1	1	1	0	1	0	1	1	1	1	1	32.2	—
2	1	1	0	1	0	1	1	1	1	1	31.3	—
3	1	1	0	1	0	1	1	1	1	1	13.0	—
4	1	1	0	1	0	1	1	1	1	1	8.0	—
5	1	1	0	1	0	1	1	1	1	1	−11.6	31
6	1	1	0	1	0	1	1	1	1	1	−50.4	31
7	1	1	1	1	0	1	1	1	1	1	−12.9	31
8	1	1	1	1	0	1	1	1	1	1	−22.0	31
9	1	1	1	1	1	1	1	1	1	1	−29.9	31
10	1	1	1	1	1	1	1	1	1	1	−29.2	31
11	1	1	1	1	1	1	1	1	1	1	−28.5	31
12	1	1	1	1	1	1	1	1	1	1	−28.5	31
13	1	1	1	1	1	1	1	1	0	0	−37.4	31
14	1	1	1	1	1	1	1	1	0	0	−98.6	31
15	1	1	1	1	1	1	1	1	0	0	−34.3	31
16	1	1	1	1	1	1	1	1	0	0	5.7	—
17	1	1	1	1	1	1	1	1	0	0	12.1	—
18	1	1	1	1	1	1	1	1	0	0	−12.0	31
19	1	1	1	1	1	1	1	1	0	0	−29.1	31
20	1	1	1	1	1	1	1	1	0	0	−37.2	31
21	1	1	1	1	1	1	0	1	0	0	−100	31
22	1	1	1	1	1	0	0	0	0	0	−39.4	31, 32
23	1	1	0	1	1	0	0	0	0	0	−54.7	31, 30
24	1	1	0	1	1	0	0	0	0	0	6.6	—

SCUC solution are transiently unstable, although they are all steady-state security. The transient stability margin and CMs are also given in Table 8.1.

In order to comprehensively illustrate the phenomenon, we select the system multi-machine rotor angle trajectories and the corresponding OMIB $P - \delta$ curve of the 5th and 1st dispatch hours, which are shown in Figure 8.5 and Figure 8.6, respectively. In Figure 8.5, there is one CM, which is unit G31, and the system losses synchronism at 0.63 second. Hence, the TDS can be early terminated at that time. In Figure 8.6, the "time to first-swing stability" is 0.4 second; at this time, the TDS can be early terminated.

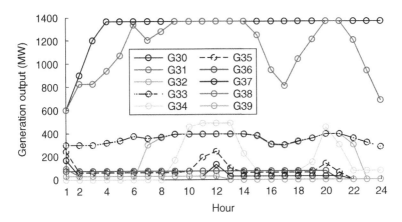

Figure 8.4 SCUC generation dispatch.

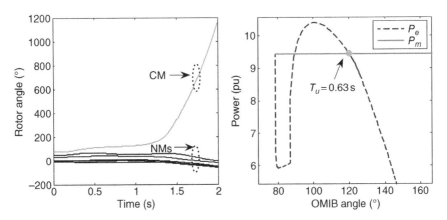

Figure 8.5 Multi-machine rotor angle trajectories (left window) and the corresponding OMIB $P - \delta$ representation (right window) – an unstable case for the New England 10-machine system at 5th dispatch.

Under C1, the TSCUC model is solved by the proposed approach. To avoid over-stabilization, the designed stability margin ε is set to a small positive value of 1.0. It should be noted that the setting of this threshold depends on practical needs regardless of larger or smaller ones. After the 1st iteration, unstable hours become stable with positive stability margins that are near the margin target except hours 7, 10, and 11. These three hours are still unstable (but their stability margins have been increased to −8.8, −8.1, and −9.8, respectively). After the 2nd iteration, all hours have become transient stable under C1.

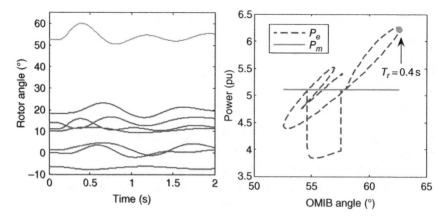

Figure 8.6 Multi-machine rotor angle trajectories (left window) and the corresponding OMIB $P-\delta$ representation (right window) – a stable case for the New England 10-machine system at 1st dispatch.

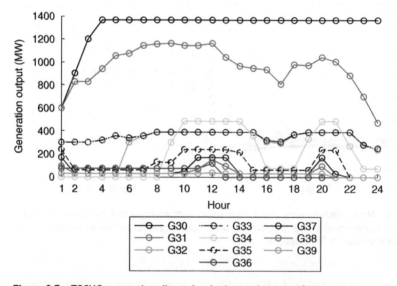

Figure 8.7 TSCUC generation dispatch–single contingency (C1).

For the obtained TSCUC solution, the total generation cost is US\$ 1 615 187.1. Only an additional 1.53% cost is required over the SCUC solution. The difference in unit status between TSCUC and SCUC solutions is shown in bold in Table 8.1, and the single-contingency TSCUC generation dispatch is shown in Figure 8.7.

In contrast to the TSCUC solution, with the SCUC solution, the generation output of G31 reduces significantly. In order to balance the generation variation, the on/off status and other generation dispatch are modified.

8.5.2.3 TSCUC Results-Multi-contingency Case

In previous work [6], only a single contingency was tested. In this chapter, a multi-contingency case is also conducted by the proposed approach. Hence, another three-phase short-circuit called C2 is applied at bus 30 and then cleared after 0.09 second. Under C2, 19 out of 24 hours are unstable, which are the hours 5–23. For common unstable hours 5–12 with C1, C2 has the same composition of CMs as C1, but C2 is more severe, which means the stability margin is smaller, in the hours 5, 7–9, and 12.

In order to stabilize C1 and C2 simultaneously, a representative contingency is selected (assuming the same instability mode) as the one with smaller stability margin for each unstable hour. Then, the multi-contingency stability constraint is composed of the representative contingency only. Consequently, the computational time can be effectively reduced.

After two iterations, a TSCUC can be obtained, which can ensure the transient stability for both C1 and C2 for each hour. The total generation cost of the multi-contingency TSCUC solution is US$ 1 616 903.5, which increases only 0.11% and 1.64% over the single-contingency TSCUC and the SCUC solution, respectively. The difference in unit status between the multi- and single-contingency TSCUC solutions is on the unit G34 at hour 8. The multi-contingency TSCUC generation dispatch is shown in Figure 8.8.

8.5.2.4 Computation Efficiency Analysis

In contrast with the prevailing TSCUC method [6], the computation efficiency of the proposed approach is much higher as it is not necessary to solve a time-consuming

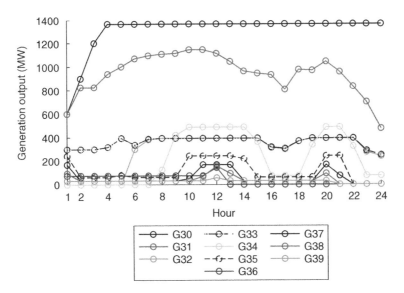

Figure 8.8 TSCUC generation dispatch–multiple contingencies (C1 and C2).

high-dimensional problem. Instead, it decomposes the whole complex problem into a set of small and tractable subproblems and then solve them by commercial solvers and TDS.

Based on the proposed computational structure, the total CPU time required for a single contingency can be roughly calculated as follows:

$$T_{total} = [T_{UC} + (T_{NSE} + T_{TDS}) \times NT] \times N_{It} + \sum_{i=1}^{N_{It}} (T_{TDS} \times Nu_i) + T_o$$

(8.18)

where T_{UC}, T_{NSE}, T_{TDS} represent the CPU time for UC solution, NSE, and TDS for a contingency, respectively, N_{It} represents the total iteration number, Nu_i denotes the number of the unstable hour for the i-th iteration, and T_o denotes all the other elapsed time during the whole computation process, including the time for reading and exporting data files, and interfacing between different software tools. In particular, the summation item in (8.18) represents that an unstable contingency requires an additional TDS to calculate the sensitivity value for deriving the stabilization cut. Table 8.2 shows the CPU time for each task of the introduced approach.

For the master problem, only 0.24 second is spent to solve the MILP by GUROBI solver. For each contingency, the NSE and generating BD cut require only 0.01 second. The major computational burden belongs to the TSA phase. Fortunately, it has been remarkably reduced by EEAC-based early termination. During the simulations, it only costs around 0.7~0.9 second for a contingency by PSS/E software. Meanwhile, the total CPU times on the New England system are 62.8 and 89.5 seconds for single- and multi-contingency TSCUC calculations, respectively. Based on the decomposition structure, parallel implementation can be applied to the proposed approach, for instance, paralleling 24 hours. Such a parallel computing platform tailored for PSS/E is employed in Ref. [23].

8.5.3 IEEE 50-Machine System

In order to further validate the proposed approach on its high computation efficiency, a larger system, IEEE 50-machine system, is used, which is derived from a representative model of a realistic power system in North America [26]. This

Table 8.2 CPU time (New England 10-machine system).

UC by MILP (T_{UC})	NSE for BD cut (T_{NSE})	TSA by IEEAC + PSSE (T_{TDS})
0.24 s	0.01 s	0.7~0.9 s

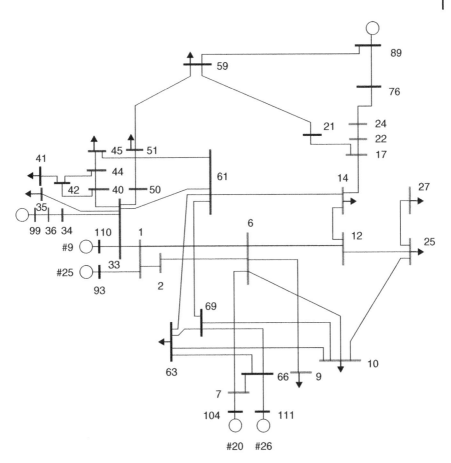

Figure 8.9 A portion of one-line diagram of the IEEE 50-machine system.

system contains 50 machines, 145 buses, and 453 branches. Figure 8.9 shows a portion of the one-line diagram of the high-voltage lines of this test system.

Original from Ref. [27], the data of the system are extended for a TSCUC study. The size of this system is similar to the IEEE 118-bus and 300-bus systems used in Ref. [6], which have 54 machines/186 branches and 69 machines/411 branches, respectively. Therefore, it is sensible to use the reported CPU time in Ref. [6] to benchmark the proposed approach.

In the system, a three-phase short-circuit fault at the line between buses 6 and 10 is investigated. A viable TSCUC solution is obtained after three iterations by the proposed method. In Figure 8.10, rotor angle trajectories and its OMIB curve for a specific hour from SCUC are indicated. It is obvious that the system is unstable under the fault and the stability margin is −18.9.

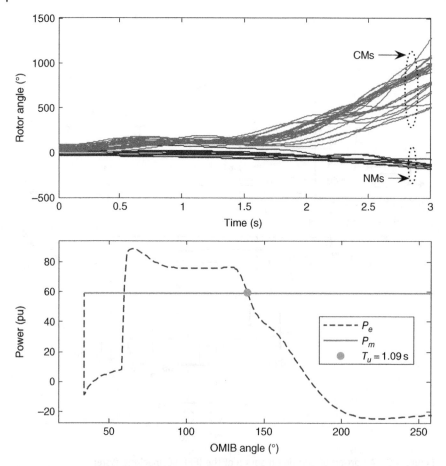

Figure 8.10 Multi-machine rotor angle trajectories (upper window) and the corresponding OMIB $P-\delta$ representation (lower window) – an unstable case for the IEEE 50-machine system.

In Figure 8.11, the trajectories are indicated for the same hour obtained from TSCUC. It can be found that the system becomes stable under the fault and the stability margin is 0.72. It should be noted that the trajectories of CMs and NMs are plotted in red and blue colors in the upper windows of these two figures, respectively.

From the lower window of Figures 8.10 and 8.11, it can be seen that they are both terminated earlier at 1.09 seconds where the instability condition is met, and 1.36 seconds where the first-swing condition is met, respectively. Therefore, the whole simulation time, such as 5 or 10 seconds, is not necessary to run, and the computational time can be greatly saved.

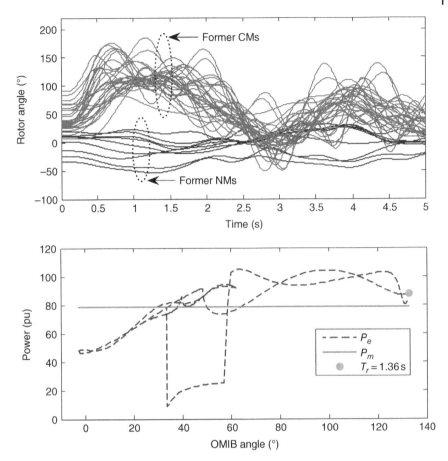

Figure 8.11 Multi-machine rotor angle trajectories (upper window) and the corresponding OMIB $P - \delta$ representation (lower window) – a stable case for the IEEE 50-machine system.

Table 8.3 CPU time (IEEE 50-machine system).

UC by MILP (T_{UC})	NSE for BD cut (T_{NSE})	TSA by EEAC + PSS/E (T_{TDS})
13.7 s	0.3 s	2.2~2.7 s

Source: Data from Ruiz-Vega and Pavella [11].

In Table 8.3, the CPU time for each task to solve TSCUC is indicated. For this test system, the total CPU time is about 319 seconds. In contrast with the prevailing approaches shown in Table 8.4, the proposed method is 140 times faster on similar test systems.

Table 8.4 Total CPU time for TSCUC computations.

50-machine 145-bus system (method of this chapter)	54-machine 118-bus system (method of [6])	69-machine 300-bus system (method of [6])
319 s	45 015 s	56 305 s

Nomenclature

Indices

i	index for units
j	index for buses
k	index for contingencies
l	index for lines
t	index for the time
C	index for critical machine
N	index for non-critical machine

Sets

NG/NT	number of units and dispatching time periods
x	sets of control variable
y	sets of states variable
s	sets of slack vector
π	*Lagrangian* multiplier vector

Variables

$\eta_{t,k}$	transient stability margin for contingency k in period t
$F_{lt,k}$	real flow on line l in period t for contingency k
$\hat{P}_{mi}(t_0)$	generation output of unit i, which is obtained from the master problem
ς_n	required generation shifting for TSC
N_{It}	total iteration number
Nu_i	number of unstable hour for the i-th iteration
$\Delta P_C, \Delta P_N$	changes in the total power of CMs and NMs

τ	approximate linear sensitivity of the stability margin with respect to generation change
τ_n	sensitivity value around the operating point n
$\Delta\eta$	required increment in stability margin
η_{us}	unstable stability margin

Parameters

D_{jt}	load demand at bus j
$PTDF^k_{lt,i}$	power transfer distribution factor of bus i to line l for contingency k at period t
ε	desired transient stability margin
M	inertia coefficient
T_{UC}	CPU time for UC solution
T_{NSE}	CPU time for NSE
T_{TDS}	CPU time for TDS
T_o	other elapsed time during the whole computation process

References

1 Wood, A.J., Wollenberg, B.F., and Sheblé, G.B. (2013). *Power Generation, Operation, and Control.* Wiley.

2 Carrion, M. and Arroyo, J.M. (2006). A computationally efficient mixed-integer linear formulation for the thermal unit commitment problem. *IEEE Transactions on Power Systems* 21 (3): 1371–1378.

3 Fu, Y., Li, Z., and Wu, L. (2013). Modeling and solution of the large-scale security-constrained unit commitment. *IEEE Transactions on Power Systems* 28 (4): 3524–3533.

4 Andersson, G., Donalek, P., Farmer, R. et al. (2005). Causes of the 2003 major grid blackouts in North America and Europe, and recommended means to improve system dynamic performance. *IEEE Transactions on Power Systems* 20 (4): 1922–1928.

5 Makarov, Y.V., Reshetov, V.I., Stroev, A., and Voropai, I. (2005). Blackout prevention in the united states, europe, and russia. *Proceedings of the IEEE* 93 (11): 1942–1955.

6 Jiang, Q., Zhou, B., and Zhang, M. (2013). Parallel augment Lagrangian relaxation method for transient stability constrained unit commitment. *IEEE Transactions on Power Systems* 28 (2): 1140–1148.

7 Pavella, M., Ernst, D., and Ruiz-Vega, D. (2012). *Transient Stability of Power Systems: A Unified Approach to Assessment and Control*. Springer Science & Business Media.

8 Pizano-Martinez, A., Fuerte-Esquivel, C.R., and Ruiz-Vega, D. (2011). A new practical approach to transient stability-constrained optimal power flow. *IEEE Transactions on Power Systems* 26 (3): 1686–1696.

9 Yong, F., Shahidehpour, M., and Zuyi, L. (2005). Security-constrained unit commitment with AC constraints. *IEEE Transactions on Power Systems* 20 (2): 1001–1013.

10 Xu, Y., Dong, Z.Y., Meng, K. et al. (2012). A hybrid method for transient stability-constrained optimal power flow computation. *IEEE Transactions on Power Systems* 27 (4): 1769–1777.

11 Ruiz-Vega, D. and Pavella, M. (2003). A comprehensive approach to transient stability control. I. Near optimal preventive control. *IEEE Transactions on Power Systems* 18 (4): 1446–1453.

12 Pizano-Martianez, A., Fuerte-Esquivel, C.R., and Ruiz-Vega, D. (2010). Global transient stability-constrained optimal power flow using an OMIB reference trajectory. *IEEE Transactions on Power Systems* 25 (1): 392–403.

13 Zarate-Minano, R., Cutsem, T.V., Milano, F., and Conejo, A.J. (2010). Securing transient stability using time-domain simulations within an optimal power flow. *IEEE Transactions on Power Systems* 25 (1): 243–253.

14 Xu, Y., Ma, J., Dong, Z.Y., and Hill, D.J. (2017). Robust transient stability-constrained optimal power flow with uncertain dynamic loads. *IEEE Transactions on Smart Grid* 8 (4): 1911–1921.

15 Xu, Y., Yin, M., Dong, Z.Y. et al. (2018). Robust dispatch of high wind power-penetrated power systems against transient instability. *IEEE Transactions on Power Systems* 33 (1): 174–186.

16 Xu, Y., Dong, Z.Y., Zhao, J. et al. (2015). Trajectory sensitivity analysis on the equivalent one-machine-infinite-bus of multi-machine systems for preventive transient stability control. *IET Generation Transmission and Distribution* 9 (3): 276–286.

17 Xue, Y., Li, W., and Hill, D.J. (2005). Optimization of transient stability control Part-I: for cases with identical unstable modes. *International Journal of Control, Automation and Systems* 3 (spc2): 334–340.

18 Xue, Y., Li, W., and Hill, D.J. (2005). Optimization of transient stability control part-II: for cases with different unstable modes. *International Journal of Control, Automation and Systems* 3 (2): 341–345.

19 CPLEX optimizer. http://www.ibm.com/us/en/ (accessed 29 August 2013).

20 GUROBI optimizer. http://www.gurobi.com/ (accessed 29 August 2013).

21 FERC (2011). Recent ISO Software Enhancements and Future Software and Modeling Plans.

22 Siemens (2011). PSS/E 33.0 program application Guide.

23 Meng, K., Dong, Z.Y., Wong, K.P. et al. Speed-up the computing efficiency of power system simulator for engineering-based power system transient stability simulations. *IET Generation Transmission and Distribution* 4 (5): 652–661. https://digital-library.theiet.org/content/journals/10.1049/iet-gtd.2009.0701.

24 Pai, M. (2012). *Energy Function Analysis for Power System Stability*. Springer Science & Business Media.

25 Kazarlis, S.A., Bakirtzis, A.G., and Petridis, V. (1996). A genetic algorithm solution to the unit commitment problem. *IEEE Transactions on Power Systems* 11 (1): 83–92.

26 IEEE. (1992). Transient stability test systems for direct stability methods. *IEEE Transactions on Power Systems* 7 (1): 37–43.

27 Richard D. Christie Power systems test case archive. http://www.ee.washington. edu/research/pstca/ (accessed 29 August 2013).

9

Transient Stability-Constrained Optimal Power Flow under Uncertainties

9.1 Introduction

While the generator dynamics are well represented in the TSC-OPF models, the load dynamics have not been fully addressed. Prevailing research always assumes the loads to be static (e.g. constant impedance [1–15]). But in practice, an actual load bus contains various static and dynamic components, such as induction motors. Such dynamic components will significantly affect the power system's short-term stability (both voltage stability and rotor angle stability, as they are strongly coupled). When subjected to a large disturbance, the induction motors may decelerate dramatically or even stall when there is not enough electrical torque that drives the mechanical load. Consequently, a very high reactive current derived from the network will dramatically reduce the voltage magnitudes and impede the voltage recovery. The power output of synchronous generators will be obstructed due to the delayed voltage recovery and/or successive voltage drops. As a result, the rotor angle separation of the synchronous generators will increase and finally lead to loss of synchronism [16]. Therefore, it is important to consider the load dynamics modeling; otherwise, the TSC-OPF may fail to maintain the transient stability if the contingency really occurs. Meanwhile, the load model parameters are uncertain and time-varying due to the stochastic nature of the load behavior. It has been widely shown that varying load composition can lead to diverse dynamic trajectories under the same disturbance [17]. Thus, it is necessary to consider the load model uncertainties for TSC-OPF. Otherwise, conservative or insecure dispatch results may be obtained. Since the implementation of smart meters is dramatically increased, it is possible to collect more load data to support power system operation and security [18].

Furthermore, the penetration of renewable energy such as wind and PV in power systems increases significantly due to its environmental and economic benefits as well as the development of the technologies. Many countries have launched

Stability-Constrained Optimization for Modern Power System Operation and Planning,
First Edition. Yan Xu, Yuan Chi, and Heling Yuan.
© 2023 The Institute of Electrical and Electronics Engineers, Inc.
Published 2023 by John Wiley & Sons, Inc.

wind power integration projects, including on-shore and off-shore wind farm constructions [19]. For instance, Denmark targets to achieve 50% wind penetration by 2025 [20]. Australia envisions a 100% renewable energy supply scenario by 2030, where wind power shares a large portion [21].

However, wind power penetration can also bring unprecedented challenges to power systems due to its uncertainties and intermittence. Specifically, wind power integration can affect the power system's dynamic security (stability) in various aspects [19], [22–24] including (i) reducing the system's effective inertia by replacing conventional synchronous generators, (ii) altering the system's dynamic characteristics due to the power electronics converter-based interfacing, (iii) fast changes of dispatch of synchronous generators in accommodating wind power variation, and (iv) fast fluctuation of magnitude and direction of power flows through the transmission network. Therefore, it becomes a significant issue to dispatch a large-scale wind-penetrated power system while considering the economy and stability for the operation stage.

Some early works start to solve optimal power flow (OPF) with the consideration of uncertain renewable generation [25, 26]. In prevailing research works, power system operation with wind power is only considered in the static security manner, such as branch overflow and bus overvoltage, which is known as security-constrained optimal power flow (SCOPF) or security-constrained unit commitment (SCUC) under wind power uncertainties [27–31]. To solve these uncertain dispatch problems, various mathematical programming algorithms have been developed since the static security constraint can be readily modeled as algebraic equations, for example, stochastic optimization [27–28], robust optimization [29–30], interval optimization [31], etc.

However, few focus on the dynamic behaviors of the power system operation when large-scale wind power is penetrated, for instance, transient stability, which is the most stringent criterion that can be lost within 150 ms subjected to a large disturbance [32]. Different from static security, the characteristics of transient stability are constructed by a large set of nonlinear differential-algebraic equations (DAEs), which are highly difficult to solve. In the current research stages, most of them concentrate on the transient stability analysis or assessment with wind power integration. In Ref. [23], the impact study of wind power is conducted via TDS and sensitivity analysis. In Ref. [33], a stochastic method to probabilistically evaluate the transient stability of power systems with wind farms is proposed. While in Ref. [34], a stochastic TDS method is proposed for transient stability assessment with stochastic parameters. In Ref. [35], an intelligent system is designed for real-time stability assessment under fast wind power variation.

Besides, current works always consider dynamic dispatch variables deterministically, which is for a conventional power system without uncertainties. They are usually known as TSC-OPF [1, 4–6, 8, 36] and TSCUC [37], which are not robust under wind power uncertainties. A comprehensive introduction to TSC-OPF and TSCUC can be found in Chapters 5 and 8, respectively.

Therefore, to model the uncertainties of the dynamic load and wind power for TSC-OPF, prevailing techniques are probabilistic programming, stochastic programming, and robust optimization. The former two depend on a probability density function (PDF) of the uncertain parameters. However, a large amount of computational burden is required by the former two methods since there are a huge number of scenarios that need to be sampled. As for robust optimization, only the range of the variation for the uncertain parameters is required, e.g. upper and lower bounds. However, its limitation is the strict construction of the model and always yields conservative solutions. For the TSC-OPF problem, it is quite difficult to construct a standard robust optimization TSC-OPF model.

Instead of employing the mathematically complicated robust optimization TSC-OPF model, this chapter uses an efficient scenario-based scheme for uncertainty modeling, which is called *Taguchi*'s orthogonal array testing (TOAT) [38]. The principle of the TOAT is to strategically select a small number of representative testing scenarios with good statistical information in the uncertainty space. TOAT is originally used for robust design, which can select a small representative set of testing scenarios to approximate all the possible combinations of uncertain variables in additive and quadratic models. Here, additive or quadratic models mean that the uncertain variables have a linear or quadratic impact on the model output. In contrast with stochastic optimization, only several deterministic scenarios are used to represent the whole uncertainty space. Hence, the computational time can be significantly reduced. Different from robust optimization, TOAT is scenario-based, does not require a strict mathematical formulation, and is less prone to conservatism. In the current research, TOAT has been used in transmission network expansion planning [39] and conventional OPF problems [40] for dealing with uncertainties coming from stochastic renewable energy generation and load demand variation. The detailed principle is introduced in the subsequent sections.

Therefore, in this chapter, TSC-OPF with uncertain dynamic loads and high-level wind power penetration is considered and modeled, respectively. The obtained solutions show that by applying the TOAT uncertainty modeling method, the system can keep transient stability against uncertain parameters.

9.2 TSC-OPF Model with Uncertain Dynamic Load Models

This section introduces a TSC-OPF model while considering the dynamics and uncertainties of the load models. The comprehensive load modeling, mathematical model, and solution approach are presented.

9.2.1 Load Modeling

9.2.1.1 Complex Load Model

During the transient process, the kernel of voltage instability is the ability of the dynamic loads to recover their power consumption within a very short time frame. As mentioned before, there are strong couplings between short-term voltage instability and transient instability. According to Ref. [41], subject to the same fault, different load models (static or dynamic) can lead to contrary transient stability statuses. Hence, considering dynamic load models in TSC-OPF is required for a reliable result.

In order to directly illustrate the problem, this section uses the complex load model "CLOD" [42]. By using this load model, all constant MVA, current, and admittance loads are integrated into a complex load containing induction motors, lighting, and other typical equipment as shown in Figure 9.1. Users can specify a minimum amount of data that describes the general character of the complex load. The model applies this data internally to establish the relative capacity of motor models for dynamic simulation and to establish typical values for the detailed parameter lists required in the detailed modeling [42].

Specifically, this complex load model is constructed by eight parameters $[p_L, p_S, p_D, p_T, p_C, K_p, R_e, X_e]$ corresponding to the percentage value of large motors (LM), small motors (SM), discharge lighting (DL), transformer exciting current (TX), constant power load (CP), the exponent of voltage-dependent real power load, and branch resistance and reactance (pu on load MW base), respectively. The model has been implemented by commercial software such as PSS/E as a standard complex load model [42].

9.2.1.2 Uncertainty Modeling

During real-time operation, load composition is stochastic and time-varying and this variation can lead to a significant impact on the system's dynamic behaviors [17], which means a single and deterministic set of load model parameters is not enough to describe the full range of dynamic response of the system. Hence, for a

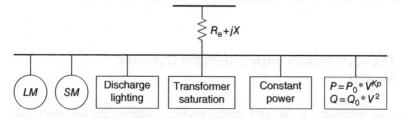

Figure 9.1 Structure of the complex load model "CLOD."

robust generation dispatch against uncertain load compositions, the load model parameters should be specified based on their statistical properties.

9.2.1.3 Taguchi's Orthogonal Array Testing (TOAT)

Provided an uncertainty space that contains M stochastic variables $\tilde{\zeta} = [\tilde{\zeta}_1, \dots \tilde{\zeta}_M]$, if each variable has B levels, for instance, the two levels can be the minimum and the maximum value of the variable, the whole uncertainty space has B^M combinations. However, the computational time will be enormous when M is large. Hence, the objective of TOAT is to select a small number of scenarios to represent the whole combination. In this connection, scenarios are determined by orthogonal arrays (OAs), where an OA is a matrix $L_H(B^M)$, which uses H combinations of levels ($H << B^M$) to represent the whole uncertainty space (H and M are the numbers of rows and columns, respectively, and B is the number of the matrix element levels) [38, 39].

For example, if a system has seven random variables and each is represented by two value levels, there will be a total of 2^7 combinations. With TOAT, an OA denoted as $L_8(2^7)$ can be constructed to represent all the combinations with only eight scenarios, as shown in Table 9.1.

In general, an OA has the following features:

1) Each testing scenario is a row in the matrix, and there are H ($H = 8$ here) scenarios in total.
2) In each OA column, every value level of a variable occurs H/B times, i.e. 1 and 2 both occur four times.

Table 9.1 Testing scenarios determined of OA $L_8(2^7)$.

Testing scenario	Variable levels						
	$\tilde{\zeta}_1$	$\tilde{\zeta}_2$	$\tilde{\zeta}_3$	$\tilde{\zeta}_4$	$\tilde{\zeta}_5$	$\tilde{\zeta}_6$	$\tilde{\zeta}_7$
1	$\zeta_1(1)$	$\zeta_2(1)$	$\zeta_3(1)$	$\zeta_4(1)$	$\zeta_5(1)$	$\zeta_6(1)$	$\zeta_7(1)$
2	$\zeta_1(1)$	$\zeta_2(1)$	$\zeta_3(1)$	$\zeta_4(2)$	$\zeta_5(2)$	$\zeta_6(2)$	$\zeta_7(2)$
3	$\zeta_1(1)$	$\zeta_2(2)$	$\zeta_3(2)$	$\zeta_4(1)$	$\zeta_5(1)$	$\zeta_6(2)$	$\zeta_7(2)$
4	$\zeta_1(1)$	$\zeta_2(2)$	$\zeta_3(2)$	$\zeta_4(2)$	$\zeta_5(2)$	$\zeta_6(1)$	$\zeta_7(1)$
5	$\zeta_1(2)$	$\zeta_2(1)$	$\zeta_3(2)$	$\zeta_4(1)$	$\zeta_5(2)$	$\zeta_6(1)$	$\zeta_7(2)$
6	$\zeta_1(2)$	$\zeta_2(1)$	$\zeta_3(2)$	$\zeta_4(2)$	$\zeta_5(1)$	$\zeta_6(2)$	$\zeta_7(1)$
7	$\zeta_1(2)$	$\zeta_2(2)$	$\zeta_3(1)$	$\zeta_4(1)$	$\zeta_5(2)$	$\zeta_6(2)$	$\zeta_7(1)$
8	$\zeta_1(2)$	$\zeta_2(2)$	$\zeta_3(1)$	$\zeta_4(2)$	$\zeta_5(1)$	$\zeta_6(1)$	$\zeta_7(2)$

3) In any two columns, combinations of two levels occur same time, i.e. "1 1," "1 2," "2 1," and "2 2" appear twice.
4) The combinations determined by OA are uniformly distributed over the uncertainty space.
5) If any two columns are exchanged or some columns are eliminated, the obtained array still satisfies the above features.

As shown in Ref. [38], all variable combinations can be represented by the testing scenarios with sufficient statistical information. OAs can be formed by mathematical methods [38] or directly looked up from OA libraries [43]. It is important to select an appropriate OA number B, which is the representative level of the random variables. Generally, if the uncertain variable has a linear effect on the system, $\tilde{\varsigma}_m$ should have two testing levels, and if $\tilde{\varsigma}_m$ follows a normal distribution, $\mu - \sigma$ and $\mu + \sigma$ should be used, where μ and σ are the mean and standard deviation of $\tilde{\varsigma}_m$, respectively; if $\tilde{\varsigma}_m$ has a quadratic effect on the system, $\tilde{\varsigma}_m$ should have three levels, and for normal distribution, $\mu - \sqrt{3/2}\sigma$, μ, and $\mu + \sqrt{3/2}\sigma$ should be applied. It should be mentioned that if there is not an OA with a desired variable number (M) in OA libraries, an OA with the next larger M can be applied by eliminating redundant columns based on feature (5).

9.2.2 Mathematical Model

Different from the conventional TSC-OPF compact model, which is introduced in Chapter 5 from Eqs. (5.1)–(5.4). The TSC-OPF compact model with uncertainties is aiming to optimize control variable $\mathbf{u}$ to make the system robustly stable to the random parameter variable of the load model $\tilde{\varsigma}_m$, namely, the system keeps transient stable against the variation of the load model parameters.

$$\min_{\mathbf{u}} \ F(\mathbf{x}, \mathbf{u}, \tilde{\varsigma}) \tag{9.1}$$

$$\text{s.t. } \mathbf{g}(\mathbf{x}, \mathbf{u}, \tilde{\varsigma}) = 0 \tag{9.2}$$

$$\mathbf{h}(\mathbf{x}, \mathbf{u}, \tilde{\varsigma}) \leq 0 \tag{9.3}$$

$$\mathbf{TSI}_k(\mathbf{x}, \mathbf{u}, \tilde{\varsigma}) \geq \varepsilon, \ \{k \in C\} \tag{9.4}$$

where $\tilde{\varsigma}$ denotes the vector of uncertain dynamic load model parameters. It should be noted that the only difference in contrast with (5.1)–(5.4) is the parameter variable. In (9.4), the parameter will have an effect on transient stability since it becomes uncertain.

The objective and the constraints $\mathbf{g}$ and $\mathbf{h}$ are the conventional OPF, which is introduced in Chapter 4 from Eqs. (4.4)–(4.9). For the transient stability constraint, the stability margin η_k is calculated by the EEAC method, which is introduced in Chapter 2.

9.2.3 Critical Uncertain Parameter Identification

As introduced before, there are eight parameters used for the complex dynamic load model here. However, not all of them will significantly affect the system's stability [44]. Therefore, it is necessary to recognize the critical model parameters, and then only these critical parameters are integrated into the TSC-OPF model. By doing this, the number of uncertainty parameters M can be reduced, and thus the number of testing scenarios can be reduced as well. Such critical parameters are also known as *well-conditioned* parameters, which means they can have a remarkable influence on the system trajectory, and the influence can be quantified in terms of trajectory sensitivities [17, 44].

Trajectory sensitivity is the time-varying sensitivity of minor changes in a parameter with respect to the system's dynamic behavior [17], the comprehensive introduction is presented in Section 2.6. It can be calculated numerically approximated through two successive TDS runs:

$$\frac{\partial \phi(x_0, t, \lambda)}{\partial x} = \frac{\phi(x_0, t, \lambda + \Delta\lambda) - \phi(x_0, t, \lambda)}{\Delta\lambda} \tag{9.5}$$

where $\partial \phi(x_0, t, \lambda)/\partial x$ represents the trajectory sensitivity with time t with the initial condition x_0. ϕ represents the dynamic trajectory of the systems. $\Delta\lambda$ denotes the small perturbation. In this section, 5% is set for the problem.

According to the concept of trajectory sensitivity, a sensitivity index (SI) can be defined to rank the influence of the load model parameters on system transient stability:

$$SI = \frac{\Delta TSI}{\Delta\varsigma} = \frac{TSI(x, u, \varsigma + \Delta\varsigma) - TSI(x, u, \varsigma)}{\Delta\varsigma} \tag{9.6}$$

Based on practical requirements, it is better to select the top several parameters as the critical load parameters. Specifically, it should exclude some parameters that have quite lower *SI* values. Moreover, when the parameters' *SI* values are very close, it is necessary to determine the number of critical parameters.

9.2.4 Solution Approach

The introduced TSC-OPF model with uncertain parameters is a large-scale nonlinear problem, which is difficult to handle. Therefore, a decomposition-based solution approach is applied to deal with it. In Figure 9.2, the principle of the approach is to divide the large-scale problem into a master problem and a series of tractable smaller subproblems, which interact with each other iteratively.

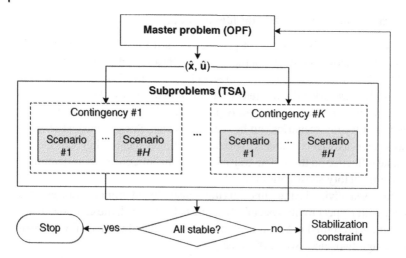

Figure 9.2 Decomposition-based solution approach.

9.2.4.1 Master Problem

The master problem is to solve the standard OPF model (9.1)–(9.3) and the output is an operating point $(\hat{x}, \hat{u})$.

As mentioned before, the OPF problem is nonconvex and nonlinear. In this chapter, the mature interior-point (IP) method is applied to solve the master problem. Although it can only ensure local optima and convergence, its speed and solution quality have been validated to be good in both academia and industry. Therefore, the solution of the whole TSC-OPF model is also locally convergent.

In the first iteration, (9.3) only contains the steady-state operational constraints (4.6)–(4.9) in Chapter 4, while the transient stability constraints are ignored at this time. After solving the subproblems, additional constraints will be generated if the stability is violated, and then they will be attached to (9.3) in the subsequent iterations to eliminate stability violations.

9.2.4.2 Subproblem

The subproblems are to assess the stability of the master problem solution under each contingency and scenario and generate stabilization constraints if the system is unstable.

For each testing scenario obtained from the TOAT, the EEAC-based TSA is performed and if (9.4) is not satisfied, the stabilization constraint, i.e. generation shifting between CMs and NMs, is generated. To simultaneously stabilize multiple contingencies/scenarios with the least number of additional constraints, the contingencies/scenarios can be grouped according to their resulting instability modes. Those whose instability modes are the same will be divided into one group.

In each group, the severest mode (the one with the smallest stability margin) will be the representative of the group. For every representative instability mode, the stability constraint is constructed below:

$$\sum_{i\in C}\hat{P}_{mi}(t_0) - \sum_{i\in C}P_{mi}(t_0) \geq \left(\frac{-\eta_{us} + \varepsilon}{\tau_n}\right)\cdot\left(\frac{M}{M_C} + \frac{M}{M_N}\right)^{-1} \tag{9.7}$$

where $\hat{P}_{mi}(t_0)$ represents the active power output of i-th generator obtained from the master problem. The detailed derivation of the stabilization constraint can be referred to in Section 8.3.

The stabilization constraint (9.7) illustrates the information about how the generation dispatch should be modified to maintain transient stability. It should be highlighted that (9.7) is a linear constraint. Hence, it can be directly added to (9.3) in the master problem. By doing this, the original problem size can be reduced to one similar to a standard OPF.

9.2.4.3 Computation Process

The computation process of the proposed method can be summarized below:

1) Prepare the computation data, including the system data (network data and dynamic models), contingency set, and the mean value μ and the standard deviation σ of the load model parameters. The load model parameters can be obtained from load composition statistics, load modeling practices [45], and advanced smart grid functions.
2) For each load model parameter, evaluate their influence on the transient stability of the system based on Eq. (9.6). Only those with high SIs are selected as uncertainty parameters whose dimension is denoted as M and the remaining load model parameters are fixed at their mean values.
3) For the uncertainty parameters, select an appropriate representative level (i.e. OA level), denoted as B. It will be shown later that $B = 2$ is sufficient as the load model parameters show a quasi-linear impact on the stability margin.
4) Given the values of M and B, construct an OA $L_H(B^M)$, which has H rows, each representing a testing scenario.
5) Solve the master problem (Eqs. (9.1)–(9.3)) by applying the IP method and then obtaining the current operating point $(\hat{x}, \hat{u})$.
6) Given the master solution, for each contingency, perform the TSA based on EEAC for each testing scenario. Given K contingencies, there will be total $K \times H$ EEAC simulation runs. The earlier termination can be applied to save the total computational time. Meanwhile, due to the decomposition nature of the proposed approach, parallel computing can be applied to speed up the solution procedure [46].

7) If all the contingencies under all the testing scenarios are transient stable, the process can be terminated; otherwise, go to Step (8).
8) For each representative instability mode, form the stability constraint (9.7) based on the transient stability control rule introduced in Section 8.3.
9) Add the generated stability constraints to the inequality constraint set (9.3), and return to Step (5).

It should be mentioned that the stability constraint derived from EEAC approximates the original multi-machine rotor angle trajectories. Since the TSC-OPF problem is nonlinear, the stabilization process may not be achieved immediately after the stability constraints are added to the master problem (similar issues are also indicated in Refs. [7, 8]); thus, the whole computation should be iteratively performed until all the contingencies/scenarios are stabilized.

9.3 Case Studies for TSC-OPF Under Uncertain Dynamic Loads

9.3.1 Simulation Settings

The numerical simulation is conducted on a 64-bit PC with 3.10 GHz CPU and 4.0 GB RAM. TDS is conducted by using the commercial software PSS/E [42], and the EEAC algorithm is integrated into the MATLAB platform. OPF is solved using MATPOWER with the embedded IP algorithm [47].

The method is validated on the standard New England 10-machine 39-bus system, which is a well-known benchmark in TSC-OPF research [5, 7, 8, 12, 13]. Three commonly studied contingencies in the literature are considered as shown in Table 9.2. The base case is the optimal power flow solution from [13] (see Table 9.3), which is the best dispatch result reported in the literature. It is assumed that the mean values μ of parameters are indicated in Table 9.4 and the standard deviation σ is 5%.

Table 9.2 Contingency set.

Contingency ID	Fault location	Fault duration	Tripped line
C1	Bus 21	0.16 s	Line 21–22
C2	Bus 4	0.25 s	Line 4–5
C3	Bus 29	0.10 s	Line 29–26

Table 9.3 Base case-initial generation dispatch (MW).

Generator	Output	Generator	Output	Generator	Output
G1	252.59	G4	652.12	G7	532.00
G2	576.20	G5	494.50	G8	552.57
G3	660.33	G6	623.50	G9	788.00
Total cost		60 971.25 $/h		G10	1005.59

Source: Adapted from Xu et al. [13].

Table 9.4 Mean value (μ) of the load model parameters.

p_L	p_S	p_D	p_T	p_C	K_p	R_e	X_e
20%	20%	10%	10%	20%	2	0	0

9.3.2 Transient Stability with Dynamic Loads

In the simulation, the base case OP is stable and subjected to all three contingencies without considering dynamic load components. However, when the dynamic load models are taken into account, the system becomes unstable with respect to all three contingencies, which is indicated in Figure 9.3.

Therefore, it is obvious that dynamic loads can have a significant influence on system stability, and the need for incorporating them in TSC-OPF is quite practical.

9.3.3 Single-contingency Case for New England 10-Machine 39-Bus System

A single contingency C1 is considered first. According to Eq. (9.6), the *SI*s are calculated for each load component, and the values are: -18.4 (*LM*), -15.9 (*SM*), -0.0004 (*TX*), 3.68 (*DL*), and -6.74 (*CP*), which illustrate that load components have various impact of degree on system stability. Hence, *LM*, *SM*, *CP*, and *DL* are selected as the uncertainty vectors, while *TX* is set to its mean value. In order to determine appropriate representative levels of the OA level *B*, the quantitative impact of the load model parameters on the transient stability margin is then studied. To achieve it, the percentage values for each load component are continuously changed and EEAC is conducted with respect to them. In Figure 9.4, the results are shown.

As Figure 9.4 indicates, the variations in the percentage of load components have a clear linear relationship with the transient stability margin. More specifically, *LM*, *SM*, and *CP* are negatively related to the transient stability margin,

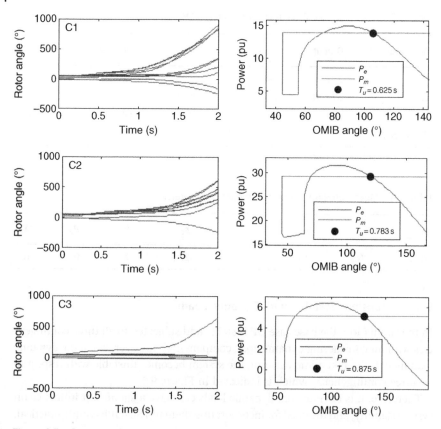

Figure 9.3 System trajectories for base case with dynamic load model.

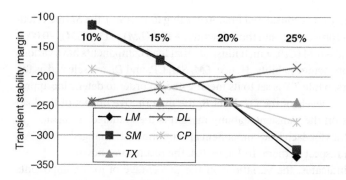

Figure 9.4 Transient stability margin with varying load component percentage.

which means increasing the percentage value will harm the system's stability, while DL is positively related to the transient stability margin, which means increasing the percentage value will benefit system stability.

Based on the TOAT theory, if the uncertain parameter has a linear impact on the system, two levels can be determined to represent the uncertain parameter space. Therefore, for an OA, B is two and M is four, which denote the number of testing levels and uncertain variables, respectively. While there is not a proper one in the OA libraries, a redundant OA $L_8(2^7)$ can be applied (see Table 9.1) with the last three columns being eliminated. Determined by $L_8(2^7)$, there are a total of eight testing scenarios, $\varsigma(1)$ and $\varsigma(2)$ represent the lower level ($\mu - \sigma$) and upper level ($\mu + \sigma$) of the uncertain variable $\tilde{\varsigma}_m$, respectively.

The TSC-OPF model with dynamic loads is solved using the proposed solution approach introduced in Section 9.2, and the re-dispatch results are given in Table 9.5. The CMs under this contingency are G4, G5, G6, and G7, and the total generation shifting amount from the CMs to NMs is 320 MW. The increased cost over the base case is 542.35 $/h or 0.89%, which is acceptable for stability improvement purposes.

9.3.4 Multi-contingency Case for New England 10-Machine 39-Bus System

For the multi-contingency case, all three contingencies are considered simultaneously for the TSCOPF model. The obtained dispatch results are shown in Table 9.6

Table 9.5 Single contingency (C1) solution results (MW).

Generator	Output	Generator	Output	Generator	Output
G1	275.05	G4	526.75	G7	465.37
G2	622.36	G5	437.88	G8	586.21
G3	701.91	G6	552.12	G9	900.00
Total cost		61 513.6 $/h		G10	1070.3

Table 9.6 Multi-contingency solution results (MW).

Generator	Output	Generator	Output	Generator	Output
G1	252.53	G4	542.14	G7	480.52
G2	536.67	G5	449.15	G8	634.88
G3	619.87	G6	570.32	G9	648.00
Total cost		62 786.9 $/h		G10	1397.33

and the system trajectories including rotor angles and OMIB power-angle trajectories under each contingency are shown in Figure 9.5. The CMs are G2, G3, G4, and G5 for C2; and G9 for C3. The increased cost over the base case is 1815.7 \$/h or 2.98%.

The obtained solution is stable for all three contingencies under all of the eight testing scenarios. In order to clearly indicate, the system trajectories under the eighth testing scenario are shown in Figure 9.5. Note that the earlier termination of the TDS can also be employed here to reduce computing time.

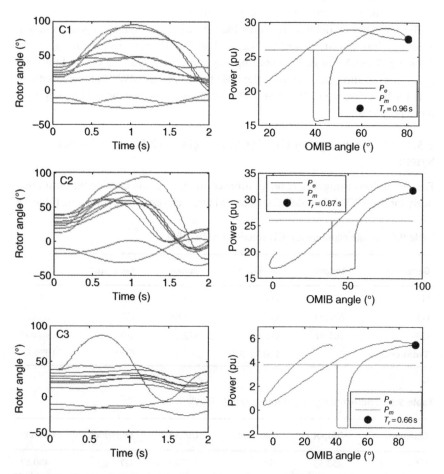

Figure 9.5 System trajectories for the multi-contingency TSCOPF solution under the eighth testing scenario.

9.3.5 Result Verification

To check the robustness of the solution against the load model uncertainty, we generate a set of random verification scenarios based on Monte Carlo simulation with assumed μ and σ of the load model parameters. Each scenario is evaluated under the three contingencies. If the system is stable, the verification scenario is considered as feasible, otherwise considered as infeasible. The robustness degree is calculated below:

$$\gamma = \frac{M_s}{M} \times 100\% \tag{9.8}$$

where M is the total number of verification scenarios (set to 100) and M_s is the number of feasible (stable) scenarios under k-th contingency.

In Table 9.7, the robustness degree of the solution obtained from the proposed model and from Ref. [13] is compared. Note that the solution reported in Ref. [13] is the most economical TSCOPF result available in the literature and the only one that considers three contingencies at the same time.

Averagely, the robustness degree of the solution from the standard TSCOPF model without dynamic load models is 16%. In contrast, the robustness degree for the proposed model is 98.67%, which indicates its well-immunized capability against the load model uncertainty.

9.3.6 Computation Efficiency Analysis

In contrast with traditional TSCOPF methods, the proposed one here makes full use of EEAC and decomposition strategy to transfer the whole large-scale problem into smaller tractable problems. In each iteration, the computation includes OPF calculation (master problem) and TSA checking for each contingency under each testing scenario. Meanwhile, for each unstable scenario, transient stability margin sensitivity is calculated to generate stabilization constraint. Thus, the total computation time can be estimated as follows:

$$T_{total} = (T_{OPF} + T_{TSA} \times K \times H) \times N_{It} + \sum_{i=1}^{N_{It}}(T_{TSA} \times N_{ui}) \tag{9.9}$$

where T_{OPF} and T_{TSA} represent the CPU time for OPF calculation and TSA for a contingency, respectively; K and H are the number of contingencies and the testing

Table 9.7 Robustness degree.

Solution	C1	C2	C3	Average
This section	99%	97%	100%	98.7%
Ref. [13]	0%	26%	22%	16%

scenarios determined by TOAT, respectively; N_{It} is the total iteration number, and N_{ui} is the number of unstable scenarios in the i-th iteration. The second term in (9.9) means to conduct an additional TDS to obtain stability margin sensitivity and generate the stability constraint under an unstable scenario.

During the simulation, the total computation time is 34.3 and 110.4 seconds for single- and multi-contingency cases, respectively, which are much faster than conventional methods such as discretizing and EA-based methods.

For contingency number K, it is 3 here. Practically, there may be a lot of contingencies in a large power system. However, only unstable contingencies are considered in the TSCOPF since a well-planned system has limited unstable contingencies. In addition, the probabilities of occurrence for different contingencies are different. Only the contingencies that have high probabilities are included in the TSCOPF.

For testing scenario H, it is 8 here. Based on OA tables [43], H can be very limited even for a large model.

For iteration number N_{It}, as the effectiveness of EEAC-based stabilization constraint is high, this number can be small. It is 2, which is consistent with other literature [6–9].

For N_{ui}, its maximum initial value is $K \times H$, but it will be gradually reduced during the iterations as the unstable scenarios will be stabilized during each iteration.

OPF calculation can be very fast by using MATPOWER. In this book, T_{OPF} is 0.1 second. Therefore, the major computation time is cost by the TSA aspect. For T_{TSA}, it is 0.9–1.1 seconds by using PSS/E software. It should be mentioned that this time can be further reduced by an early-termination application. Meanwhile, a more powerful and specific TSA engine DSATools [48] can be used, whose speed is much faster. For the same task, DSATools only costs 0.12 second for one contingency. Hence, the total computation time will become 3.8 and 12 seconds for single- and multi-contingency cases, which are acceptable for online implementation.

Moreover, it should be highlighted that the proposed method here can be parallelized as it is a decomposition structure. For example, if H CPU cores are available, the speed can be increased H times by dispersing the subproblems to the H CPU cores. This kind of parallel computing platform for PSS/E-based TSA has been employed in our previous work [46]. The parallel computing is widely used at a modern power system control center. Hence, this computing power can be fully applied to the proposed approach in this section.

9.4 TSC-OPF Model with Uncertain Wind Power Generation

This section introduces a TSCOPF model while considering large-scale wind-penetrated power systems. The comprehensive wind uncertainty modeling, mathematical model, and solution approach are presented.

9.4.1 Mathematical Model

The compact model is shown below, which is similar to (9.1)–(9.4):

$$\min_{\mathbf{u}} F(\mathbf{x}, \mathbf{u}, \tilde{\mathbf{w}}) \tag{9.10}$$

$$\text{s.t. } \mathbf{g}(\mathbf{x}, \mathbf{u}, \tilde{\mathbf{w}}) = 0 \tag{9.11}$$

$$\mathbf{h}(\mathbf{x}, \mathbf{u}, \tilde{\mathbf{w}}) \leq 0 \tag{9.12}$$

$$\text{TSI}_k(\mathbf{x}, \mathbf{u}, \tilde{\mathbf{w}}) \geq \varepsilon, \ \{k \in C\} \tag{9.13}$$

where $\tilde{\mathbf{w}}$ represents the vector of uncertain wind active power output.

The objective and the constraints $\mathbf{g}$ and $\mathbf{h}$ are the conventional OPF, which is introduced in Chapter 4 from Eq. (4.4)–(4.9). It should be mentioned that the power flow with the wind power is a little different from conventional OPF, which is shown below:

$$\begin{cases} P_{Gi} + \tilde{P}_{Wi} - P_{Di} - V_i \sum_{j=1}^{N_B} V_j \left(G_{ij} \cos \theta_{ij} + B_{ij} \sin \theta_{ij} \right) = 0 \\ Q_{Gi} + \tilde{Q}_{Wi} - Q_{Di} - V_i \sum_{j=1}^{N_B} V_j \left(G_{ij} \sin \theta_{ij} - B_{ij} \cos \theta_{ij} \right) = 0 \end{cases} \tag{9.14}$$

This section follows OMIB angle trajectory [8] to form the transient stability constraint. This constraint restricts the OMIB trajectory to be bounded by a critical value at a critical time:

$$\hat{\delta}(t_u) - \delta_{CT}(t_u) \leq 0 \tag{9.15}$$

where t_u is the instability time of the unstable OMIB, namely, the time that the system loses synchronism. $\hat{\delta}(t_u)$ and $\delta_{CT}(t_u)$ represent the targeted equivalent OMIB angle of the system and a reference critical OMIB (C-OMIB) trajectory of the same system at the instability time t_u, respectively. The OMIB equivalence and t_u are explained in Chapter 2. It should be mentioned that C-OMIB corresponds to the initial system in the critical stable situation, e.g. the one under critical clearing time (CCT).

It is important to note that this constraint is a single equality as it is only imposed at a single time step t_u. Then, trajectory sensitivity is used to convert the constraints into algebraic inequalities. By doing this, the size of the proposed optimization problem can be remarkably reduced, and a scenario-decomposition solution approach can be employed.

9.4.2 Construction of Transient Stability Constraints

Eq. (9.15) can be reformed as:

$$\hat{\delta}(t_u) = \delta_{UT}(t_u) - \Delta\delta(t_u) \leq \delta_{CT}(t_u) \tag{9.16}$$

where $\delta_{UT}(t_u)$ is the unstable OMIB angle of the original system at t_u, and $\Delta\delta(t_u)$ is the required reduced OMIB angle to be smaller than $\delta_{CT}(t_u)$.

Then, according to trajectory sensitivity, $\Delta\delta(t_u)$ can be transferred to generation changes as below:

$$
\begin{aligned}
\Delta\delta(t_u) = \sum_{i\in S^+} \Phi_i(\delta_{UT}, t_u, P_{Gi})\Delta P_{Gi} \\
+ \sum_{j\in S^-} \Phi_j(\delta_{UT}, t_u, P_{Gj})\Delta P_{Gj} \geq \delta_{UT}(t_u) - \delta_{CT}(t_u)
\end{aligned}
\tag{9.17}
$$

where ΔP_G denotes the active power output change for synchronous generators, S^+ and S^- are the sets of generators with positive and negative sensitivities, respectively (positive sensitivity means that increment in generation output will raise the OMIB angle and vice versa), $\Phi_i(\delta_{UT}, t_u, P_{Gi})$ is the OMIB angle trajectory sensitivity with respect to the active power output of generator i at the instability time t_u.

It should be noted that the generator trajectory sensitivities and OMIB angles are first obtained from TDS, then the stability constraints are formed as (9.17) and iteratively calculated through a master–slave solution process. Moreover, it should be noted that only the sensitivity at t_u is calculated. By this way, the computing time can be significantly reduced. The detailed solution process will be presented subsequently.

9.4.3 Robust Design of Wind Uncertainty

As introduced in Section 9.2.1.3, TOAT is also applied for the modeling of uncertain wind power output. It is assumed that the wind power variation from its prediction follows a normal distribution. Hence, OA scenarios can represent uncertainties. To predict wind power variation, mature techniques such as interval forecasting can be applied to obtain the mean value and its standard deviation [49]. In addition, the wind power variation has a quasi-linear relationship with the stability margin as its variation is reflected by the variation of the synchronous generator's output. Therefore, the OA level B of $\widetilde{w}_m$ can be represented by two testing levels $\mu - \sigma$ and $\mu + \sigma$.

For example, if a system has three random variables and each is represented by two levels, a total of 2^3 combinations will be determined. With TOAT, an OA $L_4(2^3)$ is constructed to represent the whole uncertainty space by only four scenarios, which is shown in Table 9.8.

9.4.4 Solution Approach

9.4.4.1 Decomposition Scheme
Similar to Section 9.2, a decomposition scheme is applied since it is structurally analogous to the proposed model in this section. According to the proposed

Table 9.8 Testing scenarios OA $L_4(2^3)$.

Testing scenarios	Variable levels		
	$\tilde{w}_1$	$\tilde{w}_2$	$\tilde{w}_3$
l_1	1	1	1
l_2	1	2	2
l_3	2	1	2
l_4	2	2	1

Where 1 and 2 represent the lower and upper levels of the corresponding uncertain variable $\tilde{w}_m$, respectively.

algebraic stability constraint (9.17) and the TOAT scenarios, the original uncertain programming problem can be decomposed into a series of deterministic problems, which motivates us to employ the decomposition scheme.

Like Bender's decomposition, the original problem is decomposed into a master problem, which represents standard OPF (base dispatch), and a set of slave problems, which represent stability checks subjected to the contingencies and wind power variation. If the stability requirement is not satisfied, stabilization constraints will be generated and added to the master problem. The whole problem is iteratively solved between the master and slave problems until all scenarios are stable under all contingencies. In Figure 9.6, the proposed decomposition scheme is indicated.

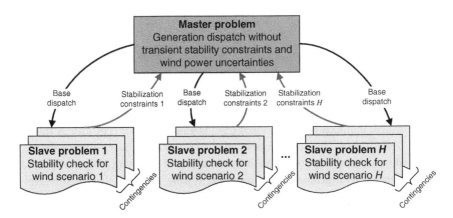

Figure 9.6 Decomposition scheme.

9.4.4.2 Master Problem

The master problem represents the steady-state generation dispatch, which is essential to solve the standard OPF model: {*min* (9.10), *subject to* (9.11)–(9.12)}.

The initial wind power output is set to the mean value, which is predicted by interval forecasting, and the uncertainties are dealt with in the slave problems. The standard OPF is solved by the IP method, which is one of the most mature algorithms. Similar to the proposed method in Section 9.2, the inequality constraint set (9.12) for the first iteration only includes the steady-state operational limits. In the subsequent iterations, additional stabilization constraints will be generated and added to (9.12) when the slave problems are solved and the stability is violated.

9.4.4.3 Slave Problem

The slave problems represent transient stability checks for each contingency under each testing scenario.

For each testing scenario selected by TOAT, TDS is run subjected to a contingency, and stability indexes including η, t_u and t_r are obtained based on EEAC. If a scenario is unstable, the trajectory sensitivities of each synchronous generator to OMIB angle at t_u are calculated as a byproduct, and the following stabilization constraint is formed:

$$
\sum_{i \in S^+} \Phi_i(\delta_{UT}, t_u, P_{Gi}) \cdot (\hat{P}_{Gi} - P_{Gi})
$$
$$
+ \sum_{j \in S^-} \Phi_j(\delta_{UT}, t_u, P_{Gj}) \cdot (\hat{P}_{Gj} - P_{Gj}) \geq \delta_{UT}(t_u) - \delta_{CT}(t_u) \tag{9.18}
$$

where $\hat{P}_G$ is the base dispatch from the master problem.

As introduced in Section 2.6, the stabilization constraint (9.18) also indicates information about how to modify the base dispatch to enhance transient stability for an unstable scenario. As the constraint (9.18) is linear, it can be directly added to the master problem.

9.4.4.4 Computation Process

In Figure 9.7, the computation flowchart is clearly indicated. The input data contain wind power forecasts, static and dynamic models of the system, and the contingency set. As introduced previously, the testing scenarios can be formed by mathematical methods [38] or indexed from OA libraries [43]. The OPF is solved by the IP algorithm. The TDS can be conducted by using industry-grade software.

In addition, instability modes such as clustering of CMs and NMs are derived from different contingencies and wind power testing scenarios. Hence, the stabilization constraints are all attached to the master problem for one solution, and if the optimization problem is feasible, stabilization constraints are satisfied simultaneously. Moreover, it is worth mentioning that when the instability mode

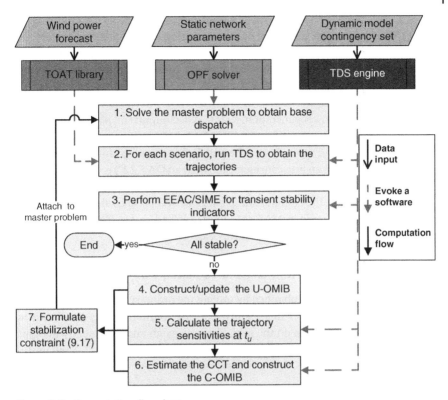

Figure 9.7 Computation flowchart.

changes, the corresponding structure of C-OMIB, t_u, and constraint (9.18) should be updated for the next iteration. Furthermore, a small perturbation should be applied in trajectory sensitivity calculation to avoid excessive stabilization. This may occur when the system is close to the stability boundary where it tends to be very sensitive to parameter variations [17]. In Ref. [50], a varying perturbation size for trajectory sensitivity calculation is proposed, which can be applied to avoid excessive power rescheduling when reaching the stability boundary.

9.5 Case Studies for TSC-OPF Under Uncertain Wind Power

9.5.1 Simulation Settings

The method is firstly validated on the New England 10-machine 39-bus system by modifications: three synchronous generators are replaced by three wind farms at buses 32, 37, and 39. The wind power output keeps the same as the original generators. The generation capacities of the remaining synchronous generators are increased by 1.5 times the initial values to let the modified system have an adequate reserve to compensate for the wind power variation. In this case, the wind power penetration level is 23.3%.

Three contingencies are tested, and the detailed information is indicated in Table 9.9. For wind power generation, its forecasted output (i.e. μ) is set to the base value and randomly varies within the standard deviation $\pm\sigma$, which is assumed to be $\pm10\%$ of its mean value. For each wind farm, two uncertain levels of output $\mu - \sigma$ and $\mu + \sigma$ are taken into consideration, then an OA $L_4(2^3)$ is constructed with four testing scenarios (see Table 9.8).

For this case study, the commercial software DSATools is used as the TDS [48], the EEAC is implemented in MATLAB, OPF is solved by MATPOWER [47], and the trajectory sensitivities are numerically calculated [17].

9.5.2 Base Dispatch for New England 10-Machine 39-Bus System

The initial base dispatch is obtained by solving the ordinary OPF without stability constraints and wind power uncertainty, and the solution results are shown in Table 9.10. TDS is then run on the initial base dispatch and the stability margins are given in Table 9.11, where the initial dispatch is stable for C1 and C2, but unstable for C3.

Table 9.9 Contingency set.

Contingency ID	Fault bus	Fault clearance	Tripped line
C1	Bus 4	0.25 s	Line 4–5
C2	Bus 21	0.16 s	Line 21–22
C3	Bus 29	0.10 s	Line 29–26

Table 9.10 Initial base dispatch (MW).

Generator	Output	Generator	Output	Generator	Output
G30	239.5	G33	624.0	G36	553.0
G31	560.9	G34	504.1	G37[a]	540.0
G32[a]	650.0	G35	644.9	G38	822.4
Total cost		39 173.9 $/h		G39[a]	1000.0

[a] DFIG wind generators.

Table 9.11 Transient stability margin for initial dispatch.

Contingency	C1	C2	C3
Stability margin	53.6	30.8	−71.6
Robustness	79.7%	82.3%	0.6%

To examine the robustness of this initial dispatch to uncertain wind power variation, a total of M wind power output scenarios are randomly generated within the variation range $[\mu - \sigma, \mu + \sigma]$. The robustness degree is evaluated as follows:

$$\gamma = \frac{M_s}{M} \times 100\% \tag{9.19}$$

where M is the total number of validation scenarios and M_s is the number of stable scenarios. In this test, M is set to 1000.

The robustness degrees for the three contingencies are also listed in Table 9.11. It is important to note that given random wind power fluctuations, C1 and C2, which are initially stable become unstable for 20.3% and 17.7% cases, respectively, while C3 is unstable for 99.4% cases. The system stability is therefore heavily impaired by the wind power uncertainty.

9.5.3 Single-contingency Case for New England 10-Machine 39-Bus System

The single contingency C1 is studied first. To illustrate the introduced approach, the stabilizing process for the first uncertain scenario l_1 is presented. For this scenario, the system is unstable, the multi-rotor angles are shown in Figure 9.8a, and the CM is G31. Based on EEAC/SIME, the OMIB equivalence is obtained and the P_e-OMIB angle plane is shown in Figure 9.9a, where P_e crosses P_m at t_u 0.39 second. The corresponding stability margin of the U-OMIB is −89.34. The corresponding U-OMIB trajectory is shown in Figure 9.10 (solid line). The trajectory sensitivities of the synchronous machine's active power output with respect to the OMIB angle are shown in Figure 9.11.

The quasi-linearity of the stability margin with respect to some key parameters is examined here. The generation output of the three wind farms and the fault clearing time are varied, and then the corresponding stability margins are calculated. As shown in Figure 9.12, a general quasi-linear relationship between these parameters can be observed. Especially, around the zero stability margin range, the linearity is quite strong. While for very stable and very unstable conditions, where the stability margin is close to −100 or 100, the linearity becomes weaker. Since the quasi-linear sensitivity of the stability margin is mainly used to estimate the CCT in this section, the accuracy is acceptable for engineering use (it is worth

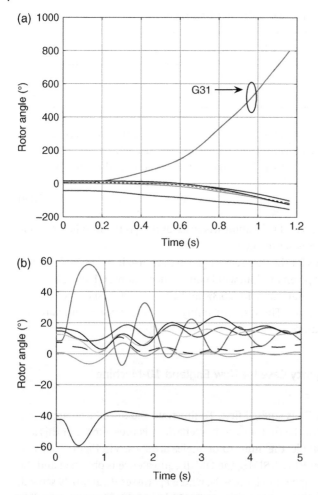

Figure 9.8 Rotor angle trajectories for C1 under l_1: (a)-unstable (under initial clearing time); (b)-stable (under CCT).

indicating that the industry-grade software packages FASTEST and DSATools both utilize this quasi-linear property to estimate the CCT).

Based on the quasi-linear relationship between the stability margin and the fault clearing time, the CCT is extrapolated by an additional EEAC run with a smaller fault clearing time. It is estimated that the CCT is 0.142 seconds. Under the CCT, the corresponding multi-rotor angles and P_e-OMIB plane, and C-OMIB trajectory are shown in Figures 9.8b, 9.9b, and 9.10 (dash line), respectively. The corresponding stability margin of the C-OMIB is 1.96 (which is very close to 0). In Figure 9.9b, it is important to observe that P_e curve starts to return at t_r 0.68 second. It is also

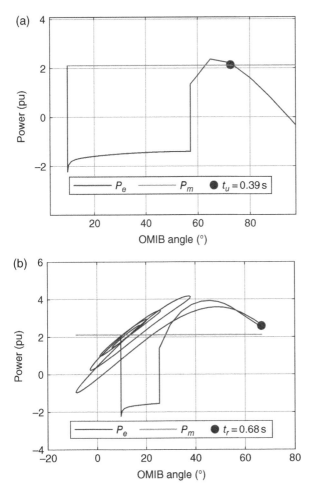

Figure 9.9 Pe-OMIB angle plane for C1 under l_1: (a)-unstable (under initial clearing time); (b)-stable (under CCT).

important to note that at t_r, P_e is very close to P_m because the stability margin is close to 0. This implies that the estimated CCT is very accurate.

In Figure 9.10, it can be observed that the U-OMIB angle excurses rapidly, especially after t_u. By contrast, the C-OMIB angle returns to a stable level after first-swing. Based on the proposed stability constraint Eq. (9.16), the U-OMIB angle at t_u should be limited to that of the C-OMIB. The trajectory sensitivities in Figure 9.11 show that the CM, G31, has positive sensitivities implying that an increase in its active output will raise the OMIB angle, and the remaining generators have negative sensitivities meaning that an increase in their active output will

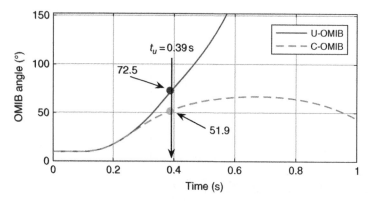

Figure 9.10 OMIB angle trajectories for C1 under l_1.

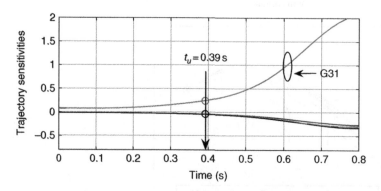

Figure 9.11 Trajectory sensitivities of synchronous machine's output to OMIB angle.

decrease the OMIB angle. Note that only the sensitivities at t_u are used for formulating the stabilization constraint Eq. (9.17). Then, the stabilization constraint is added back to the master problem for stabilizing the system. After 1st iteration, the system becomes stable for scenario l_1. The computation process repeats until all the scenarios are stable for this contingency. The final dispatch results for single contingency C1 are listed in Table 9.12. The robustness degree of this solution becomes 99.8%. It is worth noting that the total generation cost becomes only 0.46% higher, which implies the optimality of the proposed solution approach

9.5.4 Multi-contingency Case for New England 10-Machine 39-Bus System

The three contingencies C1, C2, and C3 are all considered simultaneously. Using the scenario-decomposition approach, after 4 iterations, a robust stable dispatch solution is obtained. The dispatch results are listed in Table 9.13.

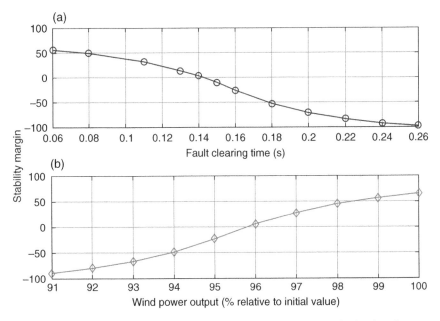

Figure 9.12 Transient stability margin versus key parameters: (a)-fault clearing time; (b)-wind power output.

Table 9.12 C1-constrained robust stability dispatch results.

Generator	Output	Generator	Output	Generator	Output
G30	251.6	G33	646.7	G36	573.6
G31	444.9	G34	519.7	G37[a]	540.0
G32[a]	650.0	G35	667.0	G38	849.0
Total cost		39 352.8 $/h		G39[a]	1000.0

[a] DFIG wind generators.

Table 9.13 Multi-contingency robust stability dispatch results.

Generator	Output	Generator	Output	Generator	Output
G30	411.7	G33	687.0	G36	523.0
G31	442.9	G34	547.2	G37[a]	540.0
G32[a]	650.0	G35	613.9	G38	721.4
Total cost		40 072.6 $/h		G39[a]	1000.0

[a] DFIG wind generators.

Table 9.14 Transient stability margin for multi-contingency dispatch.

Fault / Scenario	C1	C2	C3
l_1	12.4	57.4	16.9
l_2	87.1	57.4	3.8
l_3	87.1	44.1	4.2
l_4	90.4	41.6	7.0
Robustness	100%	99.2%	98.6%

For this dispatch solution, the stability margins under four TOAT-selected scenarios are given in Table 9.14 and the robustness degrees for 1000 randomly sampled wind power scenarios are also given. It can be seen that after satisfaction with the four TOAT selected scenarios, the dispatch becomes highly immune against random wind power variations. For C1, the system is always stable for the 1000 random scenarios, while for C2 and C3, only 11 and 14 cases are unstable, respectively. It is also worth indicating that the generation cost is increased by 2.1% to be robustly stable for the three contingencies, which is quite economical for practical implementation.

Figures 9.8 and 8.9 illustrate a single CM case that appears under C1 and scenario l_1. The illustration of multiple CMs is shown in Figures 9.13 and 8.14, which is obtained under C2 and scenario l_2. Note that Figures 9.13a and 9.14a show the angles before the dispatch, and Figures 9.13b and 9.14b show the angles after the dispatch.

9.5.5 Numerical Test on Nordic32 System

To further validate the method, a larger test system, the Nordic32 system [51], is tested in this section. This system consists of 23 generators, 41 buses, and 69 branches, and its one-line diagram is shown in Figure 9.15.

This system has been widely used for power system dynamics study in the literature. In this section, all the generation cost coefficients are set to equal values; hence, the optimal dispatch is to minimize the power loss of the network. An initial operating state is obtained by running the static optimal dispatch for the original system. Five synchronous machines (g1, g10, g17, g17b, and g18) are then replaced by wind farms modeled as equivalent DFIGs. This corresponds to a wind power penetration level of 25.2%. The mean power output of these wind farms μ is set to the initially dispatched value of the replaced synchronous machines, and it is assumed that the standard deviation σ is ±10% of its mean value. Given the five

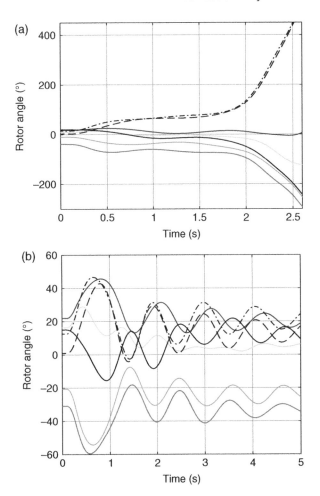

Figure 9.13 Rotor angle trajectories for C2 under l_2: (a)-unstable (before dispatch); (b)-stable (after dispatch).

uncertain variables, a total of eight TOAT testing scenarios are selected from $L_8(2^7)$.

For this system, an extremely unstable situation is studied, a three-phase short-circuit fault at bus 4022 that cleared after 0.16 seconds. TDS shows that the system is unstable for all eight wind power variation scenarios. An illustration of the extremely unstable trajectories is given in Figure 9.16.

For such an extremely unstable situation where the P_e curve does not cross P_m curve, the instability margin is measured by the shortest distance between P_e and P_m curves. For Figure 9.16, the instability margin η' is −0.48 (pu in power). Note

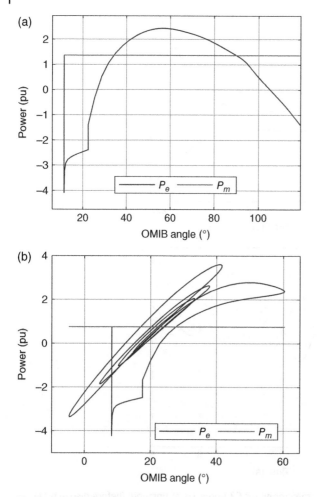

Figure 9.14 Pe-OMIB angle plane for C2 under l_2: (a)-unstable (before dispatch); (b)-stable (after dispatch).

that this is in a different unit from the normally unstable situation. The relationship between η' and fault clearing time is examined in Figure 9.17, where a very clear quasi-linearity can be seen. Note that, after the fault clearing time is smaller than 0.133 s, the system trajectories become "normally unstable" where P_e and P_m curves have an intersection, and the stability margin should be then measured by the difference between the deceleration area and the acceleration area.

The "time to instability" t_u is estimated by a backward extrapolated intersection in Figure 9.16b, which is 0.352 second. Using the DSATools, the CCT is estimated as 0.044 second.

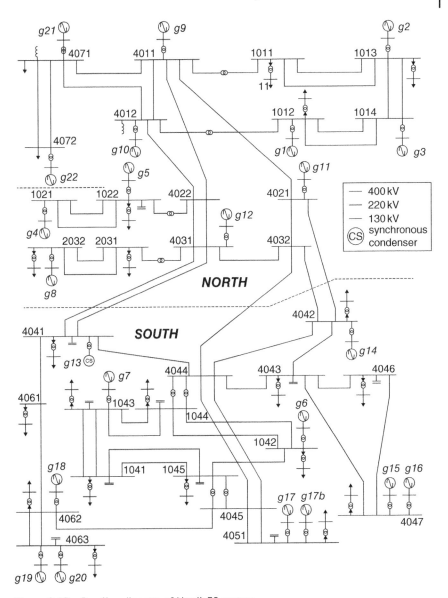

Figure 9.15 One-line diagram of Nordic32 system.

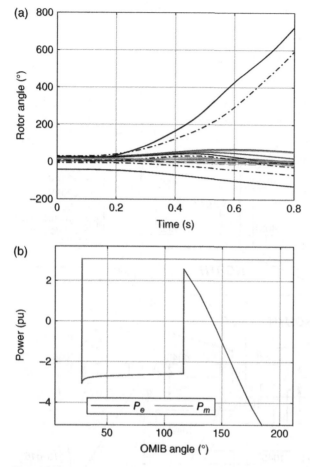

Figure 9.16 Unstable trajectories for Nordic32 system before dispatch: (a)-multi-machine angle; (b)-Pe-OMIB angle.

Using the introduced approach, a robust transient stability-constrained dispatch solution is obtained, which is stable for all eight testing scenarios. It is noted that for this extremely unstable situation, five iterations are needed. To validate it, a 1000 random scenario test is done and the robustness degree of the obtained solution is 99.7%. An illustration of the stable trajectories is given in Figure 9.18.

9.5.6 Computation Efficiency Analysis

Transient stability-constrained dispatch problem is naturally computationally demanding for its high dimensional and high nonlinear characteristics. With stochastic wind power generation being modeled, the computation burden will

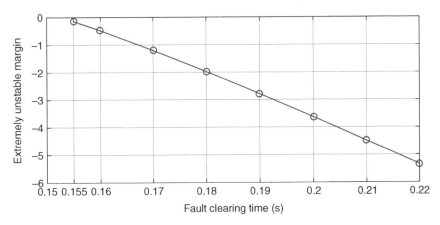

Figure 9.17 Extremely unstable margin versus fault clearing time.

increase further. In this section, we aim to minimize such burden through approximately linearized stabilization constraints and selecting a small set of representative wind power testing scenarios. It should be noted that although a large system can have a large number of contingencies, there is no need to include all of them in the dispatch process. Only the severe (unstable) and very likely (high probability) ones need to be considered. In practice, a contingency set can be selected based on historical statistics.

For the introduced computation method, only unstable ones will generate stabilization constraints and be included in the optimization process. In each iteration, the computation involves solving an OPF model (master problem) and performing TDS for each contingency under each wind power scenario (slave problems). For each unstable case, the stabilization constraint is formed based on EEAC and trajectory sensitivity. Since the OPF solution and EEAC calculation are very efficient, and the trajectory sensitivity can be obtained as a byproduct of TDS with implicit integration techniques, the total computation time is mainly spent on the TDS phase. For the proposed solution method, due to the strong effectiveness of the stabilization constraints, the computation converges within 1–5 iterations, which is consistent with results in the literature [8].

In this section, the OPF is solved using the MATPOWER, and EEAC is realized in MATLAB platform, TDS is performed using DSATools/TSAT module. The total and respective CPU times for each phase are given in Table 9.15. Note that as a state-of-the-art industry-grade TDS software package, DSATools is very fast. Therefore, with such high-performance TDS tools, the computation speed can allow for online applications.

Besides, it is important to indicate that given the decomposition structure of the proposed method, the computation can be parallelized. For example, if X CPU cores are used, the speed can be increased up to X times by distributing the

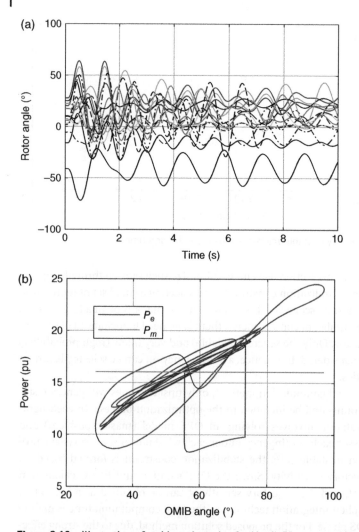

Figure 9.18 Illustration of stable trajectories for Nordic32 system after dispatch: (a)-multi-machine angle; (b)-Pe-OMIB angle.

Table 9.15 CPU time (s) for computational tasks.

Test system	OPF	EEAC	TDS	Total
New England	0.12	0.2	0.16	3.2
Nordic system	0.26	0.2	0.22	9.7

subproblems among the CPU cores. It is worth mentioning that a parallel computing platform for PSS/E-based TSA has been developed in Ref. [46], and cluster computing of trajectory sensitivities has been illustrated in Ref. [50]. Parallel computing is widely available at a modern power system control center, so the introduced approach here can make full use of such computing power for practical implementation.

9.6 Discussions and Concluding Remarks

For the EEAC/SIME paradigm, the value of the stability margin may be discontinuous with varying parameters such as fault clearing time for two reasons: (i) the OMIB structure may be changed, and (ii) under extremely unstable and normally unstable situations, the calculation principle of margin is different. However, for the proposed approach in this section, the stability margin is only used for rapidly estimating CCT. In fact, alternative methods such as binary search can also be used, which does not affect the effectiveness of the proposed method. During the iteration process, when the OMIB structure changes, the corresponding C-OMIB is updated for the next iteration.

Besides, it is important to emphasize that, for the introduced method here, the EEAC/SIME paradigm is used for determining a control reference, which includes the instability time and the corresponding OMIB angle. Based on the reference, the trajectory sensitivity is calculated to determine the required generation shifting. Note that the trajectory sensitivity is calculated for the OMIB angle, not for the stability margin. Moreover, for calculating the trajectory sensitivity, a small perturbation step should be used, especially near the stability boundary [50], to avoid the excessive re-dispatch cost.

Nomenclature

Indices

i, j indices for generators
k index for contingencies
t index for the time
W index for wind turbines
C index for critical machine
N index for non-critical machine

Sets

$\tilde{\varsigma}$	sets of uncertain dynamic load model parameters
$\mathbf{x}$	sets of state variable
$\mathbf{u}$	sets of control variable
$\tilde{\mathbf{w}}$	sets of uncertain wind active power output
S^+/S^-	sets of generators with positive and negative sensitivities

Variables

$\hat{P}_{mi}(t_0)$	active power output of i-th generator obtained from master problem
τ_n	required generation shifting for TSC
t_u	instability time of the unstable OMIB
$\hat{\delta}(t_u)$	targeted equivalent OMIB angle at the instability time t_u,
$\delta_{CT}(t_u)$	reference critical OMIB (C-OMIB) trajectory of the same system at the instability time t_u
$\delta_{UT}(t_u)$	unstable OMIB angle of the original system at t_u
$\Delta\delta(t_u)$	required reduced OMIB angle to be smaller than $\delta_{CT}(t_u)$
ΔP_G	active power output change for synchronous generators
$\Phi_i(\delta_{UT}, t_u, P_{Gi})$	OMIB angle trajectory sensitivity with respect to active power output of generator i at the instability time t_u
$\hat{P}_G$	base dispatch from the master problem
N_{It}	total iteration number
N_{ui}	number of unstable scenarios in the i-th iteration

Parameters

$\Delta\lambda$	small perturbation
M	inertia coefficient
ε	desired transient stability margin
η_{us}	unstable stability margin
T_{OPF}	CPU time for OPF calculation
T_{TSA}	CPU time for TSA for a contingency
K/H	number of contingencies and the testing scenarios determined by TOAT

References

1 Gan, D., Thomas, R.J., and Zimmerman, R.D. (2000). Stability-constrained optimal power flow. *IEEE Transactions on Power Systems* 15 (2): 535–540.

2 Chen, L., Taka, Y., Okamoto, H. et al. (2001). Optimal operation solutions of power systems with transient stability constraints. *IEEE Transactions on Circuits and Systems I: Fundamental Theory and Applications* 48 (3): 327–339.

3 Yue, Y., Kubokawa, J., and Sasaki, H. (2003). A solution of optimal power flow with multicontingency transient stability constraints. *IEEE Transactions on Power Systems* 18 (3): 1094–1102.

4 Jiang, Q. and Huang, Z. (2010). An enhanced numerical discretization method for transient stability constrained optimal power flow. *IEEE Transactions on Power Systems* 25 (4): 1790–1797.

5 Nguyen, T.B. and Pai, M.A. (2003). Dynamic security-constrained rescheduling of power systems using trajectory sensitivities. *IEEE Transactions on Power Systems* 18 (2): 848–854.

6 Ruiz-Vega, D. and Pavella, M. (2003). A comprehensive approach to transient stability control. I. near optimal preventive control. *IEEE Transactions on Power Systems* 18 (4): 1446–1453.

7 Pizano-Martinez, A., Fuerte-Esquivel, C.R., and Ruiz-Vega, D. (2011). A new practical approach to transient stability-constrained optimal power flow. *IEEE Transactions on Power Systems* 26 (3): 1686–1696.

8 Pizano-Martianez, A., Fuerte-Esquivel, C.R., and Ruiz-Vega, D. (2010). Global transient stability-constrained optimal power flow using an OMIB reference trajectory. *IEEE Transactions on Power Systems* 25 (1): 392–403.

9 Zarate-Minano, R., Cutsem, T.V., Milano, F., and Conejo, A.J. (2010). Securing transient stability using time-domain simulations within an optimal power flow. *IEEE Transactions on Power Systems* 25 (1): 243–253.

10 Tu, X., Dessaint, L.-A., and Kamwa, I. (2014). Fast approach for transient stability constrained optimal power flow based on dynamic reduction method. *IET Generation, Transmission & Distribution* 8 (7): 1293–1305. https://digital-library. theiet.org/content/journals/10.1049/iet-gtd.2013.0404.

11 Mo, N., Zou, Z.Y., Chan, K.W., and Pong, T.Y.G. (2007). Transient stability constrained optimal power flow using particle swarm optimisation. *IET Generation, Transmission & Distribution* 1 (3): 476–483.

12 Cai, H.R., Chung, C.Y., and Wong, K.P. (2008). Application of differential evolution algorithm for transient stability constrained optimal power flow. *IEEE Transactions on Power Systems* 23 (2): 719–728.

13 Xu, Y., Dong, Z.Y., Meng, K. et al. (2012). A hybrid method for transient stability-constrained optimal power flow computation. *IEEE Transactions on Power Systems* 27 (4): 1769–1777.

14 Genc, I., Diao, R., Vittal, V. et al. (2010). Decision tree-based preventive and corrective control applications for dynamic security enhancement in power systems. *IEEE Transactions on Power Systems* 25 (3): 1611–1619.

15 Xu, Y., Dong, Z.Y., Guan, L. et al. (2012). Preventive dynamic security control of power systems based on pattern discovery technique. *IEEE Transactions on Power Systems* 27 (3): 1236–1244.

16 Han, D., Ma, J., He, R., and Dong, Z. (2009). A real application of measurement-based load modeling in large-scale power grids and its validation. *IEEE Transactions on Power Systems* 24 (4): 1756–1764.

17 Hiskens, I.A. and Alseddiqui, J. (2006). Sensitivity, approximation, and uncertainty in power system dynamic simulation. *IEEE Transactions on Power Systems* 21 (4): 1808–1820.

18 Guo, Z., Wang, Z.J., and Kashani, A. (2015). Home appliance load Modeling from aggregated smart meter data. *IEEE Transactions on Power Systems* 30 (1): 254–262.

19 Dong, Z., Wong, K.P., Meng, K. et al. (2010). Wind power impact on system operations and planning. In: *IEEE PES General Meeting*, 1–5.

20 Xu, Z., Rosenberg, H., Sorensen, P.E. et al. (2009). Wind energy development in China (WED)-the Danish-Chinese collaboration project. In: *2009 IEEE Power & Energy Society General Meeting*, 1–7. IEEE.

21 Operator, A.E.M. (2013). *100 per Cent Renewables Study–Modelling Outcomes*. Australia: Melbourne.

22 Nunes, M.V.A., Lopes, J.A.P., Zurn, H.H. et al. (2004). Influence of the variable-speed wind generators in transient stability margin of the conventional generators integrated in electrical grids. *IEEE Transactions on Energy Conversion* 19 (4): 692–701.

23 Gautam, D., Vittal, V., and Harbour, T. (2009). Impact of increased penetration of DFIG-based wind turbine generators on transient and small signal stability of power systems. *IEEE Transactions on Power Systems* 24 (3): 1426–1434.

24 Margaris, I.D., Hansen, A.D., Sørensen, P., and Hatziargyriou, N.D. (2011). Dynamic security issues in autonomous power systems with increasing wind power penetration. *Electric Power Systems Research* 81 (4): 880–887.

25 Liang, J., Molina, D.D., Venayagamoorthy, G.K., and Harley, R.G. (2013). Two-level dynamic stochastic optimal power flow control for power systems with intermittent renewable generation. *IEEE Transactions on Power Systems* 28 (3): 2670–2678.

26 Liang, J., Venayagamoorthy, G.K., and Harley, R.G. (2012). Wide-area measurement based dynamic stochastic optimal power flow control for smart grids with high variability and uncertainty. *IEEE Transactions on Smart Grid* 3 (1): 59–69.

27 Wang, J., Shahidehpour, M., and Li, Z. (2008). Security-constrained unit commitment with volatile wind power generation. *IEEE Transactions on Power Systems* 23 (3): 1319–1327.

28 Zheng, Q.P., Wang, J., and Liu, A.L. (2015). Stochastic optimization for unit commitment—a review. *IEEE Transactions on Power Systems* 30 (4): 1913–1924.

29 Bertsimas, D., Litvinov, E., Sun, X.A. et al. (2013). Adaptive robust optimization for the security constrained unit commitment problem. *IEEE Transactions on Power Systems* 28 (1): 52–63.

30 Capitanescu, F., Fliscounakis, S., Panciatici, P., and Wehenkel, L. (2012). Cautious operation planning under uncertainties. *IEEE Transactions on Power Systems* 27 (4): 1859–1869.

31 Wang, Y., Xia, Q., and Kang, C. (2011). Unit commitment with volatile node injections by using interval optimization. *IEEE Transactions on Power Systems* 26 (3): 1705–1713.

32 Kundur, P., Paserba, J., Ajjarapu, V. et al. (2004). Definition and classification of power system stability IEEE/CIGRE joint task force on stability terms and definitions. *IEEE Transactions on Power Systems* 19 (3): 1387–1401.

33 Faried, S.O., Billinton, R., and Aboreshaid, S. (2009). Probabilistic evaluation of transient stability of a wind farm. *IEEE Transactions on Energy Conversion* 24 (3): 733–739.

34 Dong, Z.Y., Zhao, J.H., and Hill, D.J. (2012). Numerical simulation for stochastic transient stability assessment. *IEEE Transactions on Power Systems* 27 (4): 1741–1749.

35 Xu, Y., Dong, Z.Y., Xu, Z. et al. (2012). An intelligent dynamic security assessment framework for power systems with wind power. *IEEE Transactions on Industrial Informatics* 8 (4): 995–1003.

36 Pizano-Martínez, A., Fuerte-Esquivel, C.R., Zamora-Cárdenas, E., and Ruiz-Vega, D. (2014). Selective transient stability-constrained optimal power flow using a SIME and trajectory sensitivity unified analysis. *Electric Power Systems Research* 109: 32–44.

37 Jiang, Q., Zhou, B., and Zhang, M. (2013). Parallel augment Lagrangian relaxation method for transient stability constrained unit commitment. *IEEE Transactions on Power Systems* 28 (2): 1140–1148.

38 Peace, G.S. (1993). *Taguchi Methods: A Hands-on Approach*. Addison Wesley Publishing Company.

39 Yu, H., Chung, C.Y., and Wong, K.P. (2011). Robust transmission network expansion planning method with Taguchi's orthogonal array testing. *IEEE Transactions on Power Systems* 26 (3): 1573–1580.

40 Yu, H. and Rosehart, W.D. (2012). An optimal power flow algorithm to achieve robust operation considering load and renewable generation uncertainties. *IEEE Transactions on Power Systems* 27 (4): 1808–1817.

41 Xu, Y., Dong, Z.Y., Meng, K. et al. (2014). Multi-objective dynamic VAR planning against short-term voltage instability using a decomposition-based evolutionary algorithm. *IEEE Transactions on Power Systems* 29 (6): 2813–2822.

42 Siemens (May 2011). PSS/E 33.0 Program Application Guide.

43 Orthogonal Arrays (Taguchi Design) [Online]. www.york.ac.uk/depts/maths/tables/orthogonal.htm (accessed 29 January 2015).

44 Ma, J., Han, D., He, R. et al. (2008). Reducing identified parameters of measurement-based composite load model. *IEEE Transactions on Power Systems* 23 (1): 76–83.

45 Renmu, H., Ma, J., and Hill, D.J. (2006). Composite load modeling via measurement approach. *IEEE Transactions on Power Systems* 21 (2): 663–672.

46 Meng, K., Dong, Z.Y., Wong, K.P. et al. (2010). Speed-up the computing efficiency of power system simulator for engineering-based power system transient stability simulations. *IET Generation, Transmission & Distribution* 4 (5): 652–661. https://digital-library.theiet.org/content/journals/10.1049/iet-gtd.2009.0701.

47 Zimmerman, R.D., Murillo-Sanchez, C.E., and Thomas, R.J. (2011). MATPOWER: steady-state operations, planning, and analysis tools for power systems research and education. *IEEE Transactions on Power Systems* 26 (1): 12–19.

48 DSA *Dynamic Security Assessment Software.* http://www.dsatools.com (accessed 29 January 2015).

49 Wan, C., Xu, Z., Pinson, P. et al. (2014). Probabilistic forecasting of wind power generation using extreme learning machine. *IEEE Transactions on Power Systems* 29 (3): 1033–1044.

50 Hou, G. and Vittal, V. (2012). Cluster computing-based trajectory sensitivity analysis application to the WECC system. *IEEE Transactions on Power Systems* 27 (1): 502–509.

51 Nordic32 system, CIGRE Task Force 38.02.08 (1995). *Long Term Dynamics, Phase II: Final Report.* CIGRE.

10

Optimal Generation Rescheduling for Preventive Transient Stability Control

10.1 Introduction

Generally, according to implemented time, transient stability control (TSC) can be divided into *preventive control* and *emergency control*, where the former aims to protect the system from losing stability before the occurrence of the contingency, and the latter aims to impede the system from losing stability after the contingency has already occurred. Since transient instability can develop very fast (typically within several cycles), preventive TSC is mostly applied. Typically, the transient stability assessment (TSA) is conducted to check if the system is stable when subjected to credible contingencies, if the system is found vulnerable to a contingency, preventive actions such as generation rescheduling will be activated to change the system's operating point to withstand the contingency.

In this chapter, preventive TSC is applied, where the approaches can generally be classified into the *global* approach and *sequential* approach. For the global approach [1–3], the stability control is embedded in an optimal power flow (OPF) framework. Typically, the rotor angle DAEs are converted to a tractable form, e.g. discretized algebraic equations, and then added to the OPF model as the stability constraints [1].

By contrast, the sequential approach [4–7] includes explicit preventive TSC action derivation and stability checking via TSA in a sequential and iterative manner. If the derived control at the last iteration cannot stabilize the system, an updated and improved control action is calculated, and this process iterates until the system stability is validated by the TSA. Usually, the derived preventive TSC action is mathematically explicit and can thus be incorporated into an OPF model without increasing much computational complexity. To derive the required preventive TSC actions, trajectory sensitivity [4, 8, 9], extended equal-area criterion (EEAC) [5, 6], and knowledge extraction [7] techniques have been reported in the literature. Compared with the global approach, although it is argued that the sequential approach can only provide a near-optimal solution, it has many advantages: it can

Stability-Constrained Optimization for Modern Power System Operation and Planning,
First Edition. Yan Xu, Yuan Chi, and Heling Yuan.
© 2023 The Institute of Electrical and Electronics Engineers, Inc.
Published 2023 by John Wiley & Sons, Inc.

detect significant variables for stabilizing the system and account for the advocated control actions, which is transparent and interpretable to system operators and market participants; it is more computationally efficient (typically, it only requires $1 \sim 3$ iterations to stabilize the system [4–7]), an analysis about the time increase with more transient stability constraints is reported in Ref. [10]; and it is flexible for implementation, can be either activated directly or incorporated in an OPF model [4, 5, 7]. The sequential approach can either be used for online preventive TSC based on real-time TSA results or as a daily or hourly ahead generation dispatching tool. In the former case, the preventive TSC is to reschedule the current operating state. In the latter case, a base operating point is obtained by solving an original OPF model and followed by TSA, if the operating point is unstable with respect to the contingency, the sequential approach is applied to modify the generation dispatching results.

As one promising sequential preventive TSC approach, trajectory sensitivity analysis can obtain sensitivity metrics of the control variables with respect to system dynamic trajectories; based on these sensitivities, required control actions can be explicitly computed to restore the unstable trajectory to a stable one. Compared with other sequential approaches, this approach is mathematically rigorous, and the parameters that influence system stability can be quantitatively analyzed. However, it is also observed in previous works that the transient stability criterion and the control reference are based on heuristically selected values of maximum relative rotor angle deviation [4, 8, 9], which are however inexact and have to be tentative. Besides, the trajectory sensitivities along the whole analysis time window are calculated, which can be very time consuming.

Therefore, this chapter proposes two improved sequential approaches for preventive TSC. The trajectory sensitivity analysis is performed (i) on the OMIB equivalent angle trajectory of the multi-machine system, and exact instability time/angle is systematically determined based on the OMIB power-angle curve, the transient stability is controlled by bounding the OMIB angle excursion at the instability time within a reference value obtained from the critical-OMIB, which corresponds to the marginally stable condition of the multi-machine system; (ii) on the stability margin obtained from EEAC, and transient stability is constructed by bounding the stability margin larger than zero after wind power output variation, which is the marginally stable condition as well. The simulation is conducted on New England 10-machine 39-bus system and a 285-machine and 1648-bus system.

10.2 Trajectory Sensitivity Analysis for Transient Stability

The principle and concept of trajectory sensitivity are comprehensively introduced in Chapter 2. In the current research area, trajectory sensitivity is widely applied to transient stability enhancement and control. In the literature, applying trajectory sensitivity for preventive TSC is mainly realized by controlling the system's maximum relative angle deviation $\delta_{ij}^{\max}$ at the time (denoted as T_x) when it exceeds a pre-defined threshold, e.g. 180°, to be smaller than a reference threshold σ:

$$\delta_{ij}^{\max}(T_x) \leq \sigma \tag{10.1}$$

where generators i and j denote the most and the least advanced generators, respectively [4, 8, 9].

Based on trajectory sensitivity, the required control action can be explicitly calculated by:

$$\Delta\lambda = \frac{\sigma - \delta_{ij}^{\max}(T_x)}{\Phi(x_0, T_x, \lambda)} \tag{10.2}$$

where $\Delta\lambda$ can be generation rescheduling, load shedding, or reactive power compensation, etc. and $\Phi(x_0, T_x, \lambda)$ is the trajectory sensitivity at T_x.

The inherent merit of the trajectory sensitivity-based method is that the TSC is transparent and mathematically rigorous; besides, the impact of different variables on transient stability can be quantitatively indicated; thus, a deeper understanding of the system's dynamic characteristics can be revealed, which can support system operators and planners to make a better decision in practice; moreover, the derived stability constraints are linear and simple, which can be readily added into a standard OPF without additional computational complexity.

However, traditional trajectory sensitivity-based TSC always has two key demerits: (i) the selection of the instability angle deviation threshold $\delta_{ij}^{\max}$ is heuristic or arbitrary, which is often set to a fixed value, e.g. 180° [1, 2, 4, 8, 9], but in fact, it is system-dependent and/or OP-dependent. An empirically fixed threshold is not accurate to represent the threshold; (ii) The selection of the control reference angle σ is also difficult to define since a smaller value may lead to conservative results and higher operating costs, while a more relaxed value may result in instability even if the threshold is not exceeded. But in prevailing literatures, this value is empirically selected [4, 8, 9].

It is obvious that, without proper selection for the thresholds, $\delta_{ij}^{\max}$ and σ, the derived control action according to (10.2) is either ineffective to stabilize the system or conservative to less economical. In addition, the trajectory sensitivities at each time step need to be calculated, which is quite time-consuming [4, 8]. In the rest of this chapter, the instability time/angle, control reference angle, and stability margin are systematically and exactly determined based on the EEAC/OMIB method. Trajectory sensitivity calculation is conducted: (i) on the OMIB angle trajectory at only the instability time for deriving the required control action and (ii) based on the stability margin for deriving the required control action.

In this chapter, the control reference angle is the OMIB angle under the critical clearing time (CCT) of the fault. Hence, it is necessary to predict the CCT under a

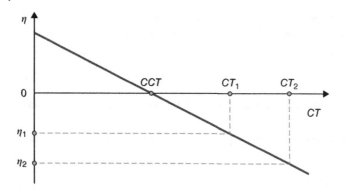

Figure 10.1 Fault clearing time (CT) versus stability margin η.

fault. In literatures [5, 11, 12], a quasi-linear relationship between the stability margin η and variations of some specific parameters has been verified, which contains fault clearing time (CT) – shown in Figure 10.1

In general, the stability margin η under CCT is zero or a very small positive value. According to this characteristic, the CCT can be extrapolated via two successive TDS runs and calculated as below:

$$CCT = \frac{\eta_1 CT_2 - \eta_2 CT_1}{\eta_1 - \eta_2} \tag{10.3}$$

Practically, there may be more than two groups of machine angles. Under this case, in order to recognize the OMIB structure, several decomposition patterns of the generators into two candidate groups of CMs and NMs are constructed according to the rotor angle deviations between adjacent machines' rotor trajectories, forming a series of candidate OMIBs. The first candidate OMIB that meets the instability condition (or the one with the least stability margin) is determined as the OMIB equivalent, which is to represent the multi-machine system [11, 12].

In order to find the accurate instability time, stability margin, and appropriate control reference angle, the EEAC theory is used here. The comprehensive introduction to EEAC can be found in Chapter 2.

10.3 Transient Stability Preventive Control Based on Critical OMIB

10.3.1 Stability Constraint Construction

The C-OMIB involves the OMIB equivalence of the multi-machine system under the CCT subjected to the fault. The C-OMIB means the system is under critically stable conditions, which can be leveraged as an exact reference for TSC.

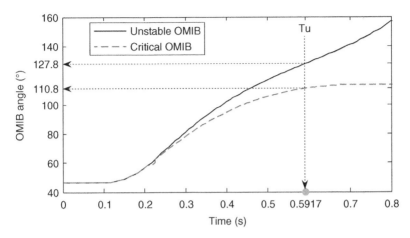

Figure 10.2 Unstable and critical OMIB trajectories – New England test system.

Particularly, bounding the unstable OMIB by the C-OMIB can be the designed control action for TSC.

According to this concept, [13] constructed a transient stability constraint to limit the angle of the unstable OMIB trajectory at instability time $\delta_{UT}(T_u)$ to be bounded by the angle of C-OMIB at the same time $\delta_{CT}(T_u)$:

$$\delta_{UT}(T_u) \leq \delta_{CT}(T_u) \tag{10.4}$$

Figure 10.2 clearly depicts the concept. The C-OMIB and the unstable OMIB angle trajectories with the same group of CM and NM are plotted.

In Ref. [13], the transient stability constraint (10.4) is incorporated in the standard TSC-OPF model, making use of the merit that an accurate instability angle can be obtained and thus the control is derived from the marginally stable trajectory, which can avoid excessive control cost; in addition, only one single and linear transient stability constraint is required, hence the computational burden will be remarkably saved based on the instability time T_u.

Based on this concept and trajectory sensitivity, this section proposes an advanced preventive TSC approach, where the trajectory sensitivities are calculated for the unstable OMIB angle trajectory at the instability time T_u, then the C-OMIB is formulated under CCT subjected to the same fault, and finally, the required stability control actions are obtained via a transient stability constraint (10.4) optimization model. The comprehensive mathematical model is introduced subsequently.

10.3.2 Mathematical Model

In this chapter, preventive generation rescheduling, which is proven to be the most potent and effective control action in practice, is used for the TSC. But it should be mentioned that other control actions, such as emergency load shedding, can be taken into consideration as well when necessary.

In previous works, the generation is empirically shifted between the generators, which have the lowest and the highest trajectory sensitivities regarding the maximum relative angle deviation [4, 8, 9]. In this chapter, according to the stability constraint (10.4), the required generation shifting to stabilize the system against a contingency is mathematically derived by solving the following linear programming problem:

$$\min_{\Delta P_G} \quad -\sum_{i \in S^+} \Delta P_{Gi} \tag{10.5}$$

$$\text{s.t.} \quad \Delta P_{Gi} \leq 0, \quad i \in S^+ \tag{10.6}$$

$$\Delta P_{Gj} \geq 0, \quad j \in S^- \tag{10.7}$$

$$\sum_{i \in S^+} \Delta P_{Gi} + \sum_{j \in S^-} \Delta P_{Gj} = 0 \tag{10.8}$$

$$\sum_{i \in S^+} \Phi_i(\delta_{UT}, T_u, P_{Gi}) \Delta P_{Gi} + \\ \sum_{j \in S^-} \Phi_j(\delta_{UT}, T_u, P_{Gj}) \Delta P_{Gj} \leq \delta_{CT}(T_u) - \delta_{UT}(T_u) \tag{10.9}$$

$$P_{Gk}^{\min} \leq P_{Gk}^0 + \Delta P_{Gk} \leq P_{Gk}^{\max}, \quad \forall k \in \{S^- \cup S^+\} \tag{10.10}$$

where ΔP_G denotes the active power output variation of dispatchable generators, which is the control variable, and subscripts i, j, k represent the generator number; P_{Gk}^0 represents the active power generation output of generator k at the base case; S^- and S^+ are the sets of generators with negative and positive trajectory sensitivities, respectively (note that positive sensitivity indicates that reducing generation output will decrease the OMIB angle); $\Phi_i(\delta_{UT}, T_u, P_{Gi})$ denotes the OMIB angle trajectory sensitivity regarding the active power output of generator i at the instability time T_u; $P_{Gk}^{\min}$ and $P_{Gk}^{\max}$ denotes the upper and lower bounds of the active power output of generator k, and the objective function (10.5) here is to minimize the total control amount of generation variation. Constraints (10.6) and (10.7) are the direction for generation shifting; (10.8) shows that the generation shifting should be balanced between different generator sets; (10.9) represents the TSC; and (10.10) bounds the generation output limits.

10.3.3 Computation Process

The overall computation flowchart of the proposed approach is depicted in Figure 10.3, and the steps are listed as follows:

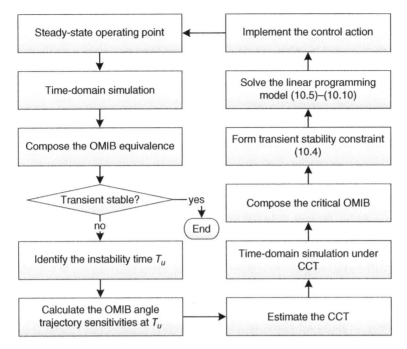

Figure 10.3 Computation flowchart of the proposed approach.

1) Given a credible contingency, perform the TDS and form the OMIB equivalence for the multi-machine trajectories based on EEAC.
2) Evaluate the transient stability of the system based on EEAC, if the system is stable, terminate the computation; otherwise, move to Step 3. Note that the earlier termination can be used during the TDS.
3) Given the unstable OMIB, identify the instability time T_u.
4) Calculate the OMIB angle trajectory sensitivities at the instability time T_u with respect to the active output of each dispatchable generator.
5) Perform an additional TDS with a different fault CT and estimate the CCT using (10.3).
6) Perform the TDS under the CCT of the contingency.
7) Compose the C-OMIB for the same composition as the unstable OMIB.
8) Find out the OMIB angles $\delta_{UT}(T_u)$ and $\delta_{CT}(T_u)$, and formulate the transient stability constraint (10.4).
9) Solve the linear programming model (10.5)–(10.10) for the TSC action.
10) Implement the solved TSC action and return to Step 1.

Since the programming problem is nonlinear and the C-OMIB only accurately indicates the maximum angle deviation at the base OP but not at the updated OP, the stabilization process may not be finished after only one iteration, namely, the generation shifting may not be adequate. Hence, it is necessary to check the updated OP to be stable; otherwise, the C-OMIB and the trajectory sensitivities should be recalculated for obtaining an updated TSC in subsequent iterations.

10.4 Case Studies of Transient Stability Preventive Control Based on the Critical OMIB

10.4.1 Simulation Setup

The proposed approach is verified on two test systems. The simulation is conducted on a 64-bit PC with 3.10 GHz CPU and 4.0 GB RAM. During the test, the TDS is run by the commercial software package PSS/E [14], and the OMIB trajectory sensitivity is calculated numerically. Other operations, including constructing the OMIB and solving the programming model, are realized in the MATLAB platform.

10.4.2 New England 10-Machine 39-Bus System

The initial operating point and the generation limit for each generator are shown in Table 10.1.

10.4.2.1 System Trajectories

The three-phase short-circuit fault is applied to bus 29 at 0.1 second and cleared after 0.12 second. The multi-machine trajectories in the center of inertia (COI) frame are shown in Figure 10.4. The maximum relative angle exceeds 180° at $T_x = 0.8167$ seconds. Previous practice is to control the system's relative angle at this time to be smaller than an empirically selected threshold, e.g. 180° or 160° [4, 8, 9].

Table 10.1 Active generation output (MW) at base operating point.

Generator	G30	G31	G32	G33	G34
P_G^0	250	572.93	650	632	508
P_G^{max}	420	780	960	900	780
Generator	G35	G36	G37	G38	G39
P_G^0	650	560	540	830	1005.7
P_G^{max}	900	900	840	1440	1080

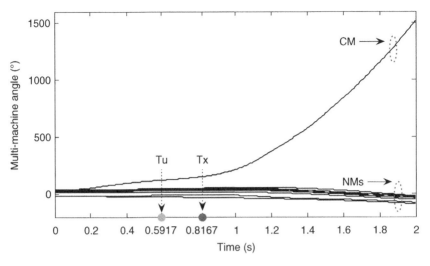

Figure 10.4 Unstable multi-machine system trajectories – New England test system.

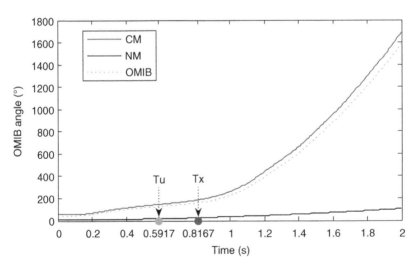

Figure 10.5 Equivalent trajectories of CM, NMs, and OMIB – New England test system.

The system's equivalent OMIB trajectory is then constructed based on EEAC. It is found that #G38 is the only CM and the remaining machines are NMs. The equivalent trajectories of CM and NMs as well as the OMIB equivalence can be found in Figure 10.5, and the power-angle curve of the OMIB can be found in Figure 10.6.

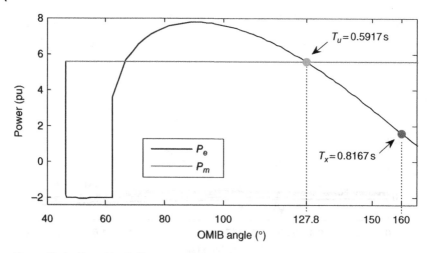

Figure 10.6 Unstable OMIB power-angle curve – New England test system.

According to Figure 10.6, P_e crosses P_m at the instability time $T_u = 0.5917$s, which is earlier than $T_x = 0.8167$s. Hence, previous practice controlling the system trajectories at T_x tends to be late. At instability time the OMIB angle $\delta_{UT}(T_u) = 127.8°$, while the corresponding maximum relative angle of the multi-machine system $\delta_{max}(T_u) = 128.1°$ which does not exceed $180°$ yet. It is important to note that the instability angle is dependent on the system and its operating condition as well as the disturbance, and can only be determined analytically; hence, the empirically selected threshold $180°$ is not appropriate.

T_u and T_x are also marked from Figures 10.4–10.6 for comparison purposes, where it should be noted that the OMIB reveals the very mechanism of the instability while the maximum relative angle-based instability criterion tends to be heuristic and hence inexact. The instability time T_u can also be used to early terminate the TDS, which means, the TDS can be stopped at 0.5917 second when the system becomes unstable. This can save a lot of computation time.

Using (10.3), the CCT is estimated, to be 0.108 second, and the C-OMIB trajectory is constructed using the same composition of the CM and NMs. The unstable OMIB and the C-OMIB trajectories can be found in Figure 10.2.

10.4.2.2 Trajectory Sensitivities

The OMIB trajectory sensitivities to the active output of the 10 generators are calculated. It is important to mention that for practical application, it is only needed to calculate the sensitivities for the instability time $T_u = 0.5917$ second rather than the whole analysis time frame; hence, a great deal of computation time can be saved. Yet in this case study, since the numerical approach is used, and for better capture of the accuracy of the method, the sensitivities along the whole trajectory (0~2 seconds) are calculated.

The time-varying trajectory sensitivity of each generator is shown in Figure 10.7, where the values at $T_u = 0.5917$ seconds of the ten generators are compared in an embedded bar chart. It should be noted that only the CM #G38 has a positive sensitivity value, which means the reduction in its output decreases the OMIB angle; on the contrary, the remaining generators (NMs) all have negative trajectory sensitivities, which means reducing their outputs will play an adverse role for stabilization (i.e. increase the OMIB angle). This is consistent with the findings in Refs. [5, 11, 12] that CMs should reduce while NMs should increase in generation output for stabilizing the system. Detailed values of the sensitivities are listed in Table 10.2. In practice, these sensitivity values provide a deeper insight into the

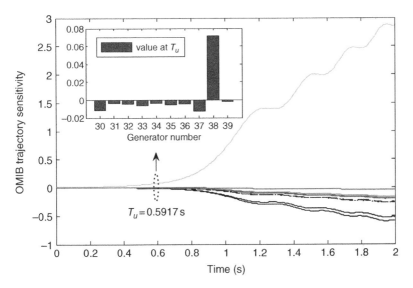

Figure 10.7 OMIB trajectory sensitivities with respect to generator active output – New England test system. *Source:* Xu et al. [15]/With permission of John Wiley & Sons.

Table 10.2 Trajectory sensitivities at $T_u = 0.5917$ second to generator active output.

Generator	G30	G31	G32	G33	G34
Sensitivity	−0.0118	−0.0039	−0.0046	−0.0060	−0.0038
Generator	G35	G36	G37	G38	G39
Sensitivity	−0.0058	−0.0044	−0.0128	0.0719	−0.0021

significant variables that affect system transient stability under the considered disturbance.

10.4.2.3 Stabilizing Results

To stabilize the system, the unstable OMIB angle at instability time $\delta_{UT}(T_u)$ is to be bounded by that of the C-OMIB at the same time $\delta_{CT}(T_u)$. According to Figure 10.2, $\delta_{UT}(T_u) = 127.8°$ and $\delta_{CT}(T_u) = 110.8°$.

By solving the linear model (10.5)–(10.10), the required generation shifting is from #G38 to #G37 by 200.7 MW, i.e. $\Delta P_{G38} = -\Delta P_{G37} = -200.7$ MW.

After this control, the system is stabilized. The resulting OMIB angle trajectory can be estimated, which is found very approximate to the C-OMIB (see Figure 10.8). This verifies that the numerically calculated trajectory sensitivities and the derived control actions are very accurate.

The system trajectories are shown in Figure 10.9. The power-angle curve of the system OMIB after control is plotted in the upper window, where it can be seen that the "time to first swing stability" $T_r = 0.7083$ second when the OMIB angle returns, hence, the TDS can be terminated at this time, saving a lot of execution time. The multi-machine system trajectories after the control are simulated and shown in the lower window, which validates the stability of the system after control.

10.4.2.4 Control Accuracy

It should be indicated that after the preventive TSC, the stability margin becomes 3.51, presenting a positive surplus against the critical value 0. This implies that the derived generation rescheduling (200.7 MW) is slightly conservative. The reason is twofold: first, the trajectory sensitivity is calculated by the numerical method with

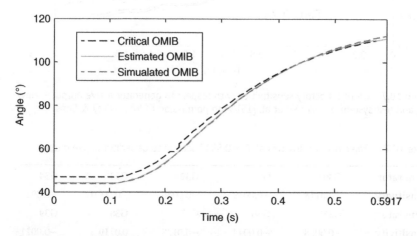

Figure 10.8 OMIB trajectories – New England test system.

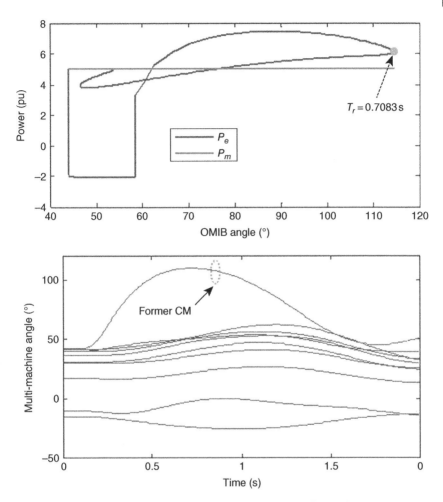

Figure 10.9 System trajectories after stability control – New England test system.

a perturbation step of 50 MW, therefore, the accuracy of the resulting generation rescheduling is on the order of 50 MW; second, the C-OMIB accurately describes the maximum angular deviations for the initial operating point, and not for the new operating point, consequently, there is an inherent mismatch between the derived generation rescheduling amount and the "critical" amount – this is also observed in previous research [13]. However, it is important to point out that this small conservation is favorable or even necessary for practical use.

For a less conservative state, i.e. a smaller yet positive stability degree, a smaller trajectory sensitivity calculation step should be used. To verify this, we then reduce

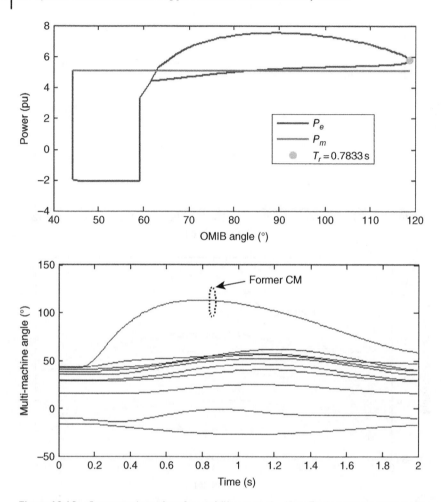

Figure 10.10 System trajectories after stability control – New England test system (using a smaller trajectory sensitivity calculation step).

the step size to 10 MW, and re-perform the proposed approach following Figure 10.3. It is found that two iterations are needed to obtain a stable operating point, and the final total generation rescheduling is 170.6 MW, and the resulting stability margin is 0.497. The corresponding system trajectories are shown in Figure 10.10, where it can be seen that $P_e(T_r)$ is closer to P_m.

Further simulations show that the derived generation rescheduling 170.6 MW is near the critical amount for stabilization. If only 160.6 MW is rescheduled, the system remains unstable – see Figure 10.11.

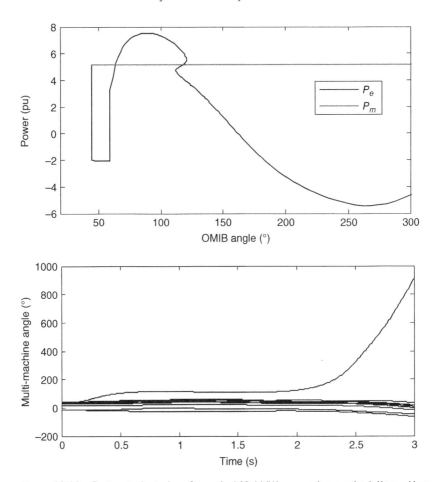

Figure 10.11 System trajectories after only 160.6 MW generation rescheduling – New England test system.

10.4.3 285-Machine 1648-Bus System

The proposed approach is also tested on a realistic large system. This system is the modified version of the example case called "bench" in the PSS/E package [14], which consists of 285 machines, 1648 buses, 2294 branches, and 55 areas.

10.4.3.1 System Trajectories

A three-phase short circuit is applied to bus 641 at 0.1 second and cleared after 0.18 second. Under this fault, the system is unstable and the resulting multi-machine trajectories are shown in Figure 10.12.

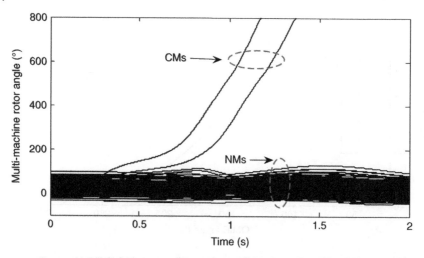

Figure 10.12 Unstable multi-machine system trajectories – the 285-machine 1648-bus system.

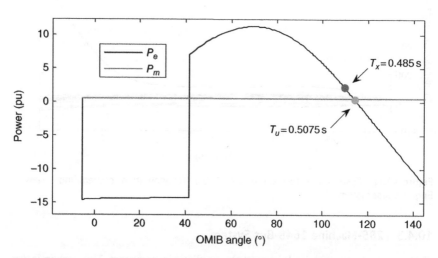

Figure 10.13 Unstable OMIB power-angle curve – the 285-machine 1648-bus system.

The OMIB equivalence is constructed, there are 2 CMs identified (#G79 and #G641), and the remaining 283 machines are NMs. The OMIB power-angle curve is shown in Figure 10.13.

Figure 10.13 also shows that the empirically determined instability time T_x is 0.485 second, while the inherent instability time T_u is 0.5075 second. Consequently, previous practice [4, 8, 9] controlling the system trajectories at T_x tends to be conservative.

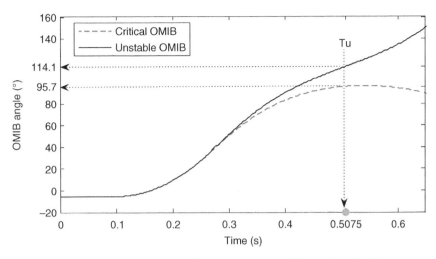

Figure 10.14 Unstable and critical OMIB trajectories – the 285-machine 1648-bus system.

For this fault, the CCT is calculated, being 0.17 seconds. Figure 10.14 shows the critical and unstable OMIB trajectories, and $\delta_{UT}(T_u) = 114.1°$, $\delta_{CT}(T_u) = 95.7°$.

10.4.3.2 Trajectory Sensitivities

The two CMs #G79 and #G641 are located in areas 5 and 32, respectively. Except for them, it is assumed that only the generators in area 32 with a capacity over 1000 MW are available for rescheduling. There are 4 such generators in area 32, see Table 10.3. The OMIB trajectory sensitivities with respect to the active power output of the 6 generators are calculated, and the sensitivity values at the instability time T_u are also listed in Table 10.3, where it can be seen that the CMs have positive sensitivities while the NMs have negative sensitivities.

10.4.3.3 Stabilizing Results

The linear programming model (10.5)–(10.10) is solved and the derived control amount is $\Delta P_{G641} = -\Delta P_{G817} = -33.8$ MW. After this control, the system

Table 10.3 Active generation output (MW) and trajectory sensitivity at T_u of the available generators.

Generator	G79	G641	G639	G817	G863	G868
P_G^0	68.8	1520	1400	1080	1520	1000
P_G^{max}	75.7	1701	1540	1188	1676	1100
Sensitivity	0.116	0.274	−0.267	−0.271	−0.266	−0.269

remains unstable, but the stability degree is increased. The second iteration is then executed, and the derived control is $\Delta P_{G641} = -\Delta P_{G817} = -17.1$ MW. After this iteration, the system is stabilized, and the corresponding OMIB and multi-machine system trajectories are shown in Figures 10.15 and 10.16, respectively.

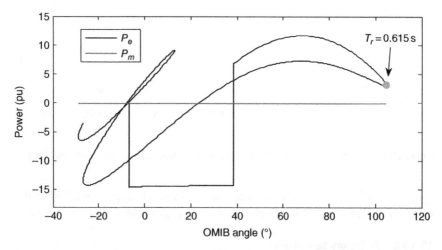

Figure 10.15 Stable OMIB power-angle curve of the system – the 285-machine 1648-bus system.

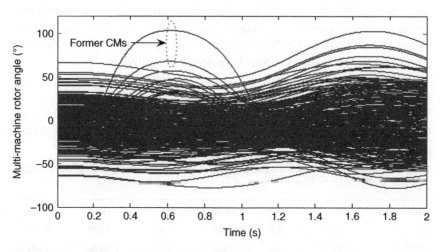

Figure 10.16 Stable multi-machine system trajectories – the 285-machine 1648-bus system.

10.4.4 Computation Efficiency Analysis

To implement the proposed approach, the following computational efforts are needed:

1) TDS: this is the basic computational component of the proposed approach. Using the commercial software PSS/E, the execution time for a fixed 2 seconds simulation time (without earlier termination) is around 1.6 seconds for the New England system and 15 seconds for the 285-machine 1648-bus system. Note that in practice, earlier termination criteria can be applied to save execution time.
2) OMIB construction: this is to apply the EEAC method to the simulated multi-machine trajectories, which are very fast. This costs only several milliseconds.
3) Calculation of the CCT: this is to obtain the C-OMIB. In this chapter, the CCT is calculated based on the EEAC, which needs an additional TDS run, see (10.3).
4) Trajectory sensitivity calculation: this is the major computational burden of the proposed approach. In this book, the numerical method is used, which requires an additional TDS run for each available generator. However, if the analytical method [8] is used, the sensitivities can be obtained as a byproduct of the TDS, minimizing the computational burden, and it is important to note that for the proposed approach, only the trajectory sensitivities at the instability time T_u is needed, this can significantly reduce the computational efforts faced by previous works [4, 8, 9].
5) Calculation of the required TSC: this is to solve the linear model (10.5)–(10.10), which is very fast. In this chapter, it takes only several milliseconds by using MATLAB optimization toolbox.

The total execution time finally depends on the number of iterations of the above steps to reach a stabilized operating equilibrium. As already reported, the computation iterates once and twice for the two test systems, respectively. This owes to the use of exact instability time and control reference. Totally, the execution time is 22 seconds for the New England system and 280 seconds for the 285-machine and 1648-bus systems.

It should be noted that compared with previous trajectory sensitivity-based approaches, the additional computation efforts required by the proposed approach are only calculating the CCT and building the OMIB, while the need for calculating the sensitivities along the whole analysis time frame is eliminated.

10.5 Transient Stability Preventive Control Based on Stability Margin

10.5.1 Stability Constraint Construction

Previously, the generation shifting was determined by trajectory sensitivities involving the OMIB angle deviation between C-OMIB (OMIB under CCT) and unstable OMIB at instability time T_u. In this section, the quantitative transient stability margin is used for generation-shifting decisions.

Based on the EEAC method, the stability margin could be calculated. The marginally stable condition is that the stability margin equals zero. Therefore, control actions can be designed to make the stability margin just larger than or equal to zero after wind power variation.

It should be mentioned that the value of zero is a critical stability margin boundary. Practically, this boundary is so critical that the system cannot be secured all the time. Consequently, a larger boundary value, such as 1 or 2, or even an arbitrary value that is larger than zero, could be used for reserving a certain stability margin. However, higher security will result in a relatively higher operating cost. In this chapter, we still use zero value for the most basic and economical purpose, i.e.

$$\eta \geq 0 \tag{10.11}$$

On the basis of trajectory sensitivities, the required generation rescheduling to stabilize an unstable case can be calculated according to Chapter 2:

$$\Delta \lambda = \frac{\Delta \eta}{\Phi(\eta, \lambda)} \tag{10.12}$$

where $\Delta \lambda$ is the form of generation rescheduling ΔP_{Gk}; $\Phi(\eta, \lambda)$ is the trajectory sensitivities of each generator.

Therefore, for generation rescheduling, the transient stability constraint (10.21) can be rewritten as:

$$\Delta \eta = \sum_{k}^{N_G} \Phi_k(\eta, P_{Gk}) \Delta P_{Gk} \tag{10.13}$$

As (10.13) indicates, the change of the stability margin after generation rescheduling can be calculated by timing generator trajectory sensitivity and the amount of the generation rescheduling.

From (10.20) and (10.13), the complete form of the inequality for transient stability constraint is shown:

$$\sum_{k}^{N_G} \Phi_k(\eta, P_{Gk}) \Delta P_{Gk} + \eta_0 \geq 0 \tag{10.14}$$

Therefore, (10.14) shows that the stability margin should be larger than zero, i.e. stable, after the generation rescheduling where $k = 1, 2, ..., N_G$; N_G is the total number of synchronous generators; and $\Phi_k(\eta, P_{Gi})$ is the trajectory sensitivity with respect to the active power output of the k-th generator.

10.5.2 Mathematical Model

The mathematical form for stabilizing the system subjected to a contingency is shown below:

$$\min_{\Delta P_{Gk}} C = \sum_{k}^{N_G} a_k P_{Gk}^2 + b_k P_{Gk} + c_k \tag{10.15}$$

$$s.t. \ P_{Gk} = P_{Gk}^0 + \Delta P_{Gk} \tag{10.16}$$

$$\Delta P_{Gi} \geq 0, \ i \in S^+ \tag{10.17}$$

$$\Delta P_{Gj} \leq 0, \ j \in S^- \tag{10.18}$$

$$\sum_{k} \Delta P_{Gk} + \Delta P_w = 0, \ \forall k \in \{S^- \cup S^+\} \tag{10.19}$$

$$\Delta \eta + \eta_0 \geq 0 \tag{10.20}$$

$$\Delta \eta = u(\Delta P_{Gk}) \tag{10.21}$$

$$0 \leq \Delta P_{Gk} \leq 50 \tag{10.22}$$

$$P_{Gk}^{\min} \leq P_{Gk}^0 + \Delta P_{Gk} \leq P_{Gk}^{\max} \tag{10.23}$$

where ΔP_{Gk} is the active power output variation, which is the control variable; i, j, and k denote the generator numbers; a_k b_k and c_k are the cost coefficient of k-th generators. S^- and S^+ are the generator set that consists of negative and positive sensitivities, respectively; the negative sensitivity means the critical machine, which denotes decreasing the generation of these generators can enhance the stability margin; η_0 is the initial stability margin of the system after the wind power variation and compensated by synchronous generators proportional to their capacities; P_{Gk}^0 is the active power output of k-th generator at base case; $P_{Gk}^{\min}$ and $P_{Gk}^{\max}$ represent the minimum and maximum values of active power output for k-th generator;

The objective function is to minimize the total cost of generation; Constraints (10.17) and (10.18) decide the direction of generation shifting; Constraint (10.19) represents that the wind variation should be balanced by the synchronous machines; Constraints (10.20) and (10.21) are the transient stability constraints. The full formulations are corresponding to (10.14) and (10.13), respectively. The constraint (10.22) is the ramp rate for each generator, and (10.23) is the active power output bound after the generation rescheduling.

Note that owning to the application of trajectory sensitivity, common advantages are also shared by the proposed approach: (i) transparent and the importance of the TSC can be reflected; (ii) linear stability constraints tend to significantly reduce the whole computational time in contrast with other nonlinear stability constraints models.

10.5.3 Computation Process

The computation process for the proposed approach is indicated in Figure 10.17. The comprehensive steps are listed below:

1) The contingency set and uncertainties are inserted as the input data.
2) EEAC is applied for OMIB construction and stability margin calculation. Trajectory sensitivity is applied for CM and NM recognition.
3) The base OP is obtained from OPF.

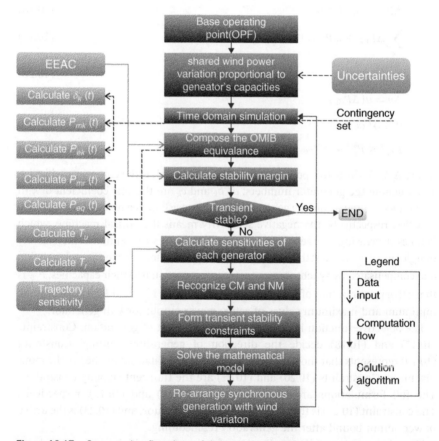

Figure 10.17 Computation flowchart of the proposed approach. *Source:* Yuan and Xu [16]/ with permission of Elsevier.

4) Input wind power variations and disperse the variation proportional to the synchronous generator's capacities.
5) Conduct TDS, apply the results to compose OMIB equivalence, and calculate stability margin
6) Check the stability of the system by stability margin. If the system is stable, terminate the procedure; otherwise, move to Step 7.
7) Calculate the stability margin sensitivities in relation to the active power of each synchronous generator and identify CM and NM.
8) Formulate the transient stability constraints based on the previous unstable stability margin
9) Solve the linear transient stability constraints model (10.15)–(10.23) for the TSC action.
10) Implement the solved TSC action and go to Step 5.

Similar to the proposed approach in Section 10.3, one iteration may not be enough for stabilizing the system. Therefore, checking the updated OP after each iteration is necessary. When the system is determined as unstable, the trajectory sensitivities should be recalculated to derive a new TSC action for the subsequent iteration. Besides, to prevent excessive stabilization, a small perturbation should be applied especially when approaching the stability boundary as it is very sensitive with respect to parameter variations [17].

10.6 Case Studies of Transient Stability Preventive Control Based on Stability Margin

10.6.1 Simulation Setup

The proposed method is performed on the New England 10-machine 39-bus system. Three synchronous generators are replaced by wind power generators at buses 32, 37, and 39, setting the same power output as the original synchronous generators. In order to guarantee the system has an adequate reserve to compensate wind power variation, the capacities of remaining synchronous generators are increased by 1.5 times of original values. The penetrated wind power is 23.3% for the base case.

The optimization and power flow are solved by YALMIP [18] and MATPOWER [19] implemented in MATLAB, respectively. Both single- and multiple-contingency cases are tested.

Three contingencies are applied in this chapter, as shown in Table 10.4. The TDS and the construction of OMIB are realized through commercial software TSAT [20]. The models of synchronous generators and wind generators are modeled

Table 10.4 Contingency set.

Contingency ID	Fault bus	Fault clearance (s)	Tripped line
C1	Bus 4	0.38	Line 4-5
C2	Bus 21	0.14	Line 21-22
C3	Bus 29	0.05	Line 29-26

Source: Yuan and Xu [16]/With permission of Elsevier.

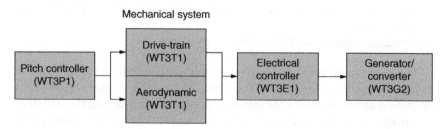

Figure 10.18 The structure of the wind generators. *Source:* Yuan and Xu [16]/With permission of Elsevier.

in PSS/E software using industry-standard models. These models can be also imported into DSAtools software. The form of "GENROU" is defined as a round rotor generator model (quadratic saturation). The excitation system uses the IEEE-type DC1A model, which is defined as "ESDC1A" in PSS/E. Wind generators are modeled as type-3 doubly fed induction generators ("WT3"). Four parts are included. The first one is a doubly fed induction generator, which is defined as "WT3G2." "WT3E1" is defined to represent the electrical control of the wind generator. The mechanical system of the wind generator is defined as "WT3T1." The pitch control of the wind generator is defined as "WT3P1." The structure of the wind generators is shown in Figure 10.18.

It should be noted that by using such detailed models of wind power generators, the dynamics of wind generators will be reflected in the time-domain simulation process. Hence, the obtained rotor angle trajectory of each synchronous generator is affected by the wind power dynamics. It should also be noted that the trajectories of wind generators are not used in the SIME approach since they are nonsynchronous trajectories. After the rotor angle trajectories are obtained, EEAC is applied to assess the transient stability of the system. The dynamics of wind as well as its impact on the system's transient stability can be reflected from the rotor angles and stability margin.

10.6.2 Single-contingency Case for New England 10-Machine 39-Bus System

The simulation is performed with the wind variation from −200 to +200 MW, which is ±9% of the base wind power output. All the wind power variations can be compensated while maintaining transient stability after iterative optimization. Some representative results are illustrated in Tables 10.5 and 10.6.

For contingency 1, the base operating point is transient stable, as shown in Table 10.5. After wind power output decreases 150 MW, all the generators compensate for the reduction of wind power proportional to their capacities, as Table 10.5 shows. However, the stability margin becomes negative, which means the system is unstable. After only one iterative optimization, the re-dispatch result is obtained in Table 10.5. The synchronous generation totally increases by 150 MW to balance wind power variation. After re-dispatch, the system becomes transiently stable at this operating point. Figure 10.19a indicates the multi-rotor angles of the system before optimization. It could be found that G31 is the critical generator. So, it illustrates that decreasing the generation of G31 (slack bus) can improve transient stability. Hence, from Table 10.5, the generation of G31 decreases by 16 MW compared with wind sharing by synchronous machines proportional to their capacity and the system becomes stable. Figure 10.19b shows the rotor angles

Table 10.5 Active generation output (MW) and wind variation (MW) at the base operating point, before optimization and after optimization with wind variation for C1.

Gens	Base	Before	After
G30	239.5	250.4	241.4
G31[a]	560.9	583.5	567.5
G32[b]	650	600	600
G33	624	647.4	654.5
G34	504.1	524.4	524.2
G35	644.9	668.4	665.9
G36	553	576.4	576.7
G37[b]	540	490	490
G38	822.4	850.5	872.4
G39[b]	1000	950	950
Wind variation	0	−150	−150
Stability margin	1.18	−22.69	12.86

[a] Slack bus generator
[b] DFIG wind generators
Source: Yuan and Xu [16]/With permission of Elsevier.

Table 10.6 Active generation output (MW) and wind variation (MW) at the base operating point, before optimization and after optimization with wind variation for C2.

Gens	Base	Before	After
G30	239.5	254.1	289.5
G31[a]	560.9	591.1	612.9
G32[b]	650	583.3	583.3
G33	624	655.2	674
G34	504.1	531.2	504.1
G35	644.9	676.2	644.9
G36	553	584.2	553
G37[b]	540	473.3	473.3
G38	822.4	859.9	872.4
G39[b]	1000	933.3	933.3
Wind variation	0	−200	−200
Stability margin	45.26	−37.8	48.85

[a] Slack bus generator.
[b] DFIG wind generators.
Source: Yuan and Xu [16]/With permission of Elsevier.

for the stable operating point. According to EEAC, the OMIB equivalence can be obtained. From Figure 10.20a, P_e crosses P_m at t_u as the circle indicates. The corresponding stability margin is −22.69. After the optimization, the corresponding stability margin is 12.86. As Figure 10.20b shows, P_e curve starts to return at $t_r = 0.751$ second before it crosses P_m.

As for contingency 2, the base operating point is stable, as Table 10.6 shows. When the wind power output decreases by 200 MW, the system becomes unstable and the stability margin is negative since all the synchronous generators increase by 200 MW proportional to their capacities for the compensation from Table 10.6. After TSC acts, the stability margin becomes positive 48.85, as Table 10.6 shows. In this case, only one iteration is performed to stabilize the system. From Figure 10.21a, the critical machines are G35 and G36, which means increasing the generation of G35 or G36 will deteriorate the transient stability. Hence, from Table 10.6, the generation of both G35 and G36 decreases by 32 MW, respectively, compared with the dispatch before the optimization. After the optimization, the system becomes stable seen from Figure 10.21b. Figure 10.22 is the OMIB angle plane for C2. It is important to note that at $t_r = 0.6985$s, P_e is a little far from P_m as Figure 10.22b indicates. This is because the stability margin is much larger than zero. It implies the optimization process is conservative. More discussion about the control accuracy will be presented later.

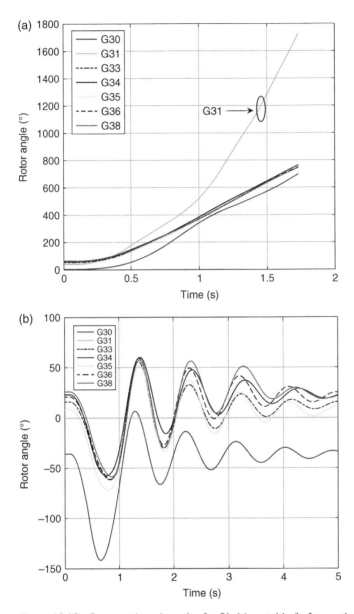

Figure 10.19 Rotor angle trajectories for C1: (a) unstable (before optimization); (b) stable (after optimization). *Source:* Yuan and Xu [16]/With permission of Elsevier.

Table 10.7 indicates the sensitivities of 7 synchronous generators for contingencies 1 and 2. As mentioned before, positive sensitivities represent noncritical machines and negative sensitivities represent critical machines. For contingency 1, it is

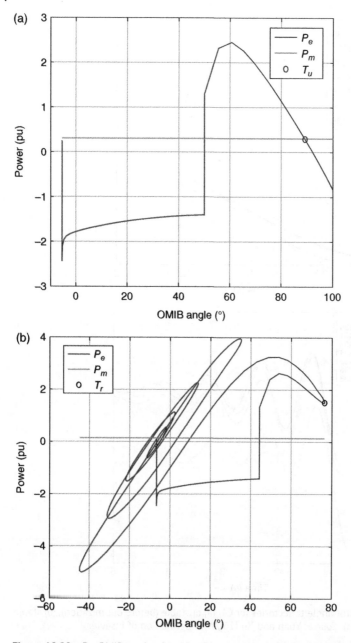

Figure 10.20 Pe-OMIB angle plane for C1: (a) unstable (before optimization); (b) stable (after optimization). *Source:* Yuan and Xu [16]/With permission of Elsevier.

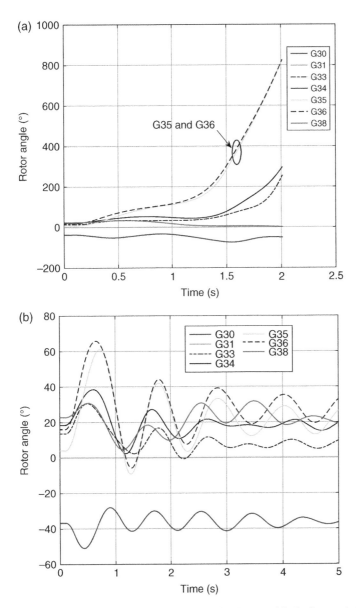

Figure 10.21 Rotor angle trajectories for C2: (a) unstable (before optimization); (b) stable (after optimization). *Source:* Yuan and Xu [16]/With permission of Elsevier.

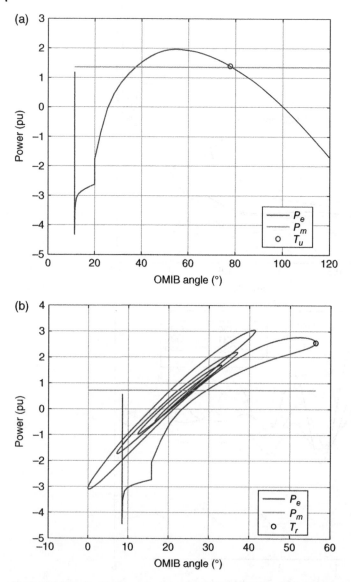

Figure 10.22 Pe-OMIB angle plane for C2: (a) unstable (before optimization); (b) stable (after optimization). *Source:* Yuan and Xu [16]/With permission of Elsevier.

obvious that all sensitivities of generators except G31 have positive values, which means G31 is a critical generator and the remaining are the noncritical generators. Meanwhile, for contingency 2, G35 and G36 have the largest negative absolute value compared with other sensitivities, which means G35 and G36 are the most critical

Table 10.7 Sensitivities of 7 synchronous generators for C1 and C2.

Generator	G30	G31	G33	G34	G35	G36	G38
C1: Sensitivity	6.359	0	1.385	1.36	1.398	1.39	1.355
C2: Sensitivity	0.2295	0	0.0145	−0.0158	−1.947	−1.3315	0.228

Source: Yuan and Xu [16]/With permission of Elsevier.

generators. Namely, these two machines can affect transient stability significantly. All these results calculated from trajectory sensitivities comply with the results calculated by time-domain simulation from Figures 10.19 and 10.21. It means that the application of trajectory sensitivity for the critical and noncritical generator recognition is corrective and can provide a quantitative sensitivity level of each generator as well.

Finally, it should be noted that the number of iterations is less than two, which is very fast and computationally efficient for practical applications.

10.6.3 Multi-contingency Case for New England 10-Machine 39-Bus System

In this case, contingencies of C1, C2, and C3 are considered concurrently. The simulation is performed with the wind variation from −150 to 150 MW, which is ±7% of the base wind power output. During this range, all the wind power variations can be compensated optimally while maintaining transient stability for the three contingencies. Some representative results are shown in Table 10.8.

Table 10.8 indicates generation dispatch while wind power output reduces 150 MW before optimization. It is obvious that C3 is stable, while C1 and C2 are unstable. After the optimization, the system is transiently stable under all of the three contingencies as Table 10.8 shows.

From Figure 10.23, the rotor angle trajectories of C1 and C2 before the optimization, it is obvious that G31, G35, and G36 are critical. Compared to the dispatch of each generator before and after the optimization, it can be found that G31, G35, and G36 decrease their generations, 103, 10.9 and 7.1 MW, separately. This action results in a transient stable operating point and can also verify that decreasing the generation of critical generators can improve stability.

10.6.4 Control Accuracy

For the single-contingency test, it should be found that after the optimization, the transient stability margin becomes 12.86 and 48.85 for C1 and C2, respectively. For C2, the value is relatively large regarding the critical value 0, which implies the conservative results of generation re-dispatch. The reason is that trajectory

Table 10.8 Active generation output (MW) and wind variation (MW) for multi-contingency before and after optimization.

Gens	Before	After
G30	250.4	298.2
G31[a]	583.5	480.5
G32[b]	600	600
G33	647.4	668.8
G34	524.44	566.85
G35	668.37	657.43
G36	576.4	569.3
G37[b]	490	490
G38	850.5	862.6
G39[b]	950	950
Wind variation	−150	−150
Stability margin	−22.7(C1)	64.19(C1)
	−5.71(C2)	24.38(C2)
	9.07(C3)	0.01(C3)

[a] Slack bus generator.
[b] DFIG wind generators.
Source: Yuan and Xu [16]/With permission of Elsevier.

sensitivities are calculated with a perturbation of 40 MW. Hence, the accuracy of the generation rescheduling is on the order of 40 MW. However, reasonable conservation is acceptable or necessary in practice since the critical stability margin is not easy to control and cannot be secure enough. But for contingency 2, the stability is quite conservative.

To achieve a less conservative result, namely, a smaller positive stability margin, the perturbation for trajectory sensitivities calculation should be smaller. Hence, the step size is reduced to 5 MW, and then the proposed approach for C2 is re-performed to verify it. For C2, the final stability margin is 36.78 after one iteration, which is much less than 48.85. From Figure 10.24, the t_r5 is 0.7402 second, which is longer than the previous 0.6985 second, t_r40. It is obvious that the P_e is closer to Γ_m at time t_r5.

An additional simulation case changing the step size from 1 to 20 MW for contingency 1 is indicated in Figure 10.25. It is found that the stability margin varies linearly against step size within a small range, which is from 1 to 6 MW. The slope represents the sensitivity of each generator. Therefore, when the step size is small, the sensitivity is unchanged, which means the optimization result is accurate by

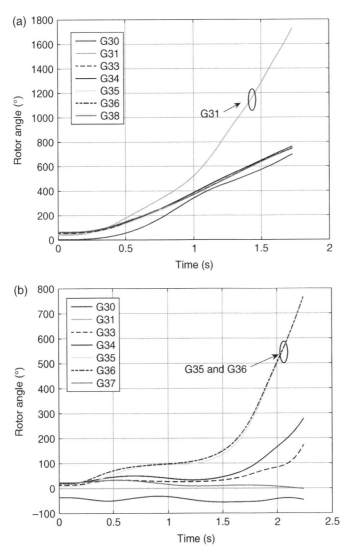

Figure 10.23 Rotor angle trajectories for multi-contingency before optimization: (a) C1; (b) C2. *Source:* Yuan and Xu [16]/With permission of Elsevier.

using trajectory sensitivity. If the step size is larger than 6 MW, the linearity cannot be guaranteed and the sensitivity varies to a larger value (the slope is steeper), which means the larger step sizes can stabilize the system fast but may over-stabilize the system. Hence, in this case, the step size can be chosen as 5 MW to obtain a reasonable stability margin after the optimization.

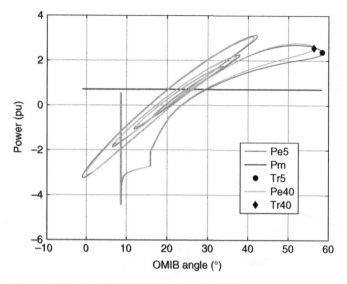

Figure 10.24 Pe-OMIB angle plane for C2 with the step size of 5 and 40 MW. *Source:* Yuan and Xu [16]/With permission of Elsevier.

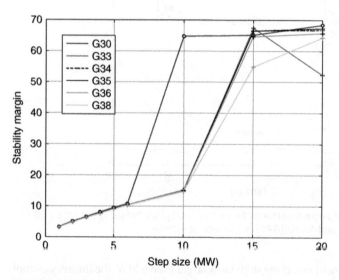

Figure 10.25 The stability margin versus step size for contingency 1. *Source:* Yuan and Xu [16]/With permission of Elsevier.

10.6.5 Computation Efficiency Analysis

The computation tasks are almost the same as before except calculation of CCT, which includes TDS, OMIB construction, trajectory sensitivity calculation, and calculation of TSC. They are simply described below:

1) TDS: In this section, TDS is realized by the commercial software TSAT. The execution time is around 0.312 second with a fixed 5 seconds simulation time (without earlier termination).
2) OMIB construction: This is the application to compose multi-machine trajectories into OMIB. The construction is implemented in the software TSAT, which is very fast and spends only several milliseconds.
3) Trajectory sensitivity calculation: Numerical method is applied for this book, which brings the main computation burden of the proposed approach since an additional TDS run is required for each available generator. In this section, the trajectory sensitivity is the system stability margin divided by the perturbation for each available generator. Compared with traditional angle trajectory sensitivity in Ref. [4], which needs to calculate the trajectory sensitivities for the whole simulation time, the proposed approach is more computationally efficient. In addition, in contrast with Ref. [15], which only considers the trajectory sensitivities at the instability time T_u, the proposed approach is more accurate and straightforward since it calculates the stability margin according to the whole trajectories.
4) Calculation of TSC: It is to solve the constructed linear model (10.15)–(10.23). In this section, the optimization toolbox YALMIP [18] is used in MATLAB. The time is not that long, since transient stability is linear, around 2.1 seconds for every optimization process.

Besides, the total execution time is dependent on the number of iterations of the optimization. As mentioned before, the iterations are all within two because of the use of stability margin and control reference. For the case study in this section, the execution time is approximately from 9 to 16 seconds for single contingency and multi-contingency, respectively. Further reduction in the computational burden for trajectory sensitivity calculation can be achieved with the application of the analytical method.

Nomenclature

Indices

i, j, k indices for generators

Sets

S^-/S^+ sets of generators with negative and positive trajectory sensitivities

Variables

$\delta_{ij}^{\max}$	maximum relative angle deviation
σ	reference threshold
$\Delta\lambda$	control action
$\Phi(x_0, T_x, \lambda)$	trajectory sensitivity
η	stability margin
T_u	instability time of the unstable OMIB
$\delta_{CT}(T_u)$	reference critical OMIB (C-OMIB) trajectory of the same system at the instability time T_u
$\delta_{UT}(T_u)$	unstable OMIB angle of the original system at T_u
ΔP_G	active power output variation of dispatchable generators
P_{Gk}^0	active power generation output of generator k at the base case
$\Phi_i(\delta_{UT},$	OMIB angle trajectory sensitivity regarding active power output of
$T_u, P_{Gi})$	generator i at the instability time T_u
$\Phi_k(\eta, P_{Gk})$	stability margin trajectory sensitivity regarding active power output of generator k
ΔP_{Gk}	active power output variation of k-th generator

Parameters

$P_{Gk}^{\min} P_{Gk}^{\max}$	upper and lower bound of active power output of generator k
a_k, b_k, c_k	cost coefficient of k-th generators
η_0	initial stability margin

References

1 Gan, D., Thomas, R.J., and Zimmerman, R.D. (2000). Stability-constrained optimal power flow. *IEEE Transactions on Power Systems* 15 (2): 535–540.

2 Chen, L., Taka, Y., Okamoto, H. et al. (2001). Optimal operation solutions of power systems with transient stability constraints. *IEEE Transactions on Circuits and Systems I: Fundamental Theory and Applications* 48 (3): 327–339.

3 Xu, Y., Dong, Z.Y., Meng, K. et al. (2012). A hybrid method for transient stability-constrained optimal power flow computation. *IEEE Transactions on Power Systems* 27 (4): 1769–1777.

4 Nguyen, T.B. and Pai, M.A. (2003). Dynamic security-constrained rescheduling of power systems using trajectory sensitivities. *IEEE Transactions on Power Systems* 18 (2): 848–854.

5 Ruiz-Vega, D. and Pavella, M. (2003). A comprehensive approach to transient stability control. I. Near optimal preventive control. *IEEE Transactions on Power Systems* 18 (4): 1446–1453.

6 Xue, Y., Li, W., and Hill, D.J. (2005). Optimization of transient stability control Part-I: For cases with identical unstable modes. *International Journal of Control, Automation and Systems* 3 (spc2): 334–340.

7 Xu, Y., Dong, Z.Y., Guan, L. et al. (2012). Preventive dynamic security control of power systems based on pattern discovery technique. *IEEE Transactions on Power Systems* 27 (3): 1236–1244.

8 Hou, G. and Vittal, V. (2012). Cluster computing-based trajectory sensitivity analysis application to the WECC system. *IEEE Transactions on Power Systems* 27 (1): 502–509.

9 Hou, G. and Vittal, V. (2013). Determination of transient stability constrained interface real power flow limit using trajectory sensitivity approach. *IEEE Transactions on Power Systems* 28 (3): 2156–2163.

10 Tang, L. and McCalley, J.D. (2012). An efficient transient stability constrained optimal power flow using trajectory sensitivity. *2012 North American Power Symposium (NAPS)*, Champaign, IL (09–11 September 2012), pp. 1–6: IEEE.

11 Xue, Y., Van Cutsem, T., and Ribbens-Pavella, M. (1988). A simple direct method for fast transient stability assessment of large power systems. *IEEE Transactions on Power Systems* 3 (2): 400–412.

12 Pavella, M., Ernst, D., and Ruiz-Vega, D. (2012). *Transient Stability of Power Systems: A Unified Approach to Assessment and Control*. Springer Science & Business Media.

13 Pizano-Martianez, A., Fuerte-Esquivel, C.R., and Ruiz-Vega, D. (2010). Global transient stability-constrained optimal power flow using an OMIB reference trajectory. *IEEE Transactions on Power Systems* 25 (1): 392–403.

14 Siemens PSS/E software (2011). PSS/E 33.0 Program Application Guide.

15 Xu, Y., Dong, Z.Y., Zhao, J. et al. (2015). Trajectory sensitivity analysis on the equivalent one-machine-infinite-bus of multi-machine systems for preventive transient stability control. *IET Generation Transmission and Distribution* 9 (3): 276–286.

16 Yuan, H. and Yan, X. (2020). Trajectory sensitivity based preventive transient stability control of power systems against wind power variation. *International Journal of Electrical Power and Energy Systems* 117: 105713.

17 Hiskens, I.A. and Alseddiqui, J. (2006). Sensitivity, approximation, and uncertainty in power system dynamic simulation. *IEEE Transactions on Power Systems* 21 (4): 1808–1820.

18 Lofberg, J. (2004). YALMIP: a toolbox for modeling and optimization in MATLAB. *2004 IEEE International Conference on Robotics and Automation (IEEE Cat. No.04CH37508)*, Taipei, Taiwan (02–04 September 2004), pp. 284–289: IEEE.

19 Zimmerman, R.D., Murillo-Sanchez, C.E., and Thomas, R.J. (2011). MATPOWER: Steady-State Operations, Planning, and Analysis Tools for Power Systems Research and Education. *IEEE Transactions on Power Systems* 26 (1): 12–19.

20 DSA. Dynamic security assessment software. http://www.dsatools.com/ (accessed 29 January 2015).

11

Preventive-Corrective Coordinated Transient Stability-Constrained Optimal Power Flow under Uncertain Wind Power

11.1 Introduction

Generally, transient stability operational control methods are divided into two categories: *preventive control* (PC) and *emergency control* (EC), depending on the implementation timing. The former is to protect the system from transient instability preventively when the contingency has not occurred yet. The latter is to prevent the system from ongoing loss of synchronism when the contingency has occurred already [1].

On the one hand, as the system quickly becomes unstable when subjected to a large disturbance, most current research always concentrates on the PC of the system. In the meantime, generation rescheduling becomes the most popular PC action to enhance transient stability, which is in the form of transient stability-constrained optimal power flow (TSC-OPF) [2–7]. A comprehensive review of the existing approaches is introduced in Chapter 5. On the other hand, with the increasing penetration of wind power in power systems, many researchers started to work on the influence of wind power on transient stability [6, 8–10]. Xu et al. [6] proposed a TSC-OPF that can stabilize the system against wind power variation on a high robustness level. In Ref. [6], a decomposition-based strategy was applied to solve the complicated optimization model. Papadopoulos and Milanović [8] assessed the transient stability by a probabilistic method for high penetration of renewables. Liu et al. [9] proposed a wind power balancing strategy and conducted a quantitative impact study on transient stability. In Ref. [10], a novel probabilistic TSC-OPF model considering the correlation of the uncertainties is proposed to enhance transient stability.

However, PC such as generation rescheduling may lead to high operation costs as it is a long-term solution but the probability of occurrence for severe contingencies is low. On the contrary, corrective control (CC) is a short-term solution to prevent transient instability. Its short-term cost may be very high, but the probability

Stability-Constrained Optimization for Modern Power System Operation and Planning, First Edition. Yan Xu, Yuan Chi, and Heling Yuan.

of acting is low, therefore its expected cost can be reasonable and acceptable. Currently, CC has been employed for the static security operation, known as corrective security-constrained optimal power flow (CSCOPF) [11, 12]. It can relax the operating region obtained by PC and thus lower the pre-contingency operating cost. Xu et al. [1] proposed an optimal coordination model between PC and CC for security-constrained optimal power flow and solved it by combing the evolutionary algorithm and interior-point methods. Similarly, in Ref. [13], a risk-based coordination bi-level model of generation rescheduling and load shedding for transient stability enhancement was proposed. However, the existing works did not consider wind power uncertainties, which is not practical in modern power systems with high-level wind power penetration.

From the above reviews, the coordination of PC and CC for transient stability considering wind power has not been fully addressed yet. Most PCs for transient stability improvement with uncertainties do not consider CC, or CC is applied without considering uncertainties of the wind power. This chapter aims to fill the gap with a preventive-corrective coordinated control method that achieves the optimal coordination cost while considering transient stability and wind power uncertainties. For the dynamic behavior of the wind, the detailed dynamic model of the wind generators is considered in time-domain simulations. For the steady state uncertain wind power output, it is modeled by selecting a small number of robust test scenarios. A risk index is proposed to quantify the probabilistic consequence of transient instability. The original DAE (differential-algebraic equations) forms of the risk constraint and stability constraint are converted into linear constraints based on trajectory sensitivity and extended equal area criterion (EEAC), which dramatically reduces the computational burden and enhances the transparency of the problem. Finally, the golden section search is applied to efficiently solve the mathematical model.

11.2 Framework of the PC–CC Coordinated TSC-OPF

Figure 11.1 indicates the framework of the proposed approach. In this chapter, the generation rescheduling and load shedding are applied for PC and CC, respectively.

To optimally coordinate the PC and CC, we propose a two-step bi-level optimization model, where the objective aims to minimize the total coordination cost (sum of PC and CC costs) in the upper level, and a two-step control actions optimization is solved in the lower level. More specifically, the first step aims to minimize the generation rescheduling cost, while the second step aims to minimize the expected load shedding cost. In the first step, the stability is contained

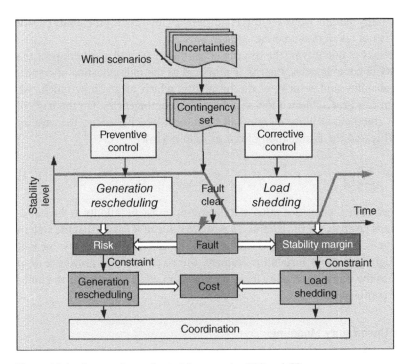

Figure 11.1 Proposed coordinated framework of PC and CC.

by a risk index threshold, which considers both the stability level and the probability of the contingency occurrence. In the meantime, a risk coordination parameter, obtained from the upper level, is proposed to adjust the risk index threshold. For the second step, the decision vector is load shedding and constrained by a stability margin boundary. If the system cannot be stabilized after the first step, the second step will act. Then, the optimal solution in the lower level is obtained and returned to the upper level. According to this, the risk coordination parameter is updated at the upper level. The whole procedure is iterative, when the termination condition satisfies, the computing will stop.

11.3 PC–CC Coordinated Mathematical Model

11.3.1 Risk Index

In order to quantitatively evaluate the transient stability, the EEAC [14, 15] is applied in this chapter. The comprehensive principle and theory of EEAC can be found in Chapter 2. By employing EEAC, the stability level can be evaluated via the stability margin. However, this is still conservative as the probability of contingency is not taken into account.

For the sake of quantifying the severity of contingencies and accounting for the probability of contingencies, the *risk* is proposed as the multiplication of contingency probability and severity of the instability, which means how much risk the system can bear subjected to a set of credible contingencies. On the basis of the EEAC method, stability margin can be gained, which assesses the severity of instability and the formulation of risk as follows [16]:

$$R = \sum_{k=1}^{N_c} p_k \times (-\eta_k) \tag{11.1}$$

As shown above, a negative value of R illustrates transient stability as the stability margin is positive. On the contrary, a positive value of R means that the system is unstable as the stability margin is negative. Therefore, a larger risk value will lead to higher transient instability. Namely, the risk index can provide a deep insight into the transient stability level of the system while considering the probability of contingencies.

11.3.2 Uncertainty Modeling

In this chapter, wind power output is integrated into the power systems. In practice, the wind power output is time-varying, intermittent, and uncertain, which may have remarkable impacts on the transient stability of power systems [6]. Therefore, deterministic wind power output is not feasible to systematically model the dynamic characteristics of the systems. The stability margin considering uncertain wind power output is indicated below:

$$\eta_k = z(\eta_{k0}) \tag{11.2}$$

where η_k is the stability margin after the wind power variation subjected to contingency k, which is a function of η_{k0}.

11.3.3 Two-step Bi-level Compact Model

It is known that the stability requirement and economic purpose are always conflicting. Namely, higher stability requirements result in higher operating costs. From the mathematical perspective, the feasible region of the solution will be reduced by high stability requirements; thus, the operating cost will be high. However, by coordinating both PC and CC, it is intuitive that PC cost will increase when the risk index threshold decreases, and then the CC cost (if activated) would reduce and vice versa (the explanation and verification are presented subsequently). Therefore, there should be a compromised optimal coordinated solution between PC and CC, which can be realized by a flexible coordination parameter to adjust the risk index threshold in PC such that the total coordination cost can be minimized. By doing this, this chapter proposes a two-step bi-level optimization model as follows:

$$\min_{x_u \in X_U, x_l \in X_L} F(x_u, x_l, \tilde{w}) \quad \text{upper level} \tag{11.3}$$

$$G_m = (x_u, x_l, \tilde{w}) \leq 0, \quad m = 1, ..., M \quad \text{upper level} \tag{11.4}$$

$$\text{s.t. } x_l \in \arg\min_{x_l \in X_L} \{f_1(x_u, x_l, \tilde{w}), f_2(x_u, x_l, \tilde{w}) : \\ g_n(x_u, x_l, \tilde{w}) \leq 0, \quad n = 1, ..., N\} \quad \text{lower level} \tag{11.5}$$

where F denotes the objective function of the upper level; f_1 and f_2 represent the two-step objective functions of the lower level. x_u and x_l denote the decision vectors of the upper level and lower level, respectively. $\tilde{w}$ are the uncertain parameters. In this chapter, wind power output is randomly changed. It should be mentioned that reactive power generated by wind turbines is not considered an uncertain parameter as it can be controlled by the inverters. G_m and g_n denote the constraints for the upper and lower levels, respectively.

11.3.4 Upper Level

The objective function and constraints of the upper level are indicated as follows:

$$\min_{\tau} \quad C = C_p + C_c \tag{11.6}$$

$$\text{s.t. } -1 \leq \tau \leq 0 \tag{11.7}$$

The objective function (11.6) is to minimize the total coordinated cost for PC and CC, which corresponds to F in (11.3) at the upper level. Constraint (11.7) is the risk coordination parameter bounds, which corresponds to G_m in (11.4).

τ represents the risk coordination parameter, which can adjust the risk index threshold such that the total coordination cost can be optimized. Rather than fix it, τ is a decision variable in the upper level, and its optimal value is searched by the golden section search method.

11.3.5 Lower Level

11.3.5.1 Preventive Control

The objective function (11.8) and constraints (11.9)–(11.15) for the first step PC in the lower level are constructed below:

$$\min_{\Delta P_{Gi}} \quad C_p = \sum_{i}^{N_G} a_i P_{Gi}^2 + b_i P_{Gi} + c_i \tag{11.8}$$

s.t. $P_{Gi} = P_{Gi}^0 + \Delta P_{Gi}$ (11.9)

$$\begin{cases} P_{Gi} + \tilde{P}_{Wi} - P_{Di} = V_i \sum_{j=1}^{N_B} V_j \left(G_{ij} \cos \theta_{ij} + B_{ij} \sin \theta_{ij} \right) \\ Q_{Gi} + \tilde{Q}_{Wi} - Q_{Di} = V_i \sum_{j=1}^{N_B} V_j \left(G_{ij} \cos \theta_{ij} - B_{ij} \sin \theta_{ij} \right) \end{cases}$$ (11.10)

$$\sum_i^{N_G} \Delta P_{Gi} = 0$$ (11.11)

$$P_{Gi}^{\min} \leq P_{Gi} \leq P_{Gi}^{\max}, \quad i = 1, 2, ..., N_G$$ (11.12)

$$\Delta R + \tau R_0 \leq 0$$ (11.13)

$$\Delta R = \sum_{k=1}^{Nc} -\Delta \eta_k \times p_k$$ (11.14)

$$\Delta \eta_k = u(\Delta P_{Gi})$$ (11.15)

The objective function (11.8) is to minimize the total generation cost after generation re-dispatch, which corresponds to f_1 in the compact model (11.5). Constraint (11.10) represents the steady-state power flow, and constraint (11.11) represents the power that should be balanced after the generation rescheduling. Constraint (11.12) denotes the power output upper and lower bounds. Meanwhile, the risk constraints are denoted by (11.13)–(11.15). All these constraints correspond to g_n in (11.5).

R_0 denotes the risk calculated by (11.1) without PC action. τ is the risk coordination parameter obtained from the upper level. The risk constraint (11.13) means that the variation of the risk after PC plus the multiplication of R_0 and τ should be less than zero. The variation of transient stability margin $\Delta \eta$ subjected to k-th contingency indicated in (11.15) is calculated based on trajectory sensitivity, which is a function of the amount of generation rescheduling. Then the variation of the risk after PC can be calculated, which is presented in (11.14).

11.3.5.2 Corrective Control
The objective function (11.16) and constraints (11.17)–(11.20) for the second step CC in the lower level are constructed below:

$$\text{for } k, \quad \min_{\Delta P_{Di}} C_c = \sum_{j=1}^{H} \sum_{i=1}^{N_D} p \times c_D \Delta P_{Di} \left(l_j \right)$$ (11.16)

$$\text{s.t. } \begin{cases} P_{ei}^t + \tilde{P}_{Wi} - P_{Di}^t = V_i^t \sum_{j=1}^{N_B} V_j^t \left(G_{ij} \cos \theta_{ij}^t + B_{ij} \sin \theta_{ij}^t \right) \\ Q_{ei}^t + \tilde{Q}_{Wi} - Q_{Di}^t = V_i^t \sum_{j=1}^{N_B} V_j^t \left(G_{ij} \cos \theta_{ij}^t - B_{ij} \sin \theta_{ij}^t \right) \end{cases}$$ (11.17)

$$P_{Di} \geq \Delta P_{Di} \geq 0$$ (11.18)

$$\Delta\eta + \eta_0^l \geq 0 \tag{11.19}$$

$$\Delta\eta = v(\Delta P_{Di}) \tag{11.20}$$

The objective function (11.16) is to minimize the total expected load shedding cost for all selected testing scenarios, which corresponds to f_2 in (11.5). Constraint (11.17) is the power flow equation during the transient state. Constraint (11.18) denotes the load shedding amount limits at i-th load bus. Constraints (11.19)–(11.20) represent transient stability constraints. $\Delta\eta$ is the variation of stability margin after load shedding, which is obtained on the basis of trajectory sensitivity. It can be presented as a function of the load shedding amount shown in (11.20). It should be noted that the variation of the stability margin after load shedding adding η_0^l should be larger than zero, i.e. be stable. All the constraints, including transient power flow, operational limits, and stability constraints, correspond to g_n in (11.5).

11.4 Solution Method for the PC–CC Coordinated Model

11.4.1 Trajectory Sensitivity-based Stabilization Constraints

According to trajectory sensitivities, the required generation rescheduling and load shedding amount to stabilize an unstable case can be formed below:

$$\Delta\eta = \Delta\lambda \cdot \Phi(\eta, \lambda) \tag{11.21}$$

where $\Delta\lambda$ denotes generation rescheduling or load shedding amount; $\Phi(\eta, \lambda)$ is the trajectory sensitivity of each generator or load bus, respectively.

Therefore, for the generation rescheduling, the risk constraint (11.15) can be constructed below:

$$\Delta\eta_k = \sum_i^{N_G} \Phi_i(\eta, P_{Gi})\Delta P_{Gi} \tag{11.22}$$

As (11.22) indicates, the variation of the stability margin after generation rescheduling can be formed by timing generator trajectory sensitivity and generation rescheduling amount.

From the Eqs. (11.13), (11.14), and (11.22), the full term of risk constraint can be formed below:

$$\sum_{k=1}^{N_c} -\left(\sum_i^{N_G} \Phi_i(\eta, P_{Gi})\Delta P_{Gi}\right)_k \times p_k + \tau R_0 \leq 0 \tag{11.23}$$

Therefore, (11.23) denotes that the variation of the risk after the generation rescheduling plus the multiplication of R_0 and τ should be less than zero.

$\Phi_i(\eta, P_{Gi})$ represents the trajectory sensitivity involving the active power output of the i-th generator.

As for load shedding, the transient stability constraint (11.20) can be reformed as follows:

$$\Delta\eta = \sum_i^{N_D} \Phi_i(\eta, P_{Di})\Delta P_{Di} \tag{11.24}$$

From (11.24), the variation of stability margin after load shedding is the multiplication of load bus trajectory sensitivity and the load shedding amount.

From (11.19) and (11.24), the full term of the inequality of transient stability constraint is constructed below:

$$\sum_i^{N_D} \Phi_i(\eta, P_{Di})\Delta P_{Di} + \eta_0^l \geq 0 \tag{11.25}$$

Therefore, (11.25) illustrates that the stability margin after the load shedding should be larger than zero, i.e. maintaining stability.

$\Phi_i(\eta, P_{Di})$ denotes the trajectory sensitivity in relation to load shedding amount at i-th load bus.

11.4.2 Taguchi's Orthogonal Array Testing (TOAT)

The comprehensive TOAT method is introduced in Section 9.2.1.3. In this chapter, with three wind power variables each represented by two levels, there will be a total of 2^3 combinations. By applying TOAT, an OA $L_4(2^3)$ is constructed to represent the whole wind uncertainty space by only four scenarios shown in Table 9.8.

By using TOAT, testing scenarios are chosen by OA, and the transient stability margin under all selected testing scenarios can be calculated as follows [16]:

$$\eta_k = \frac{1}{H}\sum_{i=1}^{H} \eta_k(l_i) \tag{11.26}$$

Then, the risk presented in Eq. (11.1) under all testing scenarios can be calculated from Eq. (11.26).

11.4.3 Golden Section Search

Since the proposed model is quite complex. In this chapter, an iterative method on the basis of the golden section search is applied to solve it.

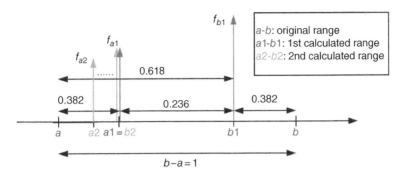

Figure 11.2 Illustration of the golden section search.

The kernel of the golden section search is to search the extremum (minimum or maximum) of a unimodal function by continuously reducing the range of values. From Figure 11.2, a and b are the initial lower and upper bounds of the function. $[a1, b1]$ is the calculated range of the function based on the golden section 0.618. f_{a1} and f_{b2} are the obtained value under $a1$ and $b1$. If f_{a1} is less than f_{b2}, the optimal value should be at the left of $a1$ and vice versa. Therefore, an updated range $[a2, b2]$ is obtained from the golden section ratio. Then, an updated optimal value can be obtained. The procedure will iteratively search the optimal value until the termination condition meets. In this chapter, the golden section search method is employed to search for an optimal risk coordination parameter, τ in the upper level. The method remains the function values for triples of points whose distances form a golden ratio. It is a direct and effective method to solve unimodal global search optimization.

11.4.4 Computation Process

In Figure 11.3, the flowchart of the whole computation process is depicted. The upper level is to search a risk coordination parameter τ by use of the golden section method. Obtained the τ from the upper level, the lower level is then solved for PC and CC (if applicable) for transient stability enhancement under wind power variation.

The comprehensive steps are listed below:

1) Run the base case and obtain the initial OP.
2) Set the lower and the upper bound for risk coordination parameter $\tau = [-1,0]$, which means $a = -1$, and $b = 0$.
3) Set two new τ by the golden section $\tau_1 = a + 0.618(b - a)$, $\tau_2 = b - 0.618(b - a)$

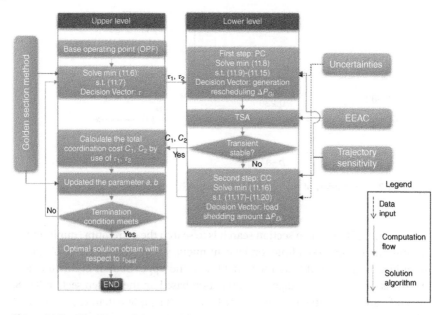

Figure 11.3 Flowchart of the solution process.

4) Solve the lower level: solve the PC model with given τ_1 and τ_2; check the stability under the selected wind power testing scenarios, if all are stable, go to Step 5 directly; otherwise, solve the CC model and then go to Step 5.

5) Calculate the total coordination costs C_1 and C_2 with given τ_1 and τ_2 for each wind power testing scenario. If there is no CC, the cost of CC is set to zero.

6) Compare C_1 and C_2, and update the current coordination parameter if $C_1 < C_2$, $\tau_{best} = \tau_1$ and $b = \tau_2$; otherwise, $\tau_{best} = \tau_2$ and $a = \tau_1$.

7) Repeat Steps 2–6 until the termination condition $|b - a| \leq \varepsilon$ is met. Thus, the optimal solution could be obtained with respect to τ_{best}.

It should be mentioned that when the optimal risk coordination parameter is obtained, the load shedding is averagely dispersed for all selected testing scenarios. However, to ensure the load shedding amount is robust against all testing scenarios, we choose the highest load shedding amount among these scenarios since it can be intuitively known that more load shedding amount will lead to a more stable system. So, the corrected expected load shedding cost is formed below:

$$\Delta P_D^c = \max(\Delta P_D(l_1), \Delta P_D(l_2), ...\Delta P_D(l_H))$$
$$C_c = p \times c_D \times \Delta P_D^c \tag{11.27}$$

From (11.27), the total coordination cost C is $C_p + C_c$.

11.5 Case Studies

11.5.1 Simulation Settings

The proposed method is tested on New England 10-machine 39-bus system [17]. The dynamic and static settings for the networks with wind farm integration are the same as that in Sections 9.5 and 10.6. The tolerance of ε is 0.001 for the termination condition.

In this chapter, three contingencies are tested and shown in Table 11.1. The probabilities of all three contingencies are set as 1% [16]. The load shedding cost is uniformly set as 100\$/MW. As for wind power generation, the base value is assumed to be the predicted mean output (i.e. μ) and the variation is set as standard deviation $\pm\sigma$, which is set to be $\pm$10%. Therefore, two uncertain levels of wind power output $\mu - \sigma$ and $\mu + \sigma$ are used for one wind farm. Based on TOAT, an OA $L_4(2^3)$ is constructed with four scenarios for three wind farms (see Table 9.8 in Chapter 9).

The TDS is conducted on commercial software, DSAtools TSAT [18]. The power flow and optimization for generation rescheduling and load shedding are solved by YALMIP [19] and MATPOWER [20], which are implemented in MATLAB, respectively.

The synchronous generators and wind generators are modeled in PSS/E software using industry-standard models, which are comprehensively introduced in Section 10.6.

11.5.2 Base Case Simulation

The base case is obtained by solving OPF without considering generation rescheduling and wind power variation. The results are indicated in Table 11.2. In the meantime, the stability margins subjected to the three contingencies are shown in Table 11.3. It is obvious that the system is unstable under all three contingencies. Moreover, the initial risk index R_0 is 1.56 considering wind power variation.

Table 11.1 Contingency set.

Contingency ID	Fault bus	Fault clearance time	Tripped line
C1	29	0.1	29-26
C2	28	0.1	28-26
C3	28	0.1	28-29

Table 11.2 Base case dispatch.

Generator	G30	G31[a]	G32[b]	G33	G34
P_G	239.5	560.9	650.0	624.0	504.1
Generator	G35	G36	G37[b]	G38	G39[b]
P_G	644.9	553.0	540	822.4	1000.0
Total cost			39 173.9 $/h		

[a] Slack bus generator.
[b] DFIG wind generator.

Table 11.3 Stability margin for base case.

Contingency	C1	C2	C3
Stability margin	−71.6	−16.5	−68.3
Robustness	0%	0%	0%

The robustness degree is to define the capability of the OP against wind power uncertainties, which is constructed as follows:

$$\text{Robustness degree: } \gamma = \frac{M_s}{M} \times 100\% \tag{11.28}$$

where M is the total randomly generated scenarios within the range $[\mu - \sigma, \mu + \sigma]$. In this chapter, M is 1000. M_s denotes the number of stable scenarios.

From Table 11.3, the robustness degree of the base case under three contingencies is 0%, which indicates that given random wind power variation, the system is completely unstable.

11.5.3 Dispatch Results

11.5.3.1 Single-contingency Case for New England 10-Machine 39-Bus System

Table 11.4 illustrates the stabilization results under three contingencies. The system can become stable with respect to the three contingencies by the coordination of generation rescheduling and load shedding. From Table 11.4, it can be seen that both PC and CC are applied to achieve an optimal coordination cost and the cost is shown in the table. Meanwhile, it can be found that different optimal risk coordination parameters τ are obtained under three contingencies, respectively, which indicates the threshold of acceptable risk index is different to find an optimal solution for different contingencies. Moreover, a comparison of the total coordination

Table 11.4 Generation rescheduling and load shedding for three contingencies.

Contingency	τ	Amount of load shedding (MW)	PC cost ($)	CC cost ($)	Total coordination cost ($)
C1	-1	$\Delta P_{D12} = 7.5$, $\Delta P_{D39} = 69$	39 256.3	77.5	39 332.8[a]
C2	-0.38	N/A	39 190.7	0	39 190.7
	-0.187	$\Delta P_{D12} = 6.0$	39 177.9	6	39 183.9[a]
C3	-0.907	$\Delta P_{D12} = 7.5$, $\Delta P_{D39} = 0.61$	39 243.9	8.1	39 252[a]

[a] Optimal coordination cost

Table 11.5 Robust generation for C2.

Generator	G30	G31[a]	G32[b]	G33	G34
P_G	235.4	552.3	650.0	639.1	513.8
Generator	G35	G36	G37[b]	G38	G39[b]
P_G	651.2	563.6	540	793.2	1000.0
Total cost			39 190.7 $/h		

[a] Slack bus generator.
[b] DFIG wind generator.

cost based on different risk coordination parameters under C2 is illustrated in Table 11.4. The system becomes stable with both -0.38 and -0.187 risk coordination parameters. For -0.38, PC is enough to stabilize the system and so there is no CC. The generation rescheduling results are shown in Table 11.5, which has a 0.043% additional cost to enhance the transient stability of the systems.

Although PC only can well stabilize the system, its long-term cost is higher compared with the coordination of PC and CC. With $\tau = -0.187$, PC and CC are coordinated and the optimal cost is US$ 39,183.9, where only an 0.025% additional cost is required. The robustness degree of the solution against wind power uncertainties is shown in Table 11.6, where the robustness for C2 increases from 0 to 93.3%. It validates the feasibility of the solution.

Table 11.6 shows the stability margins for three contingencies under four testing scenarios after controls. In contrast with Table 11.3, the stability margins under these scenarios are all stable, and the robustness with respect to uncertain wind power is all above 90% and even near 100%, which is acceptable.

Table 11.6 Transient stability margin for three contingencies under four selected scenarios.

Scenarios	Fault C1	C2	C3
l_1	14.05	13.2	15.53
l_2	1.59	0.36	0.41
l_3	2.15	1.00	1.27
l_4	5.01	3.22	3.96
Robustness	98.1%	93.3%	93%

Table 11.7 Generation rescheduling for multi-contingency (C1).

Generator	G30	G31[a]	G33	G34	G35	G36	G38
ΔP_G	−0.9	−0.2	24.54	16.2	14.8	19.4	−75.8

[a] Slack bus generator.

11.5.3.2 Multi-contingency Case for New England 10-Machine 39-Bus System

This chapter also simulates multi-contingency cases, that is, three contingencies are considered simultaneously. The results indicate that the optimal risk coordination parameter is the same as the case for single-contingency C1, i.e. −1. Therefore, the control actions and the total coordination cost are the same as single-contingency C1 shown in Table 11.4. Table 11.7 illustrates the generation rescheduling amount for the multi-contingency case (also for single-contingency C1). It can be seen that G38 reduces the most generation output, which represents this is a critical generator whose generation needs to be deceased to enhance transient stability. Table 11.8 shows the stability margin under four testing scenarios and the robustness degree under the multi-contingency case. By applying the proposed method, the robustness of C2 and C3 is 100% and that of C1 is 98%, which validates that the proposed control actions can secure the system well with three contingencies concurrently under wind power uncertainties.

11.5.4 Stabilization Results and Discussions

Figure 11.4 is the rotor angle trajectories before and after the control actions under C1. It is obvious that G38 is the critical generator as it loses synchronism with other generators, which meets the results indicated in Table 11.7. After the generation

Table 11.8 Transient stability margin for multi-contingency under four selected testing scenarios.

Fault Scenarios	C1	C2	C3
l_1	13.95	41.14	26.6
l_2	1.5	35.51	16.08
l_3	2.07	35.73	16.7
l_4	4.89	37.12	18.92
Robustness	98%	100%	100%

rescheduling and load shedding, the rotor angle trajectories become stable shown in Figure 11.4b.

According to the EEAC, the OMIB power-angle curve could be obtained. Figures 11.5 and 11.6 are the Pe-OMIB plane for C2 (with $\tau = -0.38$) and C3, respectively, where the blue line and red line indicate mechanical power output before and after generation rescheduling. The blue dash line and the green line denote the electrical power output before and after the generation rescheduling, respectively. From Figure 11.5, the system becomes stable only after generation rescheduling though it is not the optimal solution. As the green line indicates, the Pe2 curve stops its excursion and returns before crossing Pm2. In contrast with the blue and red lines, it can be found that mechanical power output becomes smaller after the generation rescheduling, which helps to decrease the accelerating area after the contingency. Then, based on the EAC criterion, once the accelerating area is less than the decelerating area, the system could maintain stable. Thus, for C2, the generation rescheduling only is enough for stabilizing the system as the accelerating area is less than the decelerating area. Nevertheless, it is not an optimal solution with respect to the coordination cost. With the optimal risk coordination parameter ($\tau = -0.187$), the system can also be stable where PC and CC are both activated.

With the optimal risk coordination parameter $\tau = -0.907$ for C3, the Pe-OMIB plane is shown in Figure 11.6. It can be observed that the system cannot maintain stable by generation rescheduling only as the green line (Pe2) crossing the red line (Pm2). In this case, load shedding should be activated. Comparing the yellow dash line (Pe3) and green line (Pe2), the electrical power output increases after load shedding, which increases the decelerating area. When the decelerating area is larger than the accelerating area, the system becomes stable since the yellow dash line returns.

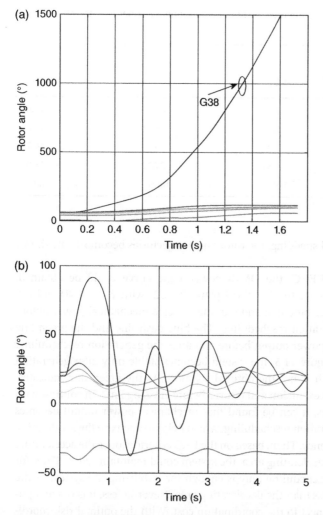

Figure 11.4 Rotor angle trajectories for C1 under base case: (a)-unstable (before control actions); (b)-stable (after control actions).

According to EAC, it is known that decreasing the accelerating area or increasing the decelerating area can both enhance transient stability. From Figures 11.5 and 11.6, it can be found that generation rescheduling contributes to decreasing the accelerating area and load shedding contributes to increasing the decelerating area. Hence, the coordination of these two control actions can help to achieve an optimal cost to improve transient stability.

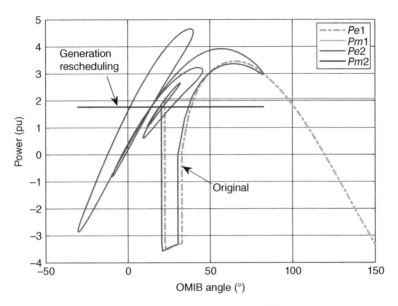

Figure 11.5 Pe-OMIB angle plane for C2 under only PC control.

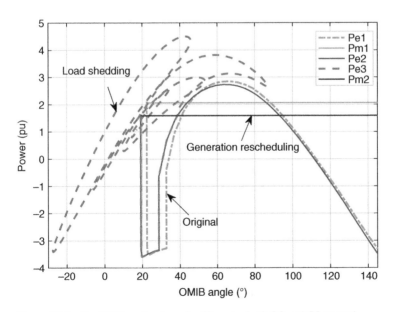

Figure 11.6 Pe-OMIB angle plane for C3 under both PC and CC control.

It should be mentioned that the risk of the system is determined by the OP, and the acceptable risk index is dependent on the risk coordination parameter τ. The proposed method is to find an optimal risk level instead of fixing it by employing a risk coordination parameter τ. It can be found that the value of the risk index determines the feasible stability region of the system in the preventive generation rescheduling control. With a large risk index, which means τ is large, the stability region becomes large. the solution for generation rescheduling is close to marginal stability and inexpensive. In the meantime, the subsequent corrective load shedding control action will be expensive. Therefore, τ plays an important coordinating role between generation rescheduling and load shedding. As for the generation rescheduling cost, it is monotonic reducing with the τ, since the PC cost will decrease with the increment of τ. On the other hand, the CC cost will increase with the increment of τ, which is a monotonic increase. Thus, the proposed method for τ is unimodal. The golden section search method can be used for it well.

The validation of the golden section search method is shown in Figure 11.7. The PC and CC costs with respect to different risk coordination parameters for C2 are indicated, respectively. It can be found that PC cost monotonously reduces. In the meantime, CC cost monotonously increases with the increment of the risk coordination parameter. This verifies the previous intuition. From the figure, the optimal coordination of PC and CC can be obtained at $\tau = -0.187$ by the golden section search.

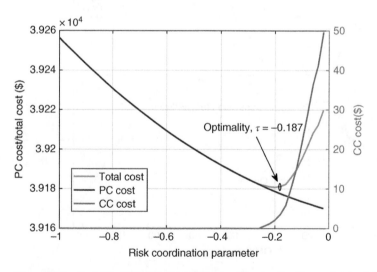

Figure 11.7 Cost function with respect to the risk coordination parameters for C2.

11.5.5 Computation Efficiency Analysis

The proposed method includes the following computation steps.

1) TDS: In this chapter, TDS is run by the commercial software TSAT. The execution time is around 0.312 second with a fixed five seconds simulation time (without earlier termination). However, the execution time can be even terminated earlier once the system is determined unstable practically.
2) OMIB construction: this is the application of the EEAC method to compose multi-machine trajectories into OMIB. The method is embedded in the software TSAT, which is very fast, and spends only several milliseconds.
3) Trajectory sensitivity calculation: in this chapter, one of the main computational burdens comes from the trajectory sensitivity calculation since the numerical method is applied. By using the numerical method, an additional TDS is required for each generator and load. Based on the EEAC method, the trajectory sensitivity is calculated from the stability margin by dividing a small disturbance for each generator and load. In contrast with Ref. [4], which calculated the trajectory sensitivities through the whole simulation time, the proposed method is efficient since it only needs to calculate one sensitivity for each generator and load. Moreover, [2] only calculated the sensitivity at the instability time, which is not accurate and straightforward since it may not represent the whole trajectories of the system. The time for calculating trajectory sensitivities for all generators and loads at a specific operating point is about 10 and 15 seconds, respectively.
4) Calculation of transient stability control: it contains two parts: PC and CC (if applicable). The mathematical models of PC and CC are both solved by the optimization toolbox YALMIP [19], which is implemented in MATLAB. Due to the linearization of the risk constraints and stability constraints for PC and CC, separately, the computation time for the transient stability control is not long, which is about 2 and 0.26 seconds for PC and CC, respectively.
5) Risk coordination parameter search: this is the most time-consuming process. In order to find an optimal risk coordination parameter, continuous iterations should be conducted. For a single-contingency case, the total computational time including all of the above processes is 1080 seconds. For the multi-contingency case, the total computation time is about 1573 seconds. The iterations of the searching process are 30 until the termination condition meets.

Nomenclature

Indices

i, j indices for generators and buses
k index for contingencies
t index for time
W index for wind turbines

Sets

x_u/x_l	sets of decision vector of upper and lower level
$\tilde{w}$	sets of uncertain wind active power output
N_G	number of synchronous generators
N_B	number of buses
N_D	number of load bus
N_c	number of defined contingencies

Variables

P_{Gi}/Q_{Gi}	generation active/reactive output of i-th generator
P_{Di}/Q_{Di}	active/reactive load of i-th load bus
ΔP_{Gi}	the amount of generation rescheduling of i-th generator
ΔP_{Di}	the amount of load shedding
V_j	the voltage magnitude of bus j
θ_{ij}	angle difference between bus i and j
R	risk
Φ	trajectory sensitivity
τ	risk coordination parameter
C_p	PC cost
C_c	CC cost
C	total coordination cost
η_k	transient stability margin of contingency k
η_{k0}	transient stability margin without wind power variation
η_0^l	initial transient stability margin before load shedding

Parameters

p_k	probability of contingency k
a_i, b_i, c_i	generation cost coefficient
c_D	load shedding cost
t	time step of the transient process
$\tilde{P}_{Wi}/\tilde{Q}_{Wi}$	active/reactive wind power output
P_{Gi}^U	generation output of i-th generators of the base case
$\Delta\lambda$	small disturbance
G/B	branch admittance
H	number of wind power output scenarios
$P_{Gi}^{min}/P_{Gi}^{max}$	lower and upper limits of the active power output of i-th generator

References

1 Xu, Y., Dong, Z.Y., Zhang, R. et al. (2014). Solving preventive-corrective SCOPF by a hybrid computational strategy. *IEEE Transactions on Power Systems* 29 (3): 1345–1355.

2 Xu, Y., Dong, Z.Y., Zhao, J. et al. (2015). Trajectory sensitivity analysis on the equivalent one-machine-infinite-bus of multi-machine systems for preventive transient stability control. *IET Generation Transmission and Distribution* 9 (3): 276–286.

3 Pizano-Martinez, A., Fuerte-Esquivel, C.R., and Ruiz-Vega, D. (2011). A new practical approach to transient stability-constrained optimal power flow. *IEEE Transactions on Power Systems* 26 (3): 1686–1696.

4 Nguyen, T.B. and Pai, M.A. (2003). Dynamic security-constrained rescheduling of power systems using trajectory sensitivities. *IEEE Transactions on Power Systems* 18 (2): 848–854.

5 Zarate-Minano, R., Cutsem, T.V., Milano, F., and Conejo, A.J. (2010). Securing transient stability using time-domain simulations within an optimal power flow. *IEEE Transactions on Power Systems* 25 (1): 243–253.

6 Xu, Y., Yin, M., Dong, Z.Y. et al. (2018). Robust dispatch of high wind power-penetrated power systems against transient instability. *IEEE Transactions on Power Systems* 33 (1): 174–186.

7 Xu, Y., Dong, Z.Y., Meng, K. et al. (2012). A hybrid method for transient stability-constrained optimal power flow computation. *IEEE Transactions on Power Systems* 27 (4): 1769–1777.

8 Papadopoulos, P.N. and Milanović, J.V. (2017). Probabilistic framework for transient stability assessment of power systems with high penetration of renewable generation. *IEEE Transactions on Power Systems* 32 (4): 3078–3088.

9 Liu, Z., Liu, C., Li, G. et al. (2015). Impact study of PMSG-based wind power penetration on power system transient stability using EEAC theory. *Energies* 8 (12).

10 Xia, S., Luo, X., Chan, K.W. et al. (2016). Probabilistic transient stability constrained optimal power flow for power systems with multiple correlated uncertain wind generations. *IEEE Transactions on Sustainable Energy* 7 (3): 1133–1144.

11 Monticelli, A., Pereira, M.V.F., and Granville, S. (1987). Security-constrained optimal power flow with post-contingency corrective rescheduling. *IEEE Power Engineering Review* PER-7 (2): 43–44.

12 Capitanescu, F. and Wehenkel, L. (2008). A new Iterative approach to the corrective security-constrained optimal power flow problem. *IEEE Transactions on Power Systems* 23 (4): 1533–1541.

13 Wang, Z., Song, X., Xin, H. et al. (2013). Risk-based coordination of generation rescheduling and load shedding for transient stability enhancement. *IEEE Transactions on Power Systems* 28 (4): 4674–4682.

14 Xue, Y., Custem, T.V., and Ribbens-Pavella, M. (1989). Extended equal area criterion justifications, generalizations, applications. *IEEE Transactions on Power Systems* 4 (1): 44–52.

15 Pavella, M., Ernst, D., and Ruiz-Vega, D. (2012). *Transient Stability of Power Systems: A Unified Approach to Assessment and Control*. Springer Science & Business Media.

16 Xu, Y., Xie, X., Dong, Z.Y. et al. (2016). Risk-averse multi-objective generation dispatch considering transient stability under load model uncertainty. *IET Generation Transmission and Distribution* 10 (11): 2785–2791.

17 Pai, M. (1989). Energy function analysis for power system stability. In: *Power Electronics and Power Systems. Norwell, MA: Kluwer* https://doi.org/10.1007/978-1-4613-1635-0.

18 DSA. Dynamic security assessment software. http://www.dsatools.com/ (accessed 29 January 2015).

19 Lofberg, J. (2004). YALMIP: a toolbox for modeling and optimization in MATLAB. *2004 IEEE International Conference on Robotics and Automation (IEEE Cat. No.04CH37508)*, Taipei, Taiwan (02–04 September 2004), pp. 284–289: IEEE.

20 Zimmerman, R.D., Murillo-Sanchez, C.E., and Thomas, R.J. (2011). MATPOWER: steady-state operations, planning, and analysis tools for power systems research and education. *IEEE Transactions on Power Systems* 26 (1): 12–19.

12

Robust Coordination of Preventive Control and Emergency Control for Transient Stability Enhancement under Uncertain Wind Power

12.1 Introduction

With the growing integration of wind power in power systems, it becomes necessary to consider the influences of uncertain wind power on transient stability. In Chapter 11, stochastic wind power output is considered via formulating a scenario-based wind power uncertainty model. Meanwhile, a bi-level model is constructed to enhance transient stability through the coordinated PC and EC. However, their scenario-based uncertainty model cannot guarantee the full robustness of the stability against uncertainties. Fortunately, in prevailing research, robust optimization has been demonstrated as an effective approach to dealing with uncertainties and ensuring robustness. It has been well applied to unit commitment problems and contingency-constrained unit commitment problems considering uncertainties [1–3]. Meanwhile, it also well performs its robustness in microgrid operation while uncertainties are considered [4–6]. Inspired by these literatures, this chapter proposes a new robust optimization approach, which aims to optimally coordinate the PC and EC, and rigorously achieve robustness under wind power uncertainty against transient instability. To the best knowledge of the authors, this is the first work that addresses the coordination of PC and EC against transient instability through the robust optimization modeling and solution process (as of year 2022). The comprehensive mathematical modeling and solution algorithm are introduced in the following sections. And the proposed method is demonstrated on the New England 39-bus system and the Nordic32 system, which show high computational efficiency and full stability robustness against uncertain wind power.

Stability-Constrained Optimization for Modern Power System Operation and Planning,
First Edition. Yan Xu, Yuan Chi, and Heling Yuan.
© 2023 The Institute of Electrical and Electronics Engineers, Inc.
Published 2023 by John Wiley & Sons, Inc.

12.2 Mathematical Formulation

This chapter proposes a two-stage robust optimization (TSRO) model that coordinates generation dispatch prior to the contingency and emergency load shedding (ELS) after the contingency to enhance the transient stability under uncertain wind power output. Different from the stochastic optimization methods, robust optimization is free from the probability distribution function of the uncertainties and, more importantly, it can reach full robustness under the worst-case uncertainty realization. The comprehensive proposed TSRO modeling is presented in this section.

12.2.1 TSRO Compact Model

The compact TSRO model is shown below:

$$\min_{x} \max_{u \in U} \min_{y \in \emptyset_k(x, u)} C_G(x) + C_{LS}(y) \tag{12.1}$$

$$\text{s.t. } g_0(x, y_0, u) = 0 \tag{12.2}$$

$$h_0(x, y_0, u) \leq 0 \tag{12.3}$$

$$\emptyset_k(x, u) = \{y : g_k(x, y_k, u) = 0, h_k(x, y_k, u) \leq 0, \eta_k(x, y_k, u) \geq \sigma\} \tag{12.4}$$

Pre-fault and post-fault states are denoted as subscripts 0 and k, respectively. Preventive generation dispatch cost and ELS cost are denoted by C_G and C_{LS}, respectively. x denotes the generation power output (preventive control), which is made before the realization of the uncertainty and is the first-stage decision variable. y is the ELS (emergency control), which is the second-stage decision variable. Besides, u denotes the uncertainty variable, and U denotes the uncertainty set.

From the compact model, a *min-max-min* structure forms the objective function (12.1). The first *min* is to minimize the generation cost by optimizing the first-stage decision variable x. The *max* is to maximize the minimization of the second-stage objective and find the worst case of the uncertainty u realization within the uncertainty set U. The second *min* is to minimize the ELS cost by optimizing the second-stage decision variable y. g_0 and h_0 represent the power flow equation and operational limits for the steady-state condition, respectively. $\emptyset_k$ denotes the constraints when the contingency occurs. g_k represents the transient power flow balance. h_k represents the bound for the second-stage decision variable operation. η_k is the stability constraint, and σ is the pre-defined stability threshold. Figure 12.1 shows the framework of the proposed model.

An uncertainty set is constructed to model the random wind power output. Within every dispatch interval, the generation dispatch decisions are implemented at the starting point. Besides, the ELS actions are stored in a decision table and will

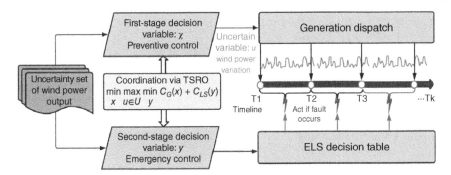

Figure 12.1 Framework of the proposed method.

be triggered immediately when the contingency is detected in this interval. Therefore, the solution results of the TSRO model contain the first-stage generation dispatch and the second-stage ELS actions under the worst case in every dispatch interval. Since the PC is optimized considering the EC under the worst case, the system should be robustly stable against any uncertain realization of wind power variation during the operating interval.

12.2.2 Detailed Model

The detailed TSRO model is constructed below:

$$\min_{P_G} \max_{\tilde{P}_w} \min_{P_L} C_G + C_{LS} \tag{12.5}$$

$$\text{s.t. } C_G = \sum_i^{N_G} a_i P_{Gi}^2 + b_i P_{Gi} + c_i \tag{12.6}$$

$$C_{LS} = \sum_i^{N_D} p_k C_{LS,O} P_{Li} \tag{12.7}$$

$$\begin{cases} P_{Gi}^t + \tilde{P}_{wi}^t - P_{Di}^t = V_i^t \sum_{j=1}^{N_B} V_j^t \left(G_{ij} \cos\theta_{ij}^t + B_{ij} \sin\theta_{ij}^t\right) & (12.8a) \\[2mm] Q_{Gi}^t + Q_{wi}^t - Q_{Di}^t = V_i^t \sum_{j=1}^{N_B} V_j^t \left(G_{ij} \sin\theta_{ij}^t - B_{ij} \cos\theta_{ij}^t\right) & (12.8b) \end{cases}$$

$$t \in [0, T]$$

$$Q_{wi}^t = \tilde{P}_{wi}^t \tan\left(\cos^{-1}\delta\right), \quad t \in [0, T] \tag{12.9}$$

$$Q_{wi}^{\min} \le Q_{wi}^t \le Q_{wi}^{\max}, \quad t \in [0, T] \tag{12.10}$$

$$\begin{cases} P_{Gi}^{\min} \le P_{Gi}^t \le P_{Gi}^{\max}, & i = 1, 2, ..., N_G \\ Q_{Gi}^{\min} \le Q_{Gi}^t \le Q_{Gi}^{\max}, & i = 1, 2, ..., N_G \\ V_i^{\min} \le V_i^t \le V_i^{\max}, & i = 1, 2, ..., N_B \\ L_i^{\min} \le L_i^t \le L_i^{\max}, & i = 1, 2, ..., N_L \end{cases} \quad t \in [0, T]$$

$$\quad (12.11a)$$
$$\quad (12.11b)$$
$$\quad (12.11c)$$
$$\quad (12.11d)$$

$$\begin{cases} P_{Di}^t = \left(P_{Di}^0 - P_{Li}\right) \\ Q_{Di}^t = \left(Q_{Di}^0 - Q_{Li}\right) \end{cases} \quad t \in [0, T]$$

$$\quad (12.12a)$$
$$\quad (12.12b)$$

$$P_{Di}^0 \ge P_{Li} \ge 0 \tag{12.13}$$

$$Q_{Li} = P_{Li} \tan\left(\cos^{-1}\vartheta\right) \tag{12.14}$$

$$O_k(P_G, P_L) \ge \sigma \tag{12.15}$$

The objective function (12.5) is to minimize the total costs of PC (generation dispatch) and EC (ELS) under the worst case of uncertain wind power, which corresponds to (12.1) in the compact model. P_G denotes the synchronous generation output corresponding to the first-stage decision variable x, while P_L denotes the load shedding amount corresponding to the second-stage decision variable y. $\tilde{P}_w$ corresponds to the uncertain variable u. Constraints (12.6) and (12.7) are the expression of the costs. When the time step $t = 0$, equality (12.8) represents the steady-state AC power flow at the pre-fault state, which corresponds to g_0 in (12.2). It should be noted that the active and reactive load demand presented in (12.8) can be separately forecasted in practice. If only a forecast of the active load demand P_{Di} is available, the reactive load demand can be calculated with a typical power factor ϑ by $Q_{Di} = P_{Di} \tan(\cos^{-1}\vartheta)$. Constraint (12.9) denotes the relationship between the active and reactive power outputs of the wind power generator, which means that the reactive power is controlled by the inverter through a constant wind power factor δ. Constraint (12.10) represents its reactive power output bounds. Inequality (12.11) represents the operational limits of the synchronous machines' real power, reactive power, voltage, and branch flow, which corresponds to h_0. When the load shedding is implemented after the contingency occurs, the power system enters the transient state, i.e. $t > 0$. Under this condition, the power flow balance equation is satisfied at each time step, which is with respect to the power flow Eq. (12.8) and corresponds to g_k. P_{Di}^t and Q_{Di}^t are the active and reactive loads of i-th load bus at t time after the load shedding, which corresponds to (12.12). In other words, the variables of the shedding loads P_{Li} and Q_{Li} are considered in (12.12) to calculate the real-time loads after load shedding, thus further impacting power flow through (12.8). It should be noted that the power flow constraints are satisfied in the time-domain simulation at $t > 0$. Constraint (12.13) represents the active power load shedding amount limits for each bus, which corresponds to h_k. Equality (12.14) denotes that the active and reactive shed loads are linked by a constant power factor ϑ. It should be noted that the reactive power load will be also shed with the same percentage as the active power load shedding, i.e. with the constant power factor. At last, constraint (12.15) is the stability constraint corresponding to η_k.

12.2.3 Uncertainty Modeling

In the compact model (12.1)–(12.4), the uncertain wind power output u is modeled as an uncertainty set as follows:

$$U_w = \left\{ \tilde{P}_{wi} \in \mathfrak{R}^{i_w} : \tilde{P}_{wi}^{\min} \leq \tilde{P}_{wi} \leq \tilde{P}_{wi}^{\max}, \mu_{w.l} \leq \frac{\sum\limits_{i \in N_w} \tilde{P}_{wi}}{\sum\limits_{i \in N_w} \tilde{P}_{wi}^{pr}} \leq \mu_{w,u} \;\; \forall i \right\} \quad (12.16)$$

The uncertainty set limits the uncertain wind power output $\tilde{P}_{wi}$ between the lower and upper bounds, where the bounds $\tilde{P}_{wi}^{\min}$ and $\tilde{P}_{wi}^{\max}$ can be obtained through interval prediction techniques [7, 8].

Moreover, the uncertainty budgets are defined with a pair of $\mu_{w,l}$ and $\mu_{w,u}$, which is to limit the overall uncertainty degree of the total wind power outputs shown in (12.16). That is to say, in contrast with the predicted output, the overall possibility of wind power output realization at different locations is constricted within the lower and upper budgets. By applying the uncertainty set (12.16), the worst case of uncertainty realization in the post-fault stage can be determined and addressed. Hence, the transient stability of the system can be maintained with respect to any possible realization within the uncertainty set.

Practically, the design rule for the budgets is based on the historical data (accuracy of forecasting tools) and robustness requirements.

On the one hand, if a high uncertainty degree is required, the uncertainty set should be enlarged by reducing $\mu_{w,l}$ and increasing $\mu_{w,u}$. However, a larger uncertainty set may result in more conservative solutions, which are more expensive. On the other hand, if the uncertainty set is designed as small to overcome the solution conservativeness, the uncertainty realization may not be covered by the uncertainty set. This will cause the solutions that are not adequately robust to maintain the system's stability for specified uncertainty realization. Therefore, a careful design for the uncertainty set is necessary to optimally compromise the solution's robustness and conservativeness [5].

12.3 Transient Stability Constraint Construction

In this chapter, based on EEAC and trajectory sensitivity calculations, the stability constraints are linearized as well. The principles of EEAC and trajectory sensitivity are briefly introduced in Chapter 2. The detailed models of stability constraints η_k in (12.4) and the linearization procedure are presented subsequently.

The stability constraint is related to the second-stage decision variable at k-th fault. Derived from the η_k in the compact model (12.4), the stability constraint can be further extended to the following constraints (12.17)–(12.19):

$$\eta_k(PC, EC, \Delta P_w) = \eta_k^{PC} + \Delta\eta_k^{\Delta P_w} + \Delta\eta_k^{EC} \geq \sigma \tag{12.17}$$

$$\Delta\eta_k^{\Delta P_w} = f_1(\Delta\tilde{P}_w) \tag{12.18}$$

$$\Delta\eta_k^{EC} = f_2(P_{Li}) \tag{12.19}$$

Equation (12.17) represents that the stability margin includes three elements: (i) η_k^{PC} denotes the initial stability margin; (ii) $\Delta\eta_k^{\Delta P_w}$ denotes the stability margin variation after the wind power variation but without the ELS, which can be expressed through a function of the wind power variation f_1 in (12.18); (iii) $\Delta\eta_k^{EC}$ represents the change of the stability margin after the ELS is subjected to a contingency, which can be expressed by a function of the ELS amount f_2 in (12.19). These two functions will be both formulated on the basis of trajectory sensitivity.

During each operating interval, the conventional synchronous generators should compensate the wind power variation to maintain the balance of generation and load. The power balance equations are shown below:

$$\sum_{i}^{N_G} \Delta P_{Gi} = -\sum_{i}^{N_W} \Delta\tilde{P}_{wi} = -\Delta\tilde{P}_w \tag{12.20}$$

$$\Delta P_{Gi} = -\kappa_i\Delta\tilde{P}_w \quad i = 1, 2, 3, ..., N_G \tag{12.21}$$

where κ_i denotes the automatic generation control (AGC) participation factor at the i-th generator. Based on the trajectory sensitivity method introduced in Chapter 2, the variation of stability margin is the multiplication of the trajectory sensitivity and the conventional generation variation. As the wind power variation is balanced by the synchronous generators regarding the AGC factor shown in (12.20) and (12.21), the variation of the stability margin is equivalent to the multiplication of the trajectory sensitivity and the wind power variation. Therefore, the expansion of f_1, which is the stability margin variation after wind power output variation, can be eventually rewritten as:

$$\Delta\eta_k^{\Delta P_w} = f_1(\Delta\tilde{P}_w) = -\sum_{i}^{N_G} \Phi_i(\eta, P_{Gi}) \kappa_i\Delta\tilde{P}_w \tag{12.22}$$

where $\Phi_i(\eta, P_{Gi})$ is the trajectory sensitivity of generator i. From (12.22), the change of the stability margin after wind power variation can be gained by multiplying generator trajectory sensitivity, wind power variation amount, and the AGC participation factor κ.

Similar to f_1, f_2 shown in (12.19), the stability margin variation after the ELS can be rewritten as below:

$$\Delta\eta_k^{EC} = f_2(P_{Li}) = \sum_i^{N_D} \Phi_i(\eta, P_{Li})P_{Li} \tag{12.23}$$

where $\Phi_i(\eta, P_{Li})$ denotes the trajectory sensitivity of load bus i, which is the stability margin variation regarding load shedding amount. As (12.23) presented, the change of the stability margin after the ELS can be calculated by timing the load bus trajectory sensitivity and the required load shedding amount.

Finally, the complete expression of the transient stability constraint indicated in (12.17) can be rewritten below:

$$\eta_k^{PC} - \sum_i^{N_G} \Phi_i(\eta, P_{Gi})\, \kappa_i \Delta\bar{P}_{wi} + \sum_i^{N_D} \Phi_i(\eta, P_{Li})P_{Li} \geq \sigma \tag{12.24}$$

As (12.24) indicates, the stability constraint is linearly formulated. With the initial stability margin, the final stability margin should be larger than the predefined threshold after the change of wind power and ELS action, that is, keeping stable.

12.4 Solution Approach

In order to solve a conventional TSRO model, the column-and-constraint generation (C&CG) algorithm [9] is often used. However, since this chapter considers the TSA process and transient stability constraints, the traditional C&CG cannot be directly used. To effectively solve the proposed model, this chapter develops a modified algorithm that integrates TSA and transient stability constraint construction in the C&CG framework.

12.4.1 Reformulation of the TSRO Model

According to the detailed model (12.5)–(12.14), (12.16), and (12.24), a TSRO matrix model can be reformulated as follows:

$$\min_x \max_{u \in U_w} \min_y a x^2 + b x + c + d^T y \tag{12.25}$$

$$\text{s.t. } Ax + Bu = b \tag{12.26}$$

$$Fx \leq f \tag{12.27}$$

$$Gy \leq g \tag{12.28}$$

$$U(x)y + V(x)u \leq e \tag{12.29}$$

$$y \geq 0 \tag{12.30}$$

$$u \in U_w \tag{12.31}$$

Constraints (12.26) and (12.27) are derived from (12.8)–(12.10) and (12.11) in the detailed model of the first stage ($t = 0$). In the meantime, constraints (12.28), (12.29), and (12.30) are derived from (12.13) and (12.24), which are the constraints of the second stage. Constraint (12.31) represents the uncertainty set (12.16).

It should be mentioned that the coefficient matrices $U(x)$ and $V(x)$ for y and u, which denote the trajectory sensitivities of the load bus and the generator, respectively, are dependent on the operating points determined by the first-stage decision variable.

12.4.2 Column-and-constraint Generation (C&CG) Framework

In practice, the TSRO problem is always solved by decomposition algorithms. The key is to decompose the TSRO into master problems and subproblems and solve them iteratively until the master problem and slave problem converge. Basically, the C&CG algorithm [9] is used in this chapter for the TSRO solution.

The TSRO problem is divided into a master problem (first stage) and a subproblem (second stage) through the C&CG framework. The master problem is presented below:

$$\min_x ax^2 + bx + c + \theta \tag{12.32}$$

$$\text{s.t. } Ax + Bu_l^* = b \tag{12.33}$$

$$Fx \leq f \tag{12.34}$$

$$\theta \geq d^T y_l \tag{12.35}$$

$$Gy_l \leq g \tag{12.36}$$

$$U(x)y_l + V(x)u_l^* \leq e \ \forall u_l^* \in S_l \tag{12.37}$$

$$y_l \geq 0 \tag{12.38}$$

In the proposed coordination model, x is the continuous decision variable. The objective is in a quadratic form and the constraints consist of power flow model, which is a nonlinear form. Then, with the fixed uncertainty variable u_l^*, which is obtained from the subproblem, the master problem can be solved by an interior point method [10]. It should be mentioned that the obtained optimal solution

(x^*, θ^*) is the current solution under the fixed uncertainty variable u_l^*. The subproblem is indicated below:

$$S(u, x^*) = \max_u \min_y d^T y \tag{12.39}$$

$$\text{s.t. } Gy \leq g \tag{12.40}$$

$$U(x^*)y + V(x^*)u \leq e \tag{12.41}$$

$$y \geq 0 \tag{12.42}$$

$$u \in U_w \tag{12.43}$$

It should be mentioned that solving this *max-min* problem is challenging with a polyhedral uncertainty set. In this chapter, to solve the problem, a strong duality method [4] and an outer approximation (OA) algorithm [3] are applied. Thus, the *max-min* problem is transferred by the strong duality into a max problem indicated as follows:

$$R(x) = \max_{u \in U, \lambda, \rho} -\lambda g + \rho^T (Vu - e) \tag{12.44}$$

$$d^T + G^T \lambda + U^T \rho \geq 0, \lambda \geq 0, \rho \geq 0 \tag{12.45}$$

where λ and ρ denote the dual variables. It can be found that the objective (12.44) consists of a non-concave bilinear term $\rho^T Vu$, which is a NP-hard problem. To overcome this problem, the OA algorithm is applied, which linearizes the bilinear term around intermediate solution points in the OA subproblem, then adds the linear forms in the OA master problem and solves these two problems iteratively. The comprehensive OA algorithm is described in [3].

After solving the *max-min* problem, the worst case is searched as an obtained value of the uncertainty variable, i.e. u^*. Meanwhile, u^* is added into a solution set *Sl* as the fixed parameters calculated in the master problem. In addition, new second-stage variables and their corresponding constraints (12.35)–(12.38) are generated and added to the master problem. Then the master problem is solved again with the worst case and its corresponding variables and constraints. By doing this, the master problem and the subproblem are solved iteratively until the lower bound (*LB*) given by the master problem and the upper bound (*UB*) given by the subproblem converge.

12.4.3 Proposed Solution Approach

In the previous sections, the classic C&CG framework for the TSRO problem was presented. However, the transient stability of the system is considered in the proposed TSRO, which will make the TSRO model cannot be solved directly by the classical C&CG algorithm. Hence, a modified algorithm is proposed in this chapter.

Firstly, the initialization of *LB* and *UB* is performed. OPF is solved. With the OPF results under the initial wind power output case, TSA is performed and trajectory sensitivities are calculated according to EEAC. Then, a linear transient stability constraint is constructed (shown in Section 12.3). Next, with the constructed stability constraint, the master problem can be solved and a current solution x_{i+1}^* is obtained so that *LB* is updated.

After solving the subproblem through the strong duality and the OA algorithm, the second-stage ELS solution y_{i+1} can be obtained. However, this may not be the accurate second-stage ELS solution as the system may not be stable at this time. Hence, the TSA is performed via EEAC to check the system's transient stability with the obtained ELS solution. If the system is still not stable under the current u_{i+1}^*, additional linear transient stability constraints (12.36)–(12.38) are generated. The objective aims to minimize the amount of ELS to obtain the second-stage ELS solution. After solving the optimization problem, a new optimal ELS solution y_{i+1} is obtained. The iteration will stop until the system is checked as stable via EEAC. Finally, the *UB* is updated by the worst case with u_{i+1}^*.

If the *UB* and *LB* do not satisfy the termination condition, the obtained *SI*, additional variable y_i, and its corresponding constraints are added to the master problem. In addition, the additional stability constraints generated in the subproblem should be added to the master problem as well.

Different from the classic C&CG, two iterative loops are implemented as the modification to the C&CG algorithm. One outer loop is the classic C&CG iteration between the master problem and the subproblem, while one inner loop is the stability checking iteration in the subproblem.

The proposed solution algorithm and its flowchart are depicted in Figure 12.2.

12.5 Case Studies

12.5.1 Numerical Simulation on New England 10-Machine 39-Bus System

The proposed coordinated method is tested on the modified New England 39-bus system, which is the same as the system applied in Section 9.5.

In this chapter, we also apply three contingencies shown in Table 12.1 to validate the proposed method. Meanwhile, three wind farms are used to replace the synchronous generators and the predicted mean values of the wind power output are assumed to be the same as the original power outputs of the replaced generators, which are 650, 540, and 1000 MW, respectively. The uncertainty budgets are set as 0.9 and 1.1 for $\mu_{w,l}$ and $\mu_{w,u}$, respectively. The cost of ELS is set as 10^4 \$/MWh. The contingency occurrence probability is set as 0.1% [11].

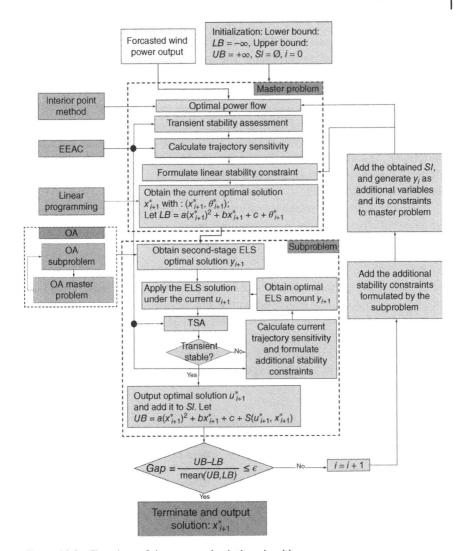

Figure 12.2 Flowchart of the proposed solution algorithm.

Table 12.1 Contingency set.

Contingency ID	Fault bus	Fault duration time	Tripped line
C1	22	0.25 s	22–23
C2	29	0.10 s	26–29
C3	26	0.15 s	26–28

During the simulation, software DSATools TSAT [12] is used to perform TDS. MATPOWER [10] is used for OPF, and the YALMIP toolbox MATLAB [13] is applied for TSRO model programming. The termination gap for the solution algorithm is 0.001. The dynamics of the wind turbines and the synchronous generators are the same as the model introduced in Section 10.6.

Through AC OPF calculation, the first-stage generation dispatch under initial mean wind power output and without the consideration of the transient stability can be obtained, which is shown in Table 12.2. With the initial dispatch, the TDS is performed, and the stability margins are obtained through EEAC shown in Table 12.3. It is obvious that the initial dispatch is stable under C1 but becomes unstable under C2 and C3.

In order to validate the robustness of the initial dispatch against wind power variation, a stability robustness index is defined below:

$$\gamma = \frac{M_s}{M} \times 100\% \tag{12.46}$$

where M denotes the total number of the generated scenarios for wind power output, which are randomly generated within the range of $\pm15\%$ of the predicted mean value. M_s denotes the number of stable scenarios. In this chapter, the total number of test scenarios is set as 1000.

Within the 1000 test scenarios, the robustness indexes under the three contingencies are presented in Table 12.3 as well. It can be found that C1, which is stable

Table 12.2 Initial generator dispatch results (MW).

Generator	Output	Generator	Output	Generator	Output
G30	239.5	G33	624.0	G36	553.0
G31	560.9	G34	504.1	G37[a]	540.0
G32[a]	650.0	G35	644.9	G38	822.4
Total cost		39 173.9$/hr		G39[a]	1000.0

[a]DFIG wind generator

Table 12.3 Transient stability margin for initial dispatch.

Contingency	C1	C2	C3
Stability margin	27.27	−21.49	−3.19
Robustness degree	91.4%	17.4%	43.3%

at initial dispatch, becomes unstable for 8.6% of all the test cases. Meanwhile, C2 and C3 have even lower robustness degrees. Hence, from the results in Table 12.3, it can be concluded that wind power uncertainty has a significant negative impact on transient stability.

In Table 12.4, the calculated load shedding amounts through (12.25)–(12.31) with 0.9–1.1 uncertainty budget are indicated. It can be found that C2 shed the most amount among the three contingencies, which means it should be the most severe contingency. This result is also consistent with the results shown in Table 12.3, where C2 owns the lowest robustness degree within the variation of ±15% of the predicted mean value.

Figure 12.3 indicates the predicted mean values and their corresponding intervals of the wind power outputs at three different locations, shown as blue circles and lines. Meanwhile, the worst-case realization of the wind power output within

Table 12.4 Load shedding amount with 0.9–1.1 uncertainty budget

Contingency ID	Amount of load shedding (MW)		
C1		$P_{L15} = 154.8$	
C2	$P_{L25} = 141.4$	$P_{L26} = 139$	$P_{L27} = 281$
C3	$P_{L25} = 72.6$	$P_{L26} = 139$	$P_{L27} = 281$

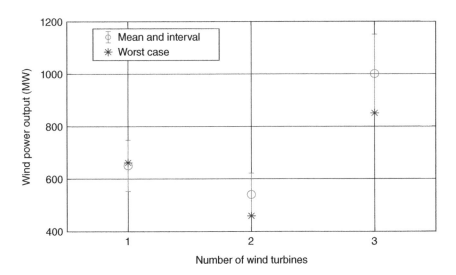

Figure 12.3 Illustration of the worst case of wind power uncertainty.

the uncertainty set is shown by the red stars at three locations, which indicate that the second and third wind farms reach the boundaries but the first one lies in the interval, fulfilling the uncertainty budgets. This implies that when some wind farms get lower wind speeds, the others at different locations get higher wind speeds, thus generating different power within the intervals. Meanwhile, it implies that the uncertain variables of the worst-case are not always at the boundaries. Staying in the interval is also possible.

Figures 12.4–12.7 indicate the generator rotor angles and the corresponding stability margins for C3. In Figure 12.4, it can be seen that the system is unstable with the initial dispatch, whose stability margin is −3.19. In Figure 12.5, the rotor angle curve under the worst-case wind power scenario without ELS is illustrated and the system is not stable, whose stability margin is −38.89 as the ELS is not acted. However, as Figure 12.6 shows, the system becomes stable since the ELS is acted and the stability margin of the system is 27. In Figure 12.7, the rotor angles with ELS under a non-worst-case of wind power output are indicated. It is obvious that the system is stable, but the stability margin is 58, which is higher than that in Figure 12.6 since it is not the worst-case.

Moreover, to comprehensively indicate the proposed method, the stability robustness degree check for the three contingencies is conducted, which are shown in Tables 12.5–12.7. In Table 12.5, three uncertainty budget pair settings with different values of $\mu_{w,\,l}$ and $\mu_{w,\,u}$ are presented. It should be noted that from uncertainty budget pairs one to three, they are enlarged by reducing the lower budgets and increasing the upper budgets, which are also applied to C2 and C3.

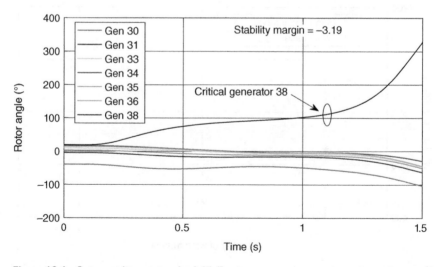

Figure 12.4 Rotor angle curve under initial case.

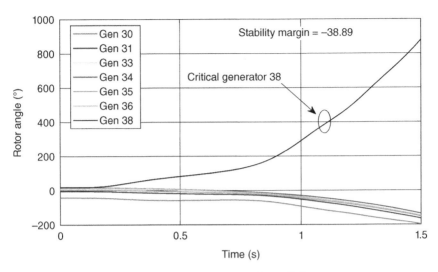

Figure 12.5 Rotor angle curve under worst-case without ELS.

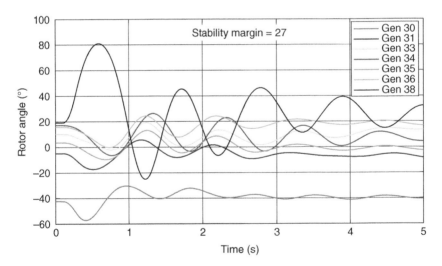

Figure 12.6 Rotor angle curve under worst-case with ELS.

In addition, three different groups of 1000 random wind power output scenarios are generated by uniform distribution-based Monte Carlo sampling (MCS). Note that from MCS Groups 1 to 3, the uncertainty realization range increases as the wind power variation ranges increase.

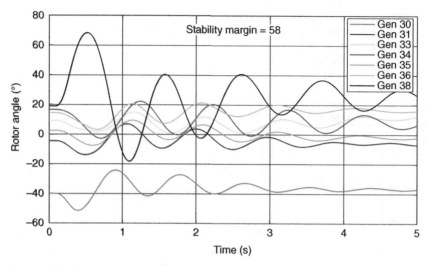

Figure 12.7 Rotor angle curve under a non-worst case with ELS.

Table 12.5 Robustness check with different uncertainty budget pair for C1.

Method	Proposed method			Deterministic
Uncertainty budget pair	1	2	3	N.A
$\mu_{w,l}$	0.95	0.9	0.85	
$\mu_{w,u}$	1.05	1.1	1.15	
Total cost under worst case ($/hr)	40 969.9	44 323	51 247	39 173.9
Stability robustness degree check (%)	MCS Group 1: $\pm 5\% \widetilde{P}_w^{pr}$			
	100%	100%	100%	100%
	MCS Group 2: $\pm 10\% \widetilde{P}_w^{pr}$			
	99.4%	100%	100%	99.4%
	MCS Group 3: $\pm 15\% \widetilde{P}_w^{pr}$			
	91.4%	97.3%	100%	91.4%

Furthermore, a deterministic optimization method is also conducted for better comparison. In this chapter, the deterministic method applies a fixed wind power output as the predicted mean values and then solves the optimal power flow with the consideration of transient stability to obtain the amount of load shedding. In

Table 12.6 Robustness check with different uncertainty budget pair for C2.

Method	Proposed method			Deterministic
Uncertainty budget pair	1	2	3	N. A
Total cost under worst case ($/hr)	44 317	48 388	53 598	40 708.91
Stability robustness degree check (%)		MCS Group 1: $\pm 5\% \widetilde{P}_w^{pr}$		
	100%	100%	100%	54.5%
		MCS Group 2: $\pm 10\% \widetilde{P}_w^{pr}$		
	97.9%	100%	100%	49.9%
		MCS Group 3: $\pm 15\% \widetilde{P}_w^{pr}$		
	86.7%	100%	100%	50.1%

Table 12.7 Robustness check with different uncertainty budget pair for C3.

Method	Proposed method			Deterministic
Uncertainty budget pair	1	2	3	N. A
Total cost under worst case ($/hr)	43 179	47 700	53 120	39 413
Stability robustness degree check (%)		MCS Group 1: $\pm 5\% \widetilde{P}_w^{pr}$		
	100%	100%	100%	52.4%
		MCS Group 2: $\pm 10\% \widetilde{P}_w^{pr}$		
	97%	100%	100%	48.9%
		MCS Group 3: $\pm 15\% \widetilde{P}_w^{pr}$		
	85.6%	100%	100%	48.9%

contrast with the deterministic method, the total costs of the proposed method with the three uncertainty budget pairs are all higher than those of the deterministic method since uncertainties are considered in the proposed method but not in the deterministic method. As for the deterministic method, many unstable scenarios exist under the three MSC groups. Especially for C2 and C3, their stability robustness degrees are only about 50% for all MSC groups. But for the proposed method, the stability robustness degrees are all higher than those of deterministic ones, which means that the proposed method can enhance the stability robustness against wind uncertainty.

For the proposed method, within a small realization range, i.e. MCS group 1, the proposed method can reach 100% robustness degrees for all three contingencies.

With the increment of the uncertainty realization, i.e. from MCS groups 2 to 3, the proposed method under uncertainty budget pairs 1 and 2 results in several unstable scenarios for the three contingencies. However, the stability robustness degree can achieve 100% under uncertainty budget pair 3. It indicates that though uncertainty budget pairs 1 and 2 can enhance the stability robustness, they cannot guarantee robustness when the uncertainty realization range becomes relatively large. As uncertainty budget pair 3 illustrates, a large uncertainty budget pair is necessary to guarantee stability robustness under wider uncertainty realizations.

More specifically, taking C1 as an instance, in Table 12.5, the stability robustness degrees are 91.4%, 97.3%, and 100% from uncertainty budget pair settings 1–3, respectively, for MCS Group 3. Namely, comparing the three budget pairs, the solution with a larger uncertainty budget pair has a higher stability robustness degree than that with a smaller pair. It illustrates that the larger uncertainty budget pair can lead to a more robustly transient stable result against more uncertainty realization scenarios. However, this result will cause a higher cost. The total costs under the worst-case are 40 969.9, 44 323, and 51 247$/hr for three uncertainty budget pairs, respectively. This is reasonable as increasing the stability and robustness of the system under high uncertainty, additional cost is required. When the uncertainty realization is within a small range, i.e. MCS Group 1, the uncertainty budget pair 1 is enough for the 100% stability requirement. In the meantime, the cost only slightly increases by 4.5%, compared with the deterministic method, which is acceptable for keeping the system stable. In practice, a larger uncertainty budget pair is suggested to be chosen as the stability is the top priority for a safe and reliable power system operation.

As shown in the three tables, the stability robustness degrees are all 100% with the widest uncertainty budget pair of 0.85 and 1.15. Namely, with this budget pair, the system is fully robustly stable against all the wind power uncertainty realization.

12.5.2 Numerical Simulation on Nordic32 System

To further validate the effectiveness of the proposed method, more case studies are conducted on a larger test system, the Nordic32 system, which has been also used for stability research in Chapter 9. The introduction and the diagram of the system are presented in Section 9.5. The forecasted mean values of the five replaced wind power farms are 600, 500, 600, 600, and 600 MW, respectively. The total penetration level of wind power in the system is about 27%.

In the case study, a critical contingency is selected, which is a three-phase short circuit fault that occurs at bus 1011 and is cleared after 0.08 second with the tripping of line 1011–4011. Two uncertainty budget pairs are set for validation: 0.95–1.05 and 0.85–1.15.

The dynamics of the synchronous generators are modeled as "GENROU" and "GENSAL" for the rotor model and "SEXS" for the excitation system. Power system stabilizers are also modeled in this system, set as "STAB2A." The wind turbine model is the same as before.

Figures 12.8–12.10 show the rotor angle trajectories under the initial case, worst-case, and worst-case with ELS, respectively. The results illustrate the same

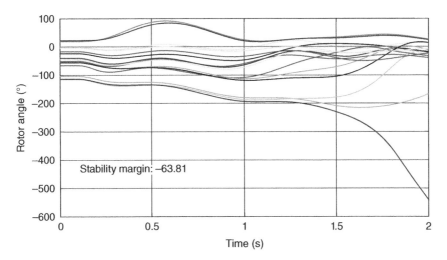

Figure 12.8 Rotor angle curve under initial case for Nordic system.

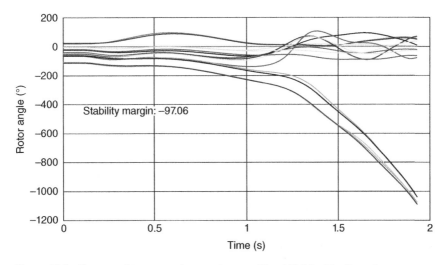

Figure 12.9 Rotor angle curve under worst-case without ELS for Nordic system.

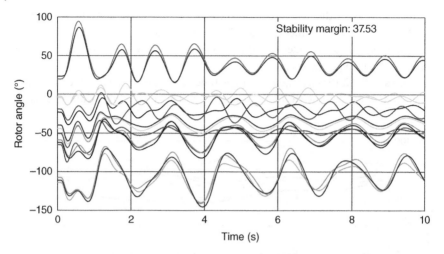

Figure 12.10 Rotor angle curve under worst-case with ELS for Nordic system.

Table 12.8 Robustness check with different uncertainty budget pair of Nordic system.

Method	Proposed method		Deterministic
Uncertainty budget pair	1	2	
$\mu_{w,l}$	0.85	0.95	N.A
$\mu_{w,u}$	1.15	1.05	
Total cost under worst case ($/hr)	246 845.54	240 553.59	235 780.33
Stability robustness degree check (%)	MCS Group 1: $\pm 5\% \widetilde{P}_w^{pr}$		
	100%	100%	62.2%
	MCS Group 2: $\pm 15\% \widetilde{P}_w^{pr}$		
	100%	99.9%	55.8%

performance as the proposed method on the New England system. With the ELS, the system can be stable under the worst-case.

Meanwhile, the robustness degree check is shown in Table 12.8. Similar to the test for the New England system, two budget pairs are tested. The results indicate the system can be fully robustly stable against wind uncertainty with the proposed method, while the deterministic method has a poor robustness degree. For the MCS group with $\pm 15\% \widetilde{P}_w^{pr}$, only the wide enough uncertainty budget pair of 0.85 and 1.15 can ensure the full robustness of the system stability.

12.5.3 Computation Efficiency Analysis

The total computation time mainly depends on how many iterations will be conducted by the proposed solution algorithm (as Figure 12.2). For different contingency cases, different iterations will be made, leading to different computational times. Besides, the total time is also impacted by other tasks such as TDS, stability margin calculation, and trajectory sensitivity calculation, which are also introduced comprehensively in Chapter 11. In this chapter, the different task is the calculation of the master problem and subproblem optimization, which are both solved by the optimization toolbox YALMIP [13] integrated into MATLAB. All in all, the time is not long, as the transient stability constraint is linearized.

For a better understanding of the computation efficiency, the time consumption for New England and Nordic systems is listed in Table 12.9, respectively.

It should be noted that the total programming time, mainly consisting of simulation time, sensitivity calculation time, and optimization solver time, depends on multiple factors.

The transient stability simulation time depends on the system size, and it only takes minority of the total time. From Table 12.9, the TDS time for the Nordic system is about two times that of New England system, which is reasonable considering their system sizes.

In this chapter, the most time-consuming tasks are the trajectory sensitivity calculations and the modified C&CG solution algorithm.

In terms of the trajectory sensitivity calculations, there are 29 (10 generators and 19 loads) sensitivities to be calculated for the New England system. As for the Nordic system, there are 45 (23 generators and 22 loads) sensitivities to be calculated. Hence, the sensitivities of the Nordic system are about 1.55 times those of the New England system. From Table 12.9, the trajectory sensitivity calculation time of the Nordic system is 1.56 times that of the New England system, which is also reasonable.

On the other hand, it should be noted that there are two loops in the whole modified C&CG solution algorithm: one is for the stability check in the subproblem, and the other is for the whole C&CG algorithm. Hence, the time cost also depends on the iteration number of the solution algorithm for different contingency cases. From Table 12.9, it can be found that the reason why the small New England system has long total time consumption of 278 seconds is that there are six iterations for C1. While for the Nordic system, only two iterations are run in the case study, such that the total time is relatively shorter in terms of its system size. Thus, it is not proper to estimate the total programming time by the system size only.

Table 12.9 Time consumption for different calculation tasks.

Calculations	Case	TDS (s)	Stability margin	Trajectory sensitivity		Solver time of master and subproblem (s)	Iteration No.	Total programming time (s)
				Gen (s)	Load (s)			
New England	C1	0.3	Negligible	25	39	2	6	278
	C2						4	176
	C3						2	101
Nordic	\	0.7	Negligible	53	47	1.3	2	170

Nomenclature

Indices

$i, j,$ indices for generators and buses
k index for contingencies
t time step before and during the transient state
W index for wind turbines

Sets

N_B number of buses
N_D number of load buses
N_G number of generators
N_W number of wind turbines
N_c number of defined contingencies

Variables

C_G total generation cost
C_{LS} total expected load shedding cost
L_i^t power flow at i-th branch
P_{Gi}^t/Q_{Gi}^t active/reactive generation power of the i-th generator at t time
P_{Li}/Q_{Li} active/reactive load shedding amount
P_{Di}^t/Q_{Di}^t active/reactive load at i-th bus at transient state $t > 0$
$\widetilde{P}_{wi}$ uncertain wind power output
$\Delta\widetilde{P}_{wi}$ variation of wind power output
ΔP_{Gi} active power variation of i-th generator
V_i^t voltage magnitude of bus j
h_k transient stability margin
Φ trajectory sensitivity
θ_{ij} angle difference of buses i and j

Parameters

a_i, b_i, c_i generation cost coefficients
$C_{LS,O}$ cost of load shedding
G/B branch admittance

$L_i^{\min}/L_i^{\max}$	lower and upper apparent power of i-th branch
P_{Di}^0/Q_{Di}^0	active/reactive load at i-th bus at pre-fault state, $t = 0$
$P_{Gi}^{\min}/P_{Gi}^{\max}$	lower and upper bounds for active power output of i-th generator
$\tilde{P}_{wi}^{\min}/\tilde{P}_{wi}^{\max}$	lower and upper bounds for the wind power output of i-th wind turbine
$\tilde{P}_{wi}^{pr}$	predicted mean wind power output of i-th wind turbine
p_k	probability of occurrence for k-th contingency
$Q_{Gi}^{\min}/Q_{Gi}^{\max}$	lower and upper bounds for reactive power output of i-th generator
U_w	uncertainty set of wind power output
$V_i^{\min}/V_i^{\max}$	lower and upper bounds for the voltage of i-th bus
$\Delta\lambda$	small perturbation
$\mu_{w,l}/\mu_{w,u}$	lower and upper budget of wind power generation
ϑ	constant power factor for the loads
δ	constant power factor for the wind turbines

References

1 Chen, Y., Zhang, Z., Chen, H., and Zheng, H. (2020). Robust UC model based on multi-band uncertainty set considering the temporal correlation of wind/load prediction errors. *IET Generation Transmission and Distribution* 14 (2): 180–190. https://doi.org/10.1049/iet-gtd.2019.1439.

2 Chen, Y., Zhang, Z., Liu, Z. et al. (2019). Robust N–k CCUC model considering the fault outage probability of units and transmission lines. *IET Generation Transmission and Distribution* 13 (17): 3782–3791. https://digital-ibrary.theiet.org/content/journals/10.1049/iet-gtd.2019.0780.

3 Bertsimas, D., Litvinov, E., Sun, X.A. et al. (2013). Adaptive robust optimization for the security constrained unit commitment problem. *IEEE Transactions on Power Systems* 28 (1): 52–63.

4 Zhang, C., Xu, Y., Dong, Z.Y., and Ma, J. (2017). Robust operation of microgrids via two-stage coordinated energy storage and direct load control. *IEEE Transactions on Power Systems* 32 (4): 2858–2868.

5 Zhang, C., Xu, Y., and Dong, Z.Y. (2020). Robustly coordinated operation of a multi energy micro-grid in grid-connected and islanded modes under uncertainties. *IEEE Transactions on Sustainable Energy* 11 (2): 640–651.

6 Zhang, C., Xu, Y., Dong, Z.Y., and Wong, K.P. (2018). Robust coordination of distributed generation and price-based demand response in microgrids. *IEEE Transactions on Smart Grid* 9 (5): 4236–4247.

7 Liu, W. and Xu, Y. (2020). Randomised learning-based hybrid ensemble model for probabilistic forecasting of PV power generation. *IET Generation Transmission and Distribution*. https://digital-library.theiet.org/content/journals/10.1049/iet-gtd.2020.0625.

8 Wan, C., Xu, Z., Pinson, P. et al. (2014). Probabilistic forecasting of wind power generation using extreme learning machine. *IEEE Transactions on Power Systems* 29 (3): 1033–1044.

9 Zeng, B. and Zhao, L. (2013). Solving two-stage robust optimization problems using a column-and-constraint generation method. *Operations Research Letters* 41 (5): 457–461.

10 Zimmerman, R.D., Murillo-Sanchez, C.E., and Thomas, R.J. (2011). MATPOWER: steady-state operations, planning, and analysis tools for power systems research and education. *IEEE Transactions on Power Systems* 26 (1): 12–19.

11 Wang, Z., Song, X., Xin, H. et al. (2013). Risk-based coordination of generation rescheduling and load shedding for transient stability enhancement. *IEEE Transactions on Power Systems* 28 (4): 4674–4682.

12 DSA. Dynamic security assessment software. http://www.dsatools.com/ (accessed 29 January 2015).

13 Lofberg, J. (2004). YALMIP : a toolbox for modeling and optimization in MATLAB. *2004 IEEE International Conference on Robotics and Automation (IEEE Cat. No.04CH37508)*, Taipei, Taiwan (02–04 September 2004). pp. 284–289: IEEE.

Part III

**Voltage Stability-Constrained Dynamic VAR
Resources Planning**

List of Acronyms

AAT	all-at-a-time
AEMO	Australian Energy Market Operator
ASR	acceptable sensitivity region
CBS	candidate bus selection
CI	crowdedness index
CLOD	complex load model
CP	constant power
CPF	continuation power flow
CC	corrective control
DE	differential evolution
DFIG	doubly-fed induction generator
DL	discharging lighting load
DR	demand response
EDLS	event-driven load shedding
EP	external population
FACTS	flexible AC transmission system
FCM	fuzzy c-mean
FSR	feasibility sensitivity region
FV	function value
FWCSR	feasibility worst case sensitivity region
GSC	grid-side converter
HVRT	high voltage ride-through
IBDR	incentive based demand response
ISGA	integral squared generator angle
LCC	life cycle cost
LDC	line-drop compensation
LHS	Latin hypercube sampling

Stability-Constrained Optimization for Modern Power System Operation and Planning, First Edition. Yan Xu, Yuan Chi, and Heling Yuan.
© 2023 The Institute of Electrical and Electronics Engineers, Inc.
Published 2023 by John Wiley & Sons, Inc.

LM	large motor
LVRT	low voltage ride-through
MOEA/D	multi-objective evolutionary algorithm based on decomposition
NERC	North American Electric Reliability Corporation
NPV	net present value
NSGA-II	non-dominated sorting genetic algorithm II
NSGA-III	non-dominated sorting genetic algorithm III
OLTC	on-load tap changer
OPF	optimal power flow
PBDR	price based demand response
PC	preventive control
PDF	probability density function
PF	Pareto front
PS	Pareto set
RDLS	response-driven load shedding
RI	robustness index
RMS	root-mean-square
RSC	rotor-side converter
RVSI	root-mean-squared voltage-dip severity index
SBX	simulated binary crossovers
SM	small motor
STATCOM	static synchronous compensator
SVC	static VAR compensator
TEP	transmission expansion planning
TOAT	Taguchi's orthogonal array testing
TS	transformer saturation
TVDI	transient voltage deviation index
TVSI	transient voltage severity index
UVLS	under-voltage load shedding
VAR	reactive power
VCPI	voltage collapse proximity indicator
VSC	voltage source converter
WCSR	worst case sensitivity region
WT	wind turbine
WTVI	transient voltage deviation index for wind turbines

13

Dynamic VAR Resource Planning for Voltage Stability Enhancement

13.1 Framework of Power System VAR Resource Planning

Conventionally, the power system VAR source planning is considered as an independent process of system operation and control. However, the so-called single-stage approach, which neglects the operational stage, usually generates conservative planning decisions with a relatively high investment cost. In this book, different measures are employed in a multi-stage planning framework as illustrated in Figure 13.1, aiming to coordinate the planning and the operation stages for voltage stability enhancement.

In this planning framework, network hardening measures, such as the upgrading of power lines and deployment of reactive power compensation devices, are used to improve both the static and short-term voltage stability of the power system in the planning stage. The operation stage consists of two sub-stages: pre-contingency stage and post-contingency stage. The pre-contingency stage aims to improve the capability of the system to withstand critical contingencies. The post-contingency stage, through control measures such as response-driven load shedding and event-driven load shedding, aims to rapidly restore the power system back to a stable state.

13.2 Mathematical Models for Optimal VAR Resource Planning

13.2.1 Single-objective Optimization

The compact mathematical model for a single-objective VAR source planning problem is as follows. This single objective is usually for the investment cost of

Stability-Constrained Optimization for Modern Power System Operation and Planning,
First Edition. Yan Xu, Yuan Chi, and Heling Yuan.
© 2023 The Institute of Electrical and Electronics Engineers, Inc.
Published 2023 by John Wiley & Sons, Inc.

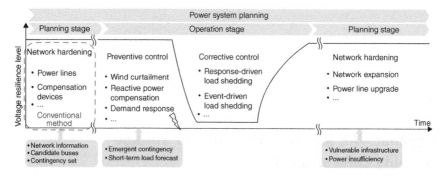

Figure 13.1 Power system VAR source planning framework for voltage stability enhancement. *Source:* Adapted from Xu et al. [1].

the network enhancement measures, while the technical performances, which are addressed as constraints, of the power system are not optimized.

$$\min_{u} \ F(x, u, \widetilde{w}, \mathcal{K}) \tag{13.1}$$

$$\text{s.t.} \ G_0(y_0, u, \widetilde{w}) = 0 \tag{13.2}$$

$$H_0^S(x_0, y_0, u, \widetilde{w}) \leq 0 \tag{13.3}$$

$$G_t(y_t, u, \widetilde{w}, \mathcal{K}) = 0 \tag{13.4}$$

$$H_t^D(x_t, y_t, u, \widetilde{w}, \mathcal{K}) \leq 0 \tag{13.5}$$

where F is the objective function to be optimized. x denotes the power system's state and y denotes the algebraic variables during the transient period. u denotes the control variables and $\widetilde{w}$ indicates the uncertain variables. $\mathcal{K}$ denotes the critical contingencies. Constraints (13.2) and (13.3) are the steady-state power balance constraints and the operational limits, respectively. Constraints (13.4) are the post-contingency power balance constraint, and constraints (13.5) represent the post-contingency inequality constraints.

13.2.2 Single-stage Multi-objective Optimization

The compact mathematical model for this multi-objective VAR planning problem is as follows, (13.6)–(13.10). In contrast to a single-objective optimization problem with a definite single solution, a multi-objective model simultaneously optimizes two or more objectives. Usually, these objectives are in conflict with each other, so there is not a single solution that can achieve a minimization of all the objectives simultaneously. Instead, a set of trade-off solutions can be obtained for the decision-makers according to the electrical engineering practices. However, the single-stage multi-objective model [1] for the power system planning problem might generate arbitrary and conservative planning decisions since only the network enhancement measures are considered while the operational control measures for voltage stability enhancement are ignored.

$$\min_u F[f_1(\boldsymbol{x}, \boldsymbol{u}, \widetilde{\boldsymbol{w}}, \mathcal{K}), f_2(\boldsymbol{x}, \boldsymbol{u}, \widetilde{\boldsymbol{w}}, \mathcal{K}), f_3(\boldsymbol{x}, \boldsymbol{u}, \widetilde{\boldsymbol{w}}, \mathcal{K})] \tag{13.6}$$

$$\text{s.t. } G_0(\boldsymbol{y}_0, u, \widetilde{w}) = 0 \tag{13.7}$$

$$H_0^S(\boldsymbol{x}_0, \boldsymbol{y}_0, \boldsymbol{u}, \widetilde{\boldsymbol{w}}) \leq 0 \tag{13.8}$$

$$G_t(\boldsymbol{y}_t, \boldsymbol{u}, \widetilde{\boldsymbol{w}}, \mathcal{K}) = 0 \tag{13.9}$$

$$H_t^D(\boldsymbol{x}_t, \boldsymbol{y}_t, \boldsymbol{u}, \widetilde{\boldsymbol{w}}, \mathcal{K}) \leq 0 \tag{13.10}$$

where F is the overall objective function, consisting of two or more sub-objectives, such as investment cost and short-term voltage stability performance metric. $\boldsymbol{x}$ denotes the power system's state and $\boldsymbol{y}$ denotes the algebraic variables during the transient period. $\boldsymbol{u}$ denotes the control variables and $\widetilde{\boldsymbol{w}}$ indicates the uncertain variables. $\mathcal{K}$ denotes the critical contingencies. Constraints (13.7) are the steady-state power balance constraints, and constraints (13.8) represent the steady-state operational limits. Constraints (13.9) are the post-contingency power balance constraint, and constraints (13.10) represent the post-contingency inequality constraints, such as rotor angle constraint, LVRT/HVRT constraint, voltage recovery constraint, and load shedding constraint. The following chapters will expand this compact mathematical model equation by equation in.

13.2.3 Multi-stage and Multi-objective Optimization

The compact mathematical model for the multi-stage and multi-objective planning problem can also be presented as (13.6)–(13.10). In contrast to the single-stage planning approach, the multi-stage approach covers a broader range of both planning and operation (Figure 13.2), as discussed in Section 13.1. The corresponding objective is also different from the one in (13.6). For instance, the total planning cost, instead of just the investment cost, should consist of the investment cost and the cost for operational controls. Although the instantaneous costs for operational controls, such as load shedding and wind generation curtailment, are neglectable compared with the investment cost of network enhancement measures, such as power lines and VAR sources, their aggregated cost over a long time is still considerable and cannot be ignored. On the other hand, since the operational control measures are adopted in the multi-stage optimization model, more

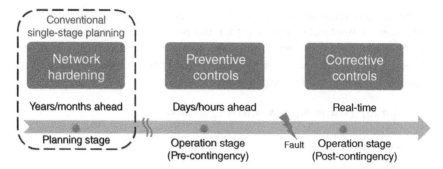

Figure 13.2 Time scales of planning stage, pre-contingency stage, and post-contingency stage. *Source:* Adapted from Mercados [2].

equity and inequity constraints should be introduced. For instance, curtailment constraint of wind power generation, load shedding constraint, and demand response constraints. In detail, the multi-stage and multi-objective optimization model will be further explained in Chapters 17–20 in detail.

13.3 Power System Planning Practices

Power system planning practices are evolving to meet growing customer expectations, guide distributed energy resources development, and ensure investments in system upgrades continue to serve customers.

In general, planning time horizons lie in one of the following ranges:

- short term (up to 1 year),
- medium term (up to 2–3 years)
- long term (between 20–30 years).

Although the practices of power system planning vary significantly from nation to nation, from utility company to utility company, they can be generally summarized as follows (Figure 13.3) [2–5]:

- study of future loads and forecasting the anticipated loads according to the planning time horizon. Multiple scenarios are usually determined based on the development of the studied area.
- evaluation of energy resources available in the future for electricity generation and the foreseeable trends in technical and economic developments.
- evaluation of the technical and economic characteristics of the current power system, including generation system, transmission system, and distribution system. Identification of the potential units for system expansion or upgrades. These characteristics include but are not limited to investment cost, operation and maintenance cost, scheduled deployment time.

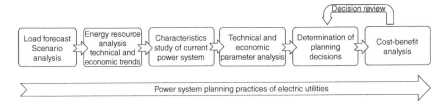

Figure 13.3 Power system planning practices of electric utilities.

- determination of the economic and technical parameters affecting planning decisions, such as reliability requirements, regulations and policies, workforce availability, and interest rates.
- determination of the optimal or strategic planning decisions within the imposed constraints.
- cost-benefit analysis followed by qualitative review of the decisions to estimate their viability on a yearly basis.

It should be noted that, as the intermittent renewable generations grow, forecasting of the system energy becomes more and more important in the planning process. The emerging practice is to include renewable energy supply early in the planning process and consider it during the energy growth forecast, which allows full integration of renewable generation into the planning process.

Specifically, there are plannings of generation system, transmission system, and distribution system. Generation planning leads to determining the capacity of units to be installed that will meet the anticipated load demand. It also defines the fuel to be used in addition to the size of units to be installed over the period. Finally, the planning should be conducted to satisfy well-defined criteria that reflect the strategies adopted within the power industry, and contribute to enhancing the security, quality, and reliability of supply at minimum cost.

Transmission and distribution (T&D) system planning objective, whether a short-term plan, targets developing a deep understanding of the existing system and preparing a roadmap for near-term and future investments required to provide services that are adequate, reliable, and economical to new and existing customers. Notably, there has been significant activity across the world related to the procurement of non-wire alternatives, such as battery storage, flow control devices, and demand response, for deferral of traditional investments in distribution and transmission systems. These efforts are at an early stage and the power industry is learning from experience gained with early projects how to address challenges associated with sourcing and implementing non-wires alternatives.

On the other hand, there are a number of different types of planning studies, focusing on different aspects of the performance of the power system, such as reliability, market efficiency, operation, and resiliency.

System planning is no longer exclusively or solely technically driven and performed strictly by utilities. Instead, it has developed into a planning process that incorporates all industry stakeholders, including but not limited to electric energy policymakers, local government, resource developers, transmission developers, environmentalists, and numerous others. All of these add up to the complexity of the planning of modern power systems. To address these issues, new approaches are adopted by utilities according to the suggestions from experienced people from multi-discipline fields. Additional planning studies or technical reports include but are not limited to the following fields [6]:

- integrated planning studies for generation and transmission systems, or (T&D) systems
- incorporating more uncertainties directly or indirectly into power system planning,
- performing power market pricing studies,
- preparing more detailed feasibility studies in addition to economic and financial analysis,
- recommending rates, tariffs, and prices after a thorough analysis.

Indices

D index for the dynamic state
o index for the original state
S index for the steady state
t index for the time

Sets

$\mathcal{K}$ set of contingencies
u set of control variables
$\tilde{w}$ set of uncertainty variables
x set of state variables
y set of algebraic variables

References

1 Xu, Y., Dong, Z.Y., Xiao, C. et al. (2015). Optimal placement of static compensators for multi-objective voltage stability enhancement of power systems. *IET Generation Transmission and Distribution* 9 (15): 2144–2151.
2 Mercados (2013). Current practices in electric transmission planning: case studies. The Global Power Best Practice 13, The Regulatory Assistance Project, Montpelier.

www.raponline.org/wp-content/uploads/2016/05/rap-globaltransmissionpractices-2013-dec.pdf (accessed 18 August 2021).

3 ENTSO-E (European Network of Transmission System Operators for Electricity) (2018). Cost benefit analysis methodology CBA 2.0 for TYNDP project assessment. European Network of Transmission System Operators for Electricity. www.entsoe. eu/publications/tyndp/#cost-benefit-analysismethodology-cba-20-for-tyndp-project-assessment (accessed 18 August 2021).

4 ENTSO-E (2014). Ten-year network development plan 2014, european network of transmission system operators for electricity. Brussels. www.entsoe.eu/publications/tyndp/tyndp-2014/ (accessed 18 August 2021).

5 Wilkerson, J., Larsen, P.H., and Barbose, G.L. (2014). Survey of western U.S. electric utility resource plans. *Energy Policy* 66: 90–103.

6 International Renewable Energy Agency (2018). *Insights on Planning for Power System Regulator*. IRENA.

14

Voltage Stability Indices

14.1 Conventional Voltage Stability Criteria

14.1.1 *P–V* Curve Method

As one of the most prevalent methods for voltage stability evaluation, the *P–V* curve (also widely called "nose curve") can be computed based on a two-bus equivalent system. As the load level of a power system is gradually increased, the power flows are re-computed until the nose of the *PV* curve (maximum loading point) is reached, and that is why it is also called the "nose curve" [1]. Then, the margin between the current operating point and the nose point is used as a voltage stability criterion. For any given pair of voltage and power factor, each curve represents a maximum system load or the loadability of the studied power system (Figure 14.1).

The *P–V* curve method maintains a good performance in an infinite bus scenario. In engineering practice, Thevenin equivalent of a power system is used, and the generations are rescheduled at each step as the load increases. It is possible that some of the generators may reach the maximum reactive power available. Consequently, the network topology may also change. Therefore, the accuracy of *P–V* curve method is compromised.

14.1.2 *V–Q* Curve Method and Reactive Power Reserve

The *V–Q* curve method [1, 2] is another widely used approach to study the voltage stability of a power system. The voltage stability of a bus is closely related to the available reactive power reserve of the bus, and the reserve can be obtained based on the *V–Q* curve of the studied bus. *V–Q* curve can be plotted using the voltage of a bus against the reactive power at that bus. Generally, a fictitious synchronous generator is installed at the bus with no active power output and no reactive power limit. Then, the power flow is computed with the bus as the generator bus for a

Stability-Constrained Optimization for Modern Power System Operation and Planning,
First Edition. Yan Xu, Yuan Chi, and Heling Yuan.
Published 2023 by John Wiley & Sons, Inc.

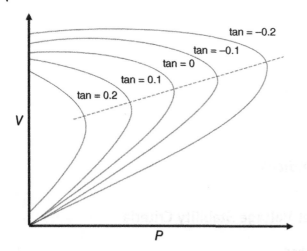

Figure 14.1 Illustration of *P–V* curve. *Source:* Kundur [1].

specific range of voltages to obtain the reactive power and the specified voltage for the *V–Q* curve.

Specifically, the reactive power margin is calculated as the reactive power difference between the current operation point and the nose point of the *V–Q* curve. Then, the stiffness of the bus can be assessed qualitatively based on the slope of the *V–Q* curve. A smaller slope indicates a stiffer bus, and the bus is less vulnerable to voltage instability. Similarly, weaker buses can also be identified based on the slope of the *V–Q* curve. The zero slope means the reach of the voltage stability limit. Therefore, the slope can be used as an index for voltage stability assessment (Figure 14.2). In practice, the *V–Q* curve method is usually used to test the

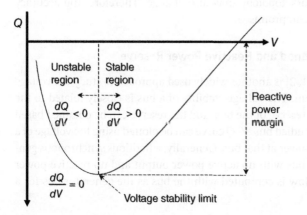

Figure 14.2 Illustration of *Q–V* curve. *Source:* Van Cutsem and Vournas [2].

strength of the bus and is helpful in determining the potential reactive power compensation requirement.

14.1.3 Modal Analysis

Based on the fact that the power flow Jacobian matrix becomes singular when voltage collapses, modal analysis [3] of the Jacobian matrix has been proposed in many studies to evaluate the voltage stability of the power system. Firstly, based on the steady-state system model, a small number of eigenvalues and the associated eigenvectors of the reduced Jacobian matrix are calculated. The eigenvalues and the eigenvectors retain Q–V relationships in the network and include the appropriate characteristics of generators, loads, reactive power compensating devices, and HVDC converters. Each eigenvalue corresponds to a specific mode of reactive power or voltage variation. The magnitude of the eigenvalues indicates an estimation of proximity to voltage instability, while the eigenvectors relate to the mechanism of voltage instability. Specifically, the power system is considered voltage stable if all the eigenvalues of the reduced Jacobian matrix are positive and the power system with at least one negative eigenvalue is voltage unstable. The system is at the edge of voltage instability if there is a zero eigenvalue. A potential voltage collapse of a stable power system can be identified based on the minimum positive eigenvalue. On the other hand, the bus participation factor, defined as the greatest contributing factor for a system to reach voltage collapse, can be used to identify the weakest bus. The modal analysis can be expressed as follows:

$$\begin{bmatrix} \Delta P \\ \Delta Q \end{bmatrix} = \begin{bmatrix} J_{11} & J_{12} \\ J_{21} & J_{22} \end{bmatrix} \begin{bmatrix} \Delta \theta \\ \Delta V \end{bmatrix} \tag{14.1}$$

By letting $\Delta P = 0$ in (14.1)

$$\Delta P = 0 = J_{11} \Delta \theta + J_{12} \Delta V \tag{14.2}$$

$$\Delta \theta = -J_{11}^{-1} J_{12} \Delta V \tag{14.3}$$

$$\Delta Q = J_{21} \Delta \theta + J_{22} \Delta V \tag{14.4}$$

Substituting (14.3) in (14.4):

$$\Delta Q = J_R \Delta V \tag{14.5}$$

$$J_R = \begin{bmatrix} J_{22} - J_{21} J_{11}^{-1} J_{12} \end{bmatrix} \tag{14.6}$$

J_R is the reduced Jacobian matrix of the system. Then, (14.5) can be written as:

$$\Delta V = J_R^{-1} \Delta Q \tag{14.7}$$

The matrix J_R represents the linearized relationship between the incremental changes in bus voltage (ΔV) and bus reactive power injection (ΔQ) for constant active power.

The eigenvalues and eigenvectors of the reduced order Jacobian matrix J_R are used for the voltage stability characteristics analysis. Voltage instability can be detected by identifying modes of the eigenvalues matrix J_R.

The magnitude of the eigenvalues provides a relative measure of proximity to instability. The eigenvectors, on the other hand, present information related to the mechanism of loss of voltage stability. According to Ref. [3], the ith modal voltage variation can be written as:

$$\Delta V_{mi} = \frac{1}{\lambda_i} \Delta Q_{mi} \tag{14.8}$$

If $\lambda_i = 0$, the ith modal voltage will collapse because any change in that modal reactive power will cause infinite modal voltage variation. If $\lambda_i < 0$, the ith modal voltage and the ith reactive power variation are in the opposite directions, indicating that the system is voltage unstable. If $\lambda_i > 0$, the ith modal voltage and the ith reactive power variation are along the same directions, indicating a stable scenario.

14.1.4 Continuation Power Flow Method

Generally, the continuation method is used to solve systems of nonlinear equations. It is a path-following methodology consisting of the following: predictor, parameterization strategy, corrector, and step length control. The numerical derivation of this method is shown in Ref. [4].

Voltage stability margin (VSM) is the MW-distance between the current operating point and the point of bifurcation. In other words, it measures how many MW can still be delivered to load buses at the base operating point along the load and generation increase direction before voltage collapse occurs. The mathematical model can be as described follows:

$$P_{Gi} = P_{Gi0} + \lambda P_{Gdi} \tag{14.9}$$

$$P_{Li} = P_{Li0} + \lambda P_{Ldi} \tag{14.10}$$

$$Q_{Li} = Q_{Li0} + \lambda Q_{Ldi} \tag{14.11}$$

where P_{Gi}, P_{Li}, and Q_{Li} denote the active power generation, active power load, and reactive power load, respectively; subscribe i denotes the ith bus or generator, 0 denotes the base operating point, subscribe d denotes the power increase direction; λ is the loading parameter. Consequently, the VSM is $\lambda \times P_{Ldi}$.

To calculate the loading parameter λ, one method is to repeatedly run a power flow program where load and generation are increased step by step. However, conventional power flow algorithms based on Newton's method usually fail to converge when the operating point approaches the bifurcation point.

To deal with the convergence problem, an alternative method known as continuation-power flow (CPF) technique has been proposed. In CPF, the loading parameter is incorporated into the power flow equations as an additional parameter.

The CPF captures this path-following feature by means of a predictor-corrector scheme that adopts locally parameterized continuation techniques to trace the power flow solution paths. Using the CPF method, the load voltage can be obtained even when the power flow Jacobian matrix is singular.

The CPF method employs a predictor-corrector scheme [5] to find a solution path of power flow equations that have been reformulated to include a load parameter. The process initiates from a known solution and is followed with a tangent predictor to estimate a subsequent solution relating to a different load parameter. The estimation is then corrected using Newton–Raphson technique. The local parameterization provides a means of integral part in avoiding singularity in the Jacobian.

14.2 Steady-State and Short-term Voltage Stability Indices

Jacobian matrix-based voltage stability indices can estimate the voltage collapse point and margin of voltage stability. However, Jacobian matrix has to be changed if there is any topological change. This issue brings a significant increase in computation time, particularly for the power system planning problem regarding placement and sizing of generators or compensation devices. On the other hand, topological change compromises the accuracy of the curve-based methods. As for the CPF method, a probable loading pattern plays a vital role in the calculation of laudability margin. It is very difficult to determine such pattern for a long-term planning problem, although it is relatively easy to obtain for the on-line operation or operational planning problems.

Conventional voltage stability indices usually overlook the fast load dynamics, which contribute directly to the short-term voltage instability phenomenon. As an alternative, the dynamic method, which is based on the time-domain simulation, can effectively model the dynamics of the system using differential-algebraic equations. The time-varying trajectories, such as voltage response after a disturbance, can be obtained by solving the differential-algebraic equations using numerical integrations. The dynamic method is necessary for the analysis of short-term voltage stability since some load components, such as induction motors, draw a significantly high reactive current within a very short timeframe after a continency. Similarly, the dynamic method can also capture the fast-changing dynamics of compensation devices for voltage stability enhancement, such as STATCOMs.

14.2.1 Steady-State Voltage Stability Index

A line-based steady-state voltage stability index, called Voltage Collapse Proximity Indicator (VCPI) [6], is used here to evaluate the steady-state voltage stability of the system. It is based on the concept of maximum power transferred through a line and can provide reliable information about the proximity to voltage collapse [6]. According to some comparison studies [6, 7], VCPI outperforms other steady-state voltage stability indices in terms of accuracy and robustness. Other favorable features of the index include simplicity, flexibility to topological changes and load modifications, and low computation burden.

VCPI has been originally derived from the idea of the voltage collapse point. The two-bus representation of a power system, as illustrated in Figure 14.3, is used to explain the voltage collapse point.

The active and reactive power at the receiving bus can be written as follows (the shunt admittances are neglected for simplicity):

$$P_r = \frac{V_r V_s \cos \cdot \delta - V_r^2 - Q_r X}{R} \tag{14.12}$$

$$Q_r = \frac{P_r X - V_r V_s \sin \cdot \delta}{R} \tag{14.13}$$

$$\delta = \delta_s - \delta_r \tag{14.14}$$

Based on (14.12) and (14.13), the fourth-degree equation can be derived as (14.15) and (14.16).

$$V_r^4 + \left(2P_r R + 2Q_r X - V_s^2\right)V_r^2 + \left(P_r^2 + Q_r^2\right)Z^2 = 0 \tag{14.15}$$

$$V_r^4 + \left(2P_r R + 2Q_r X - V_s^2\right)V_r^2 + S_r^2 Z^2 = 0 \tag{14.16}$$

At the voltage collapse point, Eq. (14.15) has two pairs of real identical roots and the Jacobian matrix is singular. Any further increase in the load makes the roots become complex with real and imaginary parts. The voltage stability requires the discriminant of this equation to be greater than or equal to zero [8].

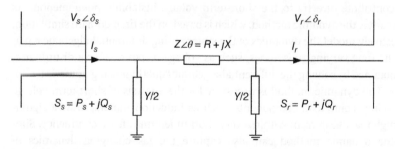

Figure 14.3 The two-bus representation of a power system.

When the boundary condition is reached, mathematically expressed as $\dfrac{Z_r}{Z_s} = 1$, the maximum active power that can be transferred to the receiving end is obtained as (14.17), where $\phi = \tan^{-1}(Q_r/P_r)$.

$$P_r^{\text{max}} = \frac{V_s^2}{Z_s} \frac{\cos \phi}{4 \cos^2 \dfrac{(\theta - \phi)}{2}} \tag{14.17}$$

Similarly, the maximum transferable reactive power Q_r^{max}, the maximum active power loss in the line P_l^{max}, and its reactive power counterpart Q_l^{max} can be accordingly obtained as follows:

$$Q_r^{\text{max}} = \frac{V_s^2}{Z_s} \frac{\sin \phi}{4 \cos^2 \dfrac{(\theta - \phi)}{2}} \tag{14.18}$$

$$P_l^{\text{max}} = \frac{V_s^2}{Z_s} \frac{\sin \theta}{4 \cos^2 \dfrac{(\theta - \phi)}{2}} \tag{14.19}$$

$$Q_l^{\text{max}} = \frac{V_s^2}{Z_s} \frac{\sin \theta}{4 \cos^2 \dfrac{(\theta - \phi)}{2}} \tag{14.20}$$

Then, VCPIs can be mathematically described as follows based on these four maximum permissible quantities P_r^{max}, Q_r^{max}, P_l^{max}, and Q_l^{max}.

$$
\begin{aligned}
VCPI(1) &= \frac{\text{active power transferred to the receiving end}}{\text{Maximum active power that can be transferred}} \\[2mm]
&= \frac{P_r}{P_r^{\text{max}}}
\end{aligned}
\tag{14.21}
$$

$$
\begin{aligned}
VCPI(2) &= \frac{\text{reactive power transferred to the receiving end}}{\text{Maximum reactive power that can be transferred}} \\[2mm]
&= \frac{Q_r}{Q_r^{\text{max}}}
\end{aligned}
\tag{14.22}
$$

$$
\begin{aligned}
VCPI(3) &= \frac{\text{active power loss in the line}}{\text{Maximum possible active power loss}} \\[2mm]
&= \frac{P_l}{P_l^{\text{max}}}
\end{aligned}
\tag{14.23}
$$

$$
\begin{aligned}
VCPI(4) &= \frac{\text{Reactive power loss in the line}}{\text{Maximum possible reactive power loss}} \\[2mm]
&= \frac{Q_l}{Q_l^{\text{max}}}
\end{aligned}
\tag{14.24}
$$

According to extensive experimental results from Ref. [9], $VCPI(1) = VCPI(2)$, and $VCPI(3) = VCPI(4)$ are always true for any loading state. Therefore, instead of using all four indices, an either active or reactive index can be used. For the steady-state voltage stability analysis, the fundamental cause of instability is either the excessive power transfer through the line or the excessive absorption of power by the line itself. Let us denote the $VCPI(1)$ and $VCPI(2)$ as $VCPI(power)$, $VCPI(1)$, and $VCPI(2)$ as $VCPI(loss)$. With the increase in load demand in the power system, the power flow also increases accordingly, which in turn results in a gradual voltage decrease at the receiving end bus. As the loading condition rises to the critical operating point, the indicators $VCPI(power)$ and $VCPI(loss)$ approach the maximum value 1.

Based on (14.17) and (14.21), the $VCPI(power)$ can be expressed as follows:

$$VCPI(power) = \frac{P_r}{P_r^{\max}} = \frac{P_r}{\dfrac{V_s^2}{Z_s} \, 4\cos^2 \dfrac{(\theta - \phi)}{2}} \cos\phi \tag{14.25}$$

The $VCPI(power)$ is computed for each line to obtain the stressed conditions of all the lines in the studied power system. So, the VCPI for the whole system (all the power lines) can be expressed as follows:

$$VCPI^T = \sum_{l \in \mathscr{L}} VCPI(power) \tag{14.26}$$

VCPI is quite simple in nature, with clear physical meaning and low computation complexity. These features make it a good candidate for steady-state voltage stability analysis of a real-world power system planning problem. Another favored advantage of the index is the identification of stressed lines.

14.2.2 Short-term Voltage Stability Index

Fast load dynamics play a vital role in short-term voltage stability analysis.

The time scale of the study is in the order of a few seconds. Differential equations of the systems are required for the analysis, and accurate dynamic modeling of loads is also necessary. Following a large voltage disturbance, induction motors decelerate dramatically by the voltage drop or may stall once the electrical torque cannot overcome the mechanical load. This draws a high reactive current, which significantly decreases the voltage magnitudes and impedes the voltage recovery.

Given the fast dynamics of the voltage response after a disturbance, short-term voltage stability analysis mainly focuses on voltage deviation, delayed voltage recovery, and potential voltage collapse in the post-contingency stage. Transient Voltage Severity Index (TVSI) [10] is a straightforward index, quantifying the transient voltage performance of the buses following the disturbance and it can be

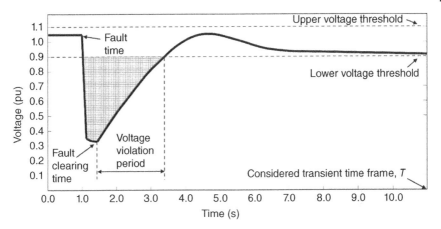

Figure 14.4 Illustration of Typical Post-contingency Scenario. *Source:* Adapted from Ajjarapu [4].

mathematically described as follows, where TVDI is defined as (14.28), based on the dynamic voltage trajectories after a disturbance (where $k \in \mathcal{K}$).

$$TVSI_k = \frac{\sum_{i \in \mathscr{B}} \sum_{t \in [T^c, T^{end}]} TVDI_{i,t,k}}{N^b} \tag{14.27}$$

$$TVDI_{i,t,k} = \begin{cases} \dfrac{|V_{i,t} - V_{i,0}|}{V_{i,0}}, & \text{if } \dfrac{|V_{i,t} - V_{i,0}|}{V_{i,0}} \geq \delta \\ 0, & \text{otherwise} \end{cases}, \forall t \in [T^c, T^{end}] \tag{14.28}$$

In practice, the threshold δ, which represents the maximum acceptable voltage deviation level used in (14.28), can be set according to specific industrial criteria [11], e.g. 20%.

As illustrated in Figure 14.4, TVSI indicates the unacceptable voltage violation of buses (the shaded area) during the transient period after a contingency or a voltage disturbance. It measures the magnitude of voltage violation and reflects the corresponding duration time of the violation. Therefore, it makes possible a quantitative comparison of the system's transient voltage performance after a voltage disturbance. A larger TVSI value means a worse voltage response, while a smaller one indicates a better transient voltage performance.

14.3 Time-Constrained Short-term Voltage Stability Index

In recent years, in addition to TVSI, there are also some indices for short-term voltage stability assessment [12–15]. The prevalent voltage stability metrics are based on the voltage thresholds and the time-varying voltage trajectories. Nonetheless, they still have limitations: (i) the steady-state voltage stability and short-term voltage stability are usually evaluated in the literature separately; (ii) there is no comprehensive evaluation of voltage resilience, and most of them only focus on one or two of the critical aspects of voltage performance, including recovery speed, duration of the deviation, and post-contingency steady-state assessment. On the other hand, given the increasing penetration level of renewable generation and their high sensitivity to not only the extent of the voltage deviation but also the time duration of the voltage deviation, more advanced metrics are needed for an accurate assessment of the dynamic voltage response of buses after a disturbance.

In this chapter, the voltage resilience performance after a contingency is evaluated by two advanced trajectory-based indices, called $TVSI^r$ and $TVSI^s$, based on the principles of TVSI. The voltage trajectories can be extracted from time-domain simulations or PMU data.

14.3.1 $TVSI^r$ for Voltage Recovery Stage Assessment

In existing studies, short-term voltage stability indices are usually calculated based on the extent of post-contingency voltage violation with a specific threshold. However, these metrics ignore the time of voltage violation or cannot fully reveal the impact of the duration time of the violation. In practice, for instance, not only the extent but also the duration of the voltage violation has a significant impact on LVRT/HVRT capability of wind turbines. Therefore, an improved index, $TVSI^r$, is proposed to evaluate the immediate voltage response of buses after a voltage disturbance. Specifically, it is defined by: (i) the extent of the deviation v^d (direct voltage deviation impact), (ii) the corresponding acceptable maximum duration t^{dl} and t^{du} (voltage recovery speed). S_1, S_2, S_3, and S_4 are used to represent different voltage violation areas formed by the time-varying voltage trajectory, upper and lower voltage limit (1.1 and 0.9 pu in this chapter), and pre-determined periods (t^{du} and t^{dl}):

S_1: voltage violation below lower voltage threshold within t^{dl}
S_2: voltage violation below lower voltage threshold beyond t^{dl}
S_3: voltage violation above upper voltage threshold within t^{du}
S_4: voltage violation above upper voltage threshold beyond t^{du}

Based on S_1, S_2, S_3, and S_4, $TVSI_i^r$ for each bus is calculated as (14.29). It consists of two parts: a quantitative assessment for a normal voltage violation and a quantitative assessment for a penalized voltage violation. Penalty parameters, α^l and α^u, are used to evaluate the impact of a delayed voltage recovery. The delayed voltage recovery is usually ignored by the conventional index. Post-contingency voltage trajectories under three typical scenarios (named as A, B, and C) are shown in

Figures 14.5–14.7, respectively. It should be noted that $TVSI_i^r$ is zero when there is no voltage violation after a disturbance.

$$TVSI_i^r = \overbrace{S_1 + S_3}^{\text{normal deviation}} + \overbrace{\alpha^l \cdot S_2 + \alpha^u \cdot S_4}^{\text{penalized deviation}}$$

$$= \int_{t^f}^{t^f + t^{dl}} \left(v_i^{dl} - v_i\right)dt + \alpha^l \cdot \int_{t^f + t^{dl}}^{t^{l2}} \left(v_i^{dl} - v_i\right)dt \qquad (14.29)$$

$$+ \int_{t^{u1}}^{t^{u1} + t^{du}} \left(v_i - v_i^{du}\right)dt + \alpha^u \cdot \int_{t^{u1} + t^{du}}^{t^{u2}} \left(v_i - v_i^{du}\right)dt$$

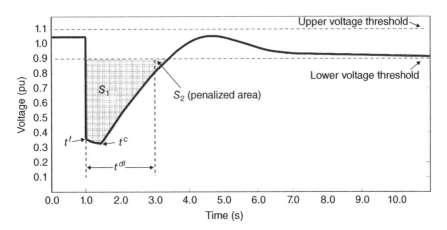

Figure 14.5 Illustration of Typical Scenario A.

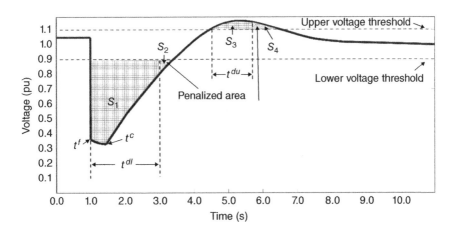

Figure 14.6 Illustration of Typical Scenario B.

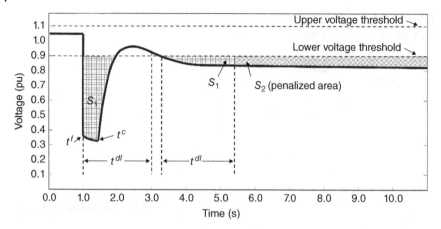

Figure 14.7 Illustration of Typical Scenario C.

As shown in Figure 14.5, Scenario A represents that only a lower voltage threshold is violated during the post-contingency stage and the voltage magnitude of the steady state after a contingency is within the boundary. In this scenario, S_3 and S_4 in Eq. (14.29) are zero. So, $TVSI_i^r$ can be calculated as (14.30).

$$TVSI_i^r = S_1 + \alpha^l \cdot S_2 = \int_{t^f}^{t^f + t^{dl}} \left(v_i^{dl} - v_i\right)dt + \alpha^l \cdot \int_{t^f + t^{dl}}^{t^{l2}} \left(v_i^{dl} - v_i\right)dt \qquad (14.30)$$

Scenario B is the case where both lower and upper voltage thresholds are violated, and the voltage magnitude of steady state after a disturbance is within the boundary. In this scenario, $TVSI_i^r$ can be calculated as (14.31).

$$\begin{aligned}
TVSI_i^r &= S_1 + \alpha^l \cdot S_2 + S_3 + \alpha^u \cdot S_4 \\
&= \int_{t^f}^{t^f + t^{dl}} \left(v_i^{dl} - v_i\right)dt + \alpha^l \cdot \int_{t^f + t^{dl}}^{t^{l2}} \left(v_i^{dl} - v_i\right)dt \\
&\quad + \int_{t^{u1}}^{t^{u1} + t^{du}} \left(v_i - v_i^{du}\right)dt + \alpha^u \cdot \int_{t^{u1} + t^{du}}^{t^{u2}} \left(v_i - v_i^{du}\right)dt
\end{aligned} \qquad (14.31)$$

Scenario C represents the case that only the lower voltage threshold is violated and the voltage magnitude of steady state after a disturbance is out of the boundary. In this case, S_2 and S_3 are zero. So, $TVSI_i^r$ can be calculated as (14.32)

$$\begin{aligned}
TVSI_i^r &= S_1 + \alpha^l \cdot S_2 \\
&= \int_{t^f}^{t^f + t^{dl}} \left(v_i^{dl} - v_i\right)dt + \int_{t^{l1}}^{t^{l1} + t^{dl}} \left(v_i^{dl} - v_i\right)dt \\
&\quad + \alpha^l \cdot \int_{t^{l1} + t^{dl}}^{t^{l2}} \left(v_i^{dl} - v_i\right)dt
\end{aligned} \qquad (14.32)$$

Similarly, for the case that the post-contingency steady state is over the upper limit voltage threshold, $TVSI_i^r$ can also be calculated according to (14.31) with t^{du}, v^{du}, t^{u1}, and t^{u2}.

The values of the weighting factors, α^l and α^u, are determined by the following two aspects:

a) Priorities of delayed upper/lower voltage recovery, which is based on the empirical engineering judgment of their importance and the specific operational requirements (for instance, wind turbines are highly sensitive to delayed voltage recovery. So, higher values of α^l and α^u are expected).

b) Sensitivity analysis against different capacities of STATCOMs. The sensitivity analysis for the weighting factors is mainly to minimize the unexpected surge in the evaluation results due to the nonlinearity of the problem and also to avoid the situation that high-priority components end up with little influence (for instance, they might be very insensitive to the capacity change) in the evaluation result. In this chapter, All-At-a-Time (AAT) method is used for the sensitivity analysis, in which output variations are induced by changing all the inputs (the capacities of all VAR compensation devices) at a time simultaneously. Therefore, both the direct impact of capacity variations and the joint impact due to interactions among different reactive power sources are considered. It should be pointed out that more advanced methods for sensitivity analysis, such as the Morris screening method [16], can be employed to determine the weights.

14.3.2 $TVSI^s$ for Post-contingency Steady-State Assessment

$TVSI^s$ is proposed in this chapter to quantitatively evaluate the voltage stabilization process after a contingency. It describes the steady state in the post-contingency stage and assesses the time that a particular bus voltage takes to restore to an acceptable voltage range. Generally, a stabilized voltage can be in one of the two post-contingency steady states: (i) stays within the magnitude limit of voltage, and (ii) out of the magnitude limit range. There are two critical variables for the calculation of $TVSI^s$: t^s and v^s. Settling time t^s is calculated based on the acceptable voltage fluctuation range $[v^{fu}, v^{fl}]$ and it indicates how fast the steady state can be reached. v^s is the average value of voltage magnitude after the t^s. As for the unstable/unacceptable case, t^s is set to be $t^{s, th}$ and v^s is calculated based on the average voltage magnitude after the settling time t^s. A general definition of $TVSI_i^s$ is presented as (14.33), and two corresponding typical scenarios are illustrated in Figures 14.8 and 14.9.

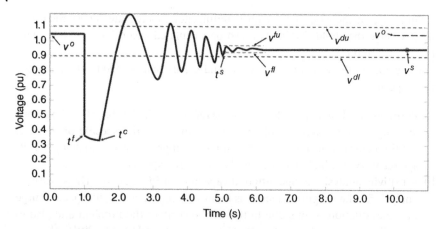

Figure 14.8 Illustration of Typical Scenario D.

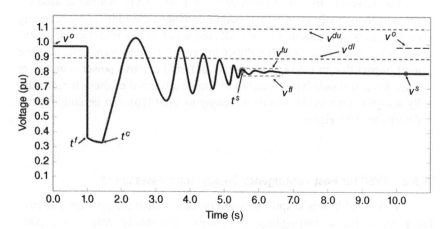

Figure 14.9 Illustration of Typical Scenario E.

$$
TVSI_i^s = \begin{cases}
1 - \dfrac{v_i^s - v^{dl}}{v_i^o - v^{dl}} \cdot \dfrac{t^s}{t^{s,th}}, & v^{dl} \leq v_i^s < v_i^o \\[3mm]
1 - \dfrac{v^{du} - v_i^s}{v^{du} - v_i^o} \cdot \dfrac{t^s}{t^{s,th}}, & v_i^o \leq v_i^s \leq v^{du} \\[3mm]
1 + \dfrac{v^{dl} - v_i^s}{v^{dl}} \cdot \dfrac{t_i^s}{t_i^{s,th}}, & 0 \leq v_i^s < v^{dl} \\[3mm]
1 + \dfrac{v_i^s - v^{du}}{v^{du}} \cdot \dfrac{t_i^s}{t_i^{s,th}}, & v^{du} < v_i^s
\end{cases}
\tag{14.33}
$$

Suppose there is sufficient reactive power support for a bus. In that case, its voltage can stay within the boundary after a short fluctuation, as illustrated in Figure 14.8. Specifically, it stabilizes within the voltage range but is lower than the original voltage value v^o. So, in this case (denoted as Typical Scenario D), $TVSI_i^s$ is calculated as (14.33), with the condition that $v^{dl} \leq v_i^s < v_i^o$.

Similarly, when the voltage stabilizes within the voltage range but is larger than the original value v^o, $TVSI_i^s$ is calculated as (14.33) with the condition that $v_i^o \leq v_i^s \leq v^{du}$.

Contrary to Typical Scenario D, when the reactive power is insufficient, the magnitude of the stabilized voltage might be inferior, as shown in Figure 14.9. The stabilized voltage is smaller than the lower threshold. So, $TVSI_i^s$ is calculated as (14.33), with the condition that $0 \leq v_i^s < v^{dl}$. Similarly, when the stabilized voltage is above the upper voltage threshold, $TVSI_i^s$ is calculated as (14.33), with the condition that $v^{du} < v_i^s$.

Generally, a smaller $TVSI_i^s$ indicates a better steady-state voltage in the post-contingency stage. For instance, the range of $TVSI_i^s$ is [0,1] for Typical Scenario D, and the range of $TVSI_i^s$ is (1,2] for Typical Scenario E. Figure 14.10 illustrates the correlation between $TVSI_i^s$ and v_i^s. $TVSI_i^s$ is designed to guarantee that any evaluation result of scenario E is always worse than that of scenario D. So, an inferior voltage recovery of a specific bus can be identified easily without an additional comparison with other buses. It is also a salient advantage of the proposed indicators not only for the planning problem but also for cases like online operational monitoring. The distinctive difference can provide direct information about the stabilized voltage after a contingency to the operators clearly. On the contrary, this feature is usually ignored by most of the existing voltage stability

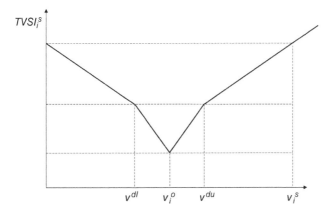

Figure 14.10 Values of $TVSI_i^s$ against the varying v_i^s.

indicators. In the literature, without comparison with others (either other bus voltage or the same bus voltage before and after compensation), the evaluation result alone hardly reveals any information about the voltage performance.

Nomenclature

Indices

i, j indices for buses
l indices for lines
r index for receiving bus
s index for sending bus
t index for the time
o index for an original state

Sets

$\mathcal{B}$ set of all buses
$\mathcal{L}$ set of all the lines
$\mathcal{K}$ set of contingencies
$\mathcal{T}$ set of time

Variables

P_r active power at receiving end
$P_l^{\max}$ maximum active power loss in the line l
Q_r reactive power at receiving end
$P_l^{\max}$ maximum active power loss in the line l
t^s settling time
V^r voltage at receiving end
V^s voltage at sending end
v^s average value of voltage magnitude after the t^s
δ_r voltage angle at receiving bus
δ_s voltage angle at sending bus
θ phase angle of line impedance

Parameters

N^b number of buses
T^c fault clearing time

T^{end} end time of set $\mathcal{T}$

t^{dl} maximum voltage deviation time (lower)

t^{du} maximum voltage deviation time (upper)

t^{end} end of time-domain simulation

t^f fault time

t^{l1} end of voltage deviation time within t^{dl} (lower)

t^{l2} end of voltage deviation time beyond t^{dl} (lower)

t^o start of time-domain simulation

$t_i^{s,th}$ time threshold for t_i^s

t^{u1} end of voltage deviation time within t^{du} (upper)

t^{u2} end of voltage deviation time beyond t^{du} (upper)

v^{dl} lower voltage threshold

v^{du} upper voltage threshold

α^l penalty parameter for voltage violation (lower)

α^u penalty parameter for voltage violation (upper)

δ maximum acceptable voltage deviation

References

1 Kundur, P. (1994). *Power System Stability and Control*. New York, NY: McGraw-Hill.

2 Cutsem, T.V. and Vournas, C. (1998). *Voltage Stability of Electric Power Systems*. New York, NY: Springer.

3 Gao, B., Morison, G., and Kundur, P. (1992). Voltage stability evaluation using modal analysis. *IEEE Transactions on Power Apparatus and Systems* 7 (4): 1423–1543.

4 Ajjarapu, V. (2010). *Computational Techniques for Voltage Stability Assessment and Control* (ed. V. Ajjarapu). New York, NY: Springer.

5 Ajjarapu, V. and Christy, C. (1992). The continuation power flow: a tool for steady state voltage stability analysis. *IEEE Transactions on Power Apparatus and Systems* 7 (1): 416–423.

6 Cupelli, M., Doig Cardet, C., and Monti, A. (2012). Comparison of line voltage stability indices using dynamic real time simulation. *2012 3rd IEEE PES Innovative Smart Grid Technologies Europe (ISGT Europe)*, Berlin, Germany (14-17 October 2012): IEEE.

7 Ismail, N.A.M., Zin, A.A.M., Khairuddin, A., and Khokhar, S. (2014). A comparison of voltage stability indices. *2014 IEEE 8th International Power Engineering and*

Optimization Conference (PEOCO2014), Langkawi, Malaysia (24–25 March 2014): IEEE.

8 Lim, Z.J., Mustafa, M.W., and bt Muda, Z. (2012). Evaluation of the effectiveness of voltage stability indices on different loadings. *2012 IEEE International Power Engineering and Optimization Conference*, Melaka, Malaysia (06–07 June 2012): IEEE.

9 Moghavvemi, M. and Faruque, O. (1998). Real-time contingency evaluation and ranking technique. *IEE Proceedings – Generation, Transmission and Distribution* 145 (5): 517.

10 Xu, Y., Dong, Z.Y., Xiao, C. et al. (2015). Optimal placement of static compensators for multi-objective voltage stability enhancement of power systems. *IET Generation Transmission and Distribution* 9 (15): 2144–2151.

11 Shoup, D.J., Paserba, J.J., and Taylor, C.W. (2005). A survey of current practices for transient voltage dip/sag criteria related to power system stability. *IEEE PES Power Systems Conference and Exposition*, New York, NY, USA (10–13 October 2004): IEEE.

12 Tiwari, A. and Ajjarapu, V. (2011). Optimal allocation of dynamic VAR support using mixed integer dynamic optimization. *IEEE Transactions on Power Apparatus and Systems* 26 (1): 305–314.

13 Han, T., Chen, Y., Ma, J. et al. (2018). Surrogate modeling-based multi-objective dynamic VAR planning considering short-term voltage stability and transient stability. *IEEE Transactions on Power Apparatus and Systems* 33 (1): 622–633.

14 Deng, Z., Liu, M., Ouyang, Y. et al. (2018). Multi-objective mixed-integer dynamic optimization method applied to optimal allocation of dynamic var sources of power systems. *IEEE Transactions on Power Apparatus and Systems* 33 (2): 1683–1697.

15 Tahboub, A.M., El Moursi, M.S., Woon, W.L., and Kirtley, J.L. (2018). Multiobjective dynamic VAR planning strategy with different shunt compensation technologies. *IEEE Transactions on Power Apparatus and Systems* 33 (3): 2429–2439.

16 Iooss, B. and Lemaître, P. (2015). A review on global sensitivity analysis methods. In: *Uncertainty Management in Simulation-Optimization of Complex Systems* (ed. G. Dellino and C. Meloni), 101–122. Boston, MA: Springer US.

15

Dynamic VAR Resources

15.1 Fundamentals of Dynamic VAR Resources

Capacitor banks have been widely deployed to compensate reactive power for static voltage regulation and power factor correction. However, as a mechanical device, its responding speed is relatively low, and its lifetime would degrade with frequent switch actions. Besides, the reactive power provided by capacitor banks is proportional to the square of the terminal voltage, which is unfavorable during low voltage conditions. As a result, conventional static capacitor banks alone are inadequate to mitigate short-term voltage instability issues. As a superior alternative, static var compensator (SVC) and static synchronous compensator (STATCOM) are able to provide rapid and dynamic reactive power support against the voltage depression following a large disturbance, thus can more effectively counteract the short-term voltage instability.

SVC and STATCOM are among the prevalent dynamic VAR sources in the FACTS family. SVC, based on high-power thyristor technology, firstly appeared in the 1970s, while STATCOM, based on gate turn-off thyristors (GTO) came into the field in the 1980s. Later, with the emergence of high-power insulated gate bipolar transistor (IGBT), STATCOM based on this technology platform became commercially available around 1990.

SVC or STATCOM should be the natural choice where the following qualities are required or desirable: rapid dynamic response, the ability for frequent variations in output, and output that is smoothly adjustable. Each in its own particular way, SVC and STATCOM offer benefits such as:

a) Grid voltage control under normal and contingency conditions
b) Fast response reactive power following contingencies
c) Preventing/reducing the risk of voltage collapses in the grid

Stability-Constrained Optimization for Modern Power System Operation and Planning,
First Edition. Yan Xu, Yuan Chi, and Heling Yuan.
© 2023 The Institute of Electrical and Electronics Engineers, Inc.
Published 2023 by John Wiley & Sons, Inc.

d) Preventing over-voltages at the loss of load
e) Boosting voltage during under-voltage disturbances and faults
f) Damping active power oscillations
g) Increased power transfer capability by stabilizing the voltage in weak and/or heavily loaded points in the grid
h) Power quality improvement by load balancing, flicker mitigation, and harmonic filtering

SVCs are shunting connected static generators/absorbers whose outputs are varied so as to control the voltage of the electric power systems. SVC is similar to a synchronous condenser but without any rotating part, since it is used to supply or absorb reactive power. It uses the equivalent of an automatic voltage regulator system to set and maintain a target voltage level. Figure 15.1 shows a typical thyristor-based SVC. The total susceptance of SVC is controlled by controlling the firing angle of thyristors. However, like fixed capacitors or fixed inductors, SVCs have the maximum and minimum limits. SVC can be built using a variety of designs, and the controllable elements used in most systems are similar. Thyristor-controlled reactor (TCR), thyristor-switched capacitor (TSC), and thyristor-switched reactor (TSR) are commonly used controllable elements of SVC.

STATCOM is a solid-state synchronous voltage source that is analogous to an ideal synchronous machine without any rotating part, which is able to generate a balanced set of sinusoidal voltages at the fundamental frequency with rapidly controllable amplitude and phase angle. Figure 15.1 shows a typical STATCOM, which is the voltage-source converter, converting a DC input voltage into an AC output voltage to compensate for the reactive power.

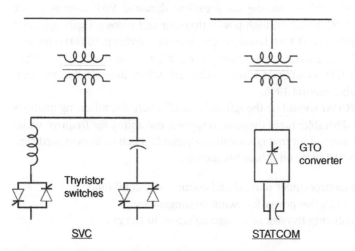

Figure 15.1 Thyristor-based SVC and STATCOM.

The VAR output of a STATCOM is the product of its current injection and the voltage magnitude of the connected bus. In response to a sudden voltage dip, the VAR output of the STATCOM decreases linearly with the bus voltage drop, which is superior to its competitor-SVC, whose VAR output falls squarely with the bus voltage drop.

Figure 15.2 illustrates a comparison between SVC and STATCOM, in terms of voltage–current characteristics. It shows that STATCOM provides constant reactive current even when the voltage is low. For example, at 70% voltage level, the SVC output becomes 70% reactive current or 49% reactive power, while the reactive current output of STATCOM remains 100% and the reactive power output of STATCOM is 70%. This performance difference is critical in supporting the re-acceleration of motor loads, which is a major driving force for short-term voltage instability. In addition, compared with SVC, STATCOM can respond in a shorter time and has a smaller size. All these advantages make it a better candidate for providing dynamic VAR support, particularly after a contingency [1, 2].

When a dynamic event occurs in the power system, such as a short-circuit fault, STATCOMs can support the bus voltage by providing rapid and dynamic reactive power injection within the system. The impact of STATCOM on bus voltage before and after a disturbance is illustrated in Figure 15.3. In pre-contingency stage (0–0.1 second), the output of STATCOM is 10 MVAr reactive powers (broken red line) and the bus voltage is maintained at the normal level (0.9–1.0 pu). When the short-circuit fault occurs at 0.1 second, the bus voltage drops significantly to 0.6 pu while the reactive power output of STATCOM only decreases slightly. Between the fault time and fault clearing time (0.1–0.2 second), there is a rapid response from STATCOM of which reactive power output increases to over

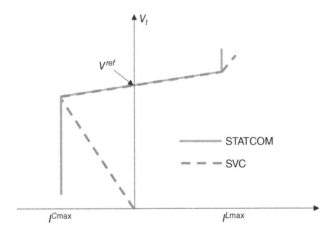

Figure 15.2 Voltage–current characteristics of SVC and STATCOM.

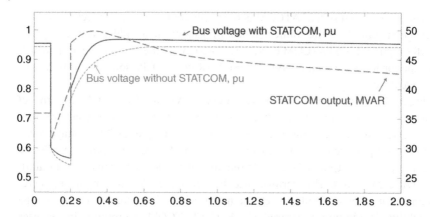

Figure 15.3 STATCOM reactive power output during a contingency.

40 MVAr to support the bus voltage. Once the fault has been cleared at 0.2 second, the bus voltage jumps up and the output of STATCOM significantly increases thanks to the increased voltage magnitude. Then, with variable STATCOM reactive power support to the nominal value, the bus voltage gradually restores to the stable pre-contingency level. The case without the STATCOM is also illustrated in Figure 15.3 as a dashed green line for comparison. It can be seen that the voltage recovers at a much slower rate, and the final voltage level is also lower than that of the case with STATCOM.

Both SVC and STATCOM can effectively increase the steady-state voltage stability margin and power transfer capability of a power system. However, in general, STATCOM outperforms SVC, in terms of loss reduction and voltage profile. A detailed comparison between SVC and STATCOM [2–6] is summarized in Table 15.1. Although the initial deployment cost of STATCOM is higher than that of SVC, the maintenance cost and complete life cycle cost make it a more economical and competitive alternative than SVC. Other issues like harmonics and losses should be examined for each scenario for an optimum investment.

15.2 Dynamic Models of Dynamic VAR Resources

The dynamic model of STATCOM used here is the generic VSC-based "SVMO3" model. It is developed by the Western Electricity Coordinating Council (WECC) [7] and has been implemented in several commercial software packages, including Siemens PTI PSS®E and GE PLSF for STATCOM modeling [7, 8]. The model has

Table 15.1 Comparison between SVC and STATCOM.

Criteria	SVC	STATCOM
Operating principle	Controlled shunt impedance	Controlled voltage source
Deployment space	Large	Medium
Dependency of the current output from the voltage level in the point of connection	Linear	Nonlinear
Dependency of the reactive power output from the voltage level in the point of connection	Quadratic	Linear
Reactive power generating during the 3-phase short circuit scenario	Low	Moderate
Overload capability	No	Yes
Response time	Fast	Faster than SVC
Loss	Step-down transformer	Converter bridge
Harmonics	Harmonic current source (major harmonic generation is at low frequency)	Harmonic voltage source (major harmonic generation is at high frequency)
Maintenance and service requirements	High	Low

also been validated with several recorded digital fault recorder traces from actual static VAr system installations through WECC task force efforts [7]. It will also be utilized in the later numerical tests of other chapters in this book.

Its block diagram is illustrated in Figure 15.4, and the parameters used in this chapter are listed in Table 15.2.

The main features of this model are primary control and secondary control. The primary control is for a proportional-integral primary voltage regulation loop with proportional gain K_p and integral gain K_i. The secondary control consists of three parts: slow-reset control, nonlinear droop control, and deadband control. It should be noted that the deadband control, slow-reset control, and nonlinear droop are all intended for the same purpose – maintaining the STATCOM at a low steady-state output when the system voltage is within a given bandwidth. These three control strategies achieve this in quite different ways and only one of the three should be used in the control strategy.

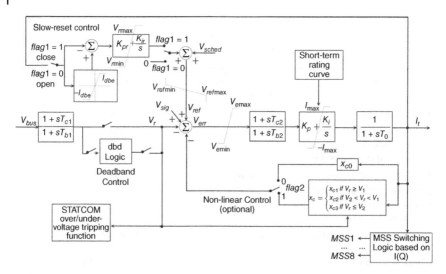

Figure 15.4   Block diagram of the SVSMO3 STATCOM model [8].

Table 15.2   SVSMO3U1 model parameters.

Parameters	Definitions	Values
X_{c0}	Linear droop	0.01
T_{c1}	Voltage measurement lead time constant (s)	0
T_{b1}	Voltage measurement lag time constant (s)	0.1
K_p	Proportional gain	4
K_i	Integral gain	25
V_{emax}	maximum voltage error (pu)	0.5
V_{emin}	Minimum voltage error (pu)	−0.5
T_0	Firing sequence control delay (s)	0.005
I_{max1}	Maximum continuous current rating (pu on STBASE)	10
dbd	Deadband range for voltage control (pu)	0
K_{dbd}	Ratio of outer to inner deadband	10
T_{dbd}	Deaband time (s)	0.1
K_{pr}	Proportional gain for slow-reset control	0
K_{ir}	Integral gain for slow-reset control	0.001
I_{dbd}	Deadband range for slow-reset control (pu on STBASE)	0.01

Table 15.2 (Continued)

Parameters	Definitions	Values
V_{rmax}	Maximum limit on slow-reset control output (pu)	0.1
V_{rmin}	Minimum limit on slow-reset control output (pu)	−0.1
I_{shrt}	Maximum short-term current rating multiplier of maximum continuous current	1.4
UV_1	Voltage at which STATCOM limit starts to be reduced linearly (pu)	0.5
UV_2	Voltage at which STATCOM is blocked (pu)	0.2
OV_1	Voltage above which STATCOM limit linearly drops (pu)	1.08
OV_2	Voltage above which STATCOM blocks (pu)	1.2
V_{trip}	Voltage above which STATCOM trips after time delay T_{delay2} (pu)	1.3
T_{delay1}	Short-term rating time (s)	1
T_{delay2}	Trip time for $V > V_{trip}$ (s)	0.08
V_{sched}	Voltage reference (pu)	1
V_{refmax}	Maximum limit on voltage reference (pu)	1.05
V_{refmin}	Minimum limit on voltage reference (pu)	0.95
T_{c2}	Lead time constant (s)	0
T_{b2}	Lag time constant (s)	0
I_{2t}	I_{2t} limit	0
$Reset$	Reset rate for I_{2t} limit	0
$hyst$	Width of hysteresis loop for I_{2t} limit	0
X_{c1}	Nonlinear droop slope 1	0.01
X_{c2}	Nonlinear droop slope 2	1
X_{c3}	Nonlinear droop slope 3	0.01
V_1	Nonlinear droop upper voltage (pu)	1.025
V_2	Nonlinear droop lower voltage (pu)	0.975
T_{mssbrk}	Time for MSS breaker to operate (s)	0.1
T_{out}	Time MSS should be out before switching back in (s)	300
T_{delLC}	Time delay for switching MSSs (pu on STBASE)	0.5
I_{upr}	Upper threshold for switching MSSs (pu on STBASE)	0.3
I_{lwr}	lower threshold for switching MSSs (pu on STBASE)	−0.3

References

1 Sapkota, B. and Vittal, V. (2010). Dynamic var planning in a large power system using trajectory sensitivities. *IEEE Transactions on Power Systems* 25 (1): 461–469.

2 Noroozian, M., Petersson, N.A., Thorvaldson, B. et al. (2003). Benefits of SVC and STATCOM for electric utility application. *2003 IEEE PES Transmission and Distribution Conference and Exposition* (IEEE Cat. No. 03CH37495), Dallas, TX, USA (07–12 September 2003), Vol. 3, pp. 1143–1150: IEEE.

3 Qi, J., Zhao, W., and Bian, X. (2020). Comparative study of svc and statcom reactive power compensation for prosumer microgrids with dfig-based wind farm integration. *IEEE Access* 8: 209878–209885.

4 Tan, Y.L. (1999). Analysis of line compensation by shunt-connected FACTS controllers: a comparison between SVC and STATCOM. *IEEE Power Engineering Review* 19 (8): 57–58.

5 Wessels, C., Hoffmann, N., Molinas, M., and Fuchs, F.W. (2013). StatCom control at wind farms with fixed-speed induction generators under asymmetrical grid faults. *IEEE Transactions on Industrial Electronics* 60 (7): 2864–2873.

6 Shahnia, F., Rajakaruna, S., and Ghosh, A. (2015). *Static Compensators (STATCOMs) in Power Systems*. Singapore: Springer.

7 Pourbeik, P., Sullivan, D.J., Bostrom, A. et al. (2012). Generic model structures for simulating static Var systems in power system studies – a WECC task force effort. *IEEE Transactions on Power Systems* 27 (3): 1618–1627.

8 Siemens Power Technologies International (2013). *PSS®E 33.4 Model Library*. New York: Siemens.

16

Candidate Bus Selection for Dynamic VAR Resource Allocation

16.1 Introduction

Given its highly nonlinear and non-convex features, the VAR planning problems considering short-term voltage stability issues are naturally with high computational complexity. To alleviate the computation burden of the planning problems, candidate buses for the reactive power source deployment can be determined before the optimization of the capacity. On the other hand, candidate bus selection (CBS) is also a critical factor influencing the optimality of the final planning results. Generally, steady-state [1–3] and short-term [4–6] voltage stability metrics are prevalent in assessing the voltage performance of different buses, and the evaluation results are usually used to rank the buses. As a steady-state evaluation, load margins of branches or voltage violations are often used as indicators to show the effectiveness of a specific reactive power source installation. As for the assessment of short-term voltage stability, it is mainly based on post-contingency voltage responses (like time-varying voltage trajectories). There are also industrial standards [7–9] that can be utilized for the stability assessment, but their evaluation results are designed for the operational scenarios focusing on a binary judgment: stable or unstable. These binary assessment results cannot be used to rank the candidate buses.

On the other hand, an accurate index is not the only concern of a CBS problem. The straightforward ranking method will inevitably lead to abundant candidate buses in the same area because, for geographically adjacent weak buses, they tend to have similar evaluation results in terms of voltage stability. If several adjacent buses are selected as candidate buses, the efficiency of the compensation cannot be guaranteed. To avoid this redundancy, zoning-based CBS methods are proposed in Ref. [10, 11] by categorizing the buses into different zones. The zoning-based approach is effective, but the existing studies either neglect the short-term voltage

Stability-Constrained Optimization for Modern Power System Operation and Planning,
First Edition. Yan Xu, Yuan Chi, and Heling Yuan.
© 2023 The Institute of Electrical and Electronics Engineers, Inc.
Published 2023 by John Wiley & Sons, Inc.

performance [10] or do not consider the capacity sensitivity of the reactive power source [11].

Uncertainties of wind power generation and load variations are hardly considered in the CBS stage [4, 5, 10–13]. In this context, there is a high risk that the selected typical operation profile cannot fully reveal the potential impact of variations in load and renewable generation level. Considering the varying load demand and the inherent stochasticity of renewable energy, results from conventional methods are not robust enough. The optimality of the candidate bus might be compromised when subjected to different operation scenarios.

Stochastic correlations exist between loads and renewable power generation, and these correlations are nonlinear and non-Gaussian [14]. Conventional random sampling methods, which considers the operational uncertainty separately, such as Monte Carlo method with independent distribution, cannot reflect the correlation or dependence among these uncertainties. Even for the method [15] that takes a joint distribution of wind and load into consideration, the simplified linear correlation cannot fully reveal the dependence between uncertainties or cannot correctly reflect the correlation. Copula theory [16], though not directly utilized for CBS, has been widely used in power system planning and voltage stability problems [14, 17, 18] to represent the dependence or correlation between different uncertainties. Specifically, a Multivariate Gaussian copula is used in these studies. It is regarded as an effective copula family [14] to represent correlations between load demand and renewable generations, such as photovoltaic generation, and wind power generation. In practice, the specific copula family used in the power system applications heavily depends on the availability of historical data and a case-by-case trial and error method. Therefore, the accuracy of copula families needs to be verified, and a more systematic approach is required to select the copula families.

To address the inadequacies of the previous works, two novel CBS methods for dynamic VAR planning are introduced. The first one is a zoning-based CBS, and the second one focuses on modeling of correlated operational uncertainties.

16.2 General Framework of Candidate Bus Selection

A conventional three-stage framework consisting of three typical procedures is illustrated in Figure 16.1, with a summary of its limitations in each step. First, historical data or empirical judgment is utilized for the selection of contingencies, and the high-severity contingencies are selected. Then, the general voltage performance of the system before and after contingencies is evaluated on post-contingency voltage stability index. The final step is to rank the buses based on the results of performance evaluation and select the required number of buses from top to bottom.

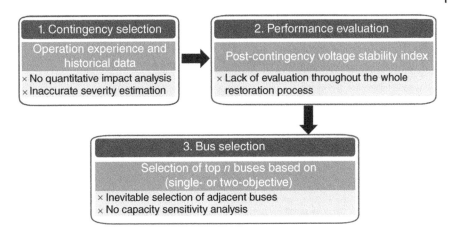

Figure 16.1 General framework of the conventional candidate bus selection method. *Source:* Chi and Xu (2020)/With permission of John Wiley & Sons.

For the conventional three-stage framework, the limitations can be summarized as follows:

a) **Contingency selection**: lack of quantitative impact analysis of contingencies, which leads to an inaccurate severity estimation.
b) **Performance evaluation**: lack of evaluation throughout the whole restoration process
c) **Bus selection**: inevitable selection of adjacent buses and lack of capacity sensitivity analysis.

16.3 Zoning-based Candidate Bus Selection Method

Considering the drawbacks of conventional methods, a multi-objective CBS scheme based on fuzzy clustering for STATCOMs to improve the voltage resilience of the wind-penetrated power system is proposed in this chapter. The proposed method also adopts a conventional three-stage framework, as illustrated in Figure 16.2, and all three stages are improved to address the limitations of conventional methods. The advantages of the proposed method can be summarized in the following three aspects. First, instead of using historical data or empirical judgment, the severity of different contingencies is evaluated quantitatively based on three criteria, including steady-state voltage stability, short-term voltage stability, and rotor angle stability. Then, the general voltage resilience of the system before and after different contingencies is evaluated on steady-state voltage stability, post-contingency voltage

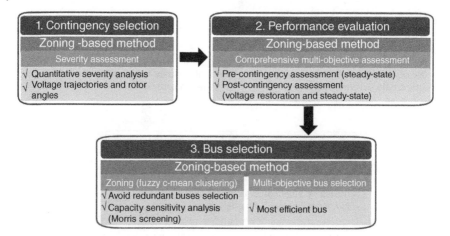

Figure 16.2 General framework of the zoning-based candidate bus selection method. *Source:* Chi and Xu (2020)/With permission of John Wiley & Sons.

restoration performance, post-contingency steady-state, and rotor angle deviations. To avoid the selection of redundant buses, the final stage of the proposed CBS scheme is divided into two steps: (i) categorizing the buses into zones based on their similarity of voltage impacts on the power system; (ii) selecting the most effective bus from each zone. Specifically, the selection can be carried out once (one bus from each zone) or several times iteratively according to the practical needs.

16.3.1 Performance Evaluation Indices for Candidate Bus Selection

The steady-state voltage stability is evaluated by the line-based voltage collapse proximity indicator (VCPI), and voltage resilience performance after a contingency is evaluated by two improved transient voltage stability indices (TVSIs), $TVSI^r$ and $TVSI^s$, based on the voltage trajectories (the trajectories data can be obtained by time-domain simulation, or from PMU data). Mathematical details of VCPI, $TVSI^r$, and $TVSI^s$ can be found in Sections 14.2 and 14.3. Specifically, the adoption of $TVSI^r$ and $TVSI^s$ can accurately reflect the severity of a large disturbance. The rotor angle stability of the system is assessed by integral squared generator angle (ISGA), which will be elaborated as follows;

When there is a contingency in a power system, rotor angle stability and voltage stability are associated with each other. In this chapter, rotor angle stability is introduced as an indicator for the evaluation of post-contingency resilience performance. Conventionally, the rotor angle is only considered as a binary constraint in VAR planning problems [12, 13]. However, this binary judgment abandons all the information about the extent of stability/instability. ISGA [19] is employed in this chapter to measure the cumulated deviation and relative severity of the incidents. ISGA is calculated and normalized as (16.1) based on a weighted center of inertia angle, and the weighted center is calculated with generator inertia constants and rotor angles.

$$ISGA = \sum_{k \in \mathcal{K}} \frac{1}{M^{TOT} T} \int_{t^0}^{t^{end}} \sum_{i \in \mathcal{G}_s} M_i \left(\delta_i - \delta^{COA} \right)^2 dt \cdot p_k \tag{16.1}$$

$$M^{TOT} = \sum_{i \in \mathcal{G}_s} M_i \tag{16.2}$$

$$\delta^{COA} = \frac{\sum_{i \in \mathcal{G}_s} M_i \cdot \delta_i}{M^{TOT}} \tag{16.3}$$

VCPI is used in the chapter to quantitatively compare the steady-state voltage stability performance of different CBS results with various capacities installed, showing the extent of improvement or deterioration.

$$VCPI = \sum_{k \in \mathcal{K}} \sum_{l \in \mathcal{L}} VCPI(\text{power})_l \cdot p_k \tag{16.4}$$

As for post-contingency voltage performance evaluation, a root-mean-square method [20] is utilized to improve the original TVSI further. The original TVSI adopts a direct averaging method, which cannot fully reveal a slow voltage recovery of a specific bus if the recovery speed of others is relatively fast. To improve this shortcoming, A root-mean-square method is used in this chapter as (16.5) and (16.6), with the consideration of probabilities of different contingencies. A smaller $TVSI^r$ is favorable since it indicates less voltage violation and faster voltage recovery. Similarly, $TVSI^s$ follows the same pattern.

$$TVSI^r = \sum_{k \in \mathcal{K}} \sqrt{\frac{\sum_{i \in \mathcal{B}} \left(TVSI_i^r \right)^2}{N^b}} \cdot p_k \tag{16.5}$$

$$TVSI^s = \sum_{k \in \mathcal{K}} \sqrt{\frac{\sum_{i \in \mathcal{B}} \left(TVSI_i^s \right)^2}{N^b}} \cdot p_k \tag{16.6}$$

16.3.2 Capacity Sensitivity Analysis

In the literature, zoning-based CBS methods [10, 11] only use one predetermined value of capacity in the whole CBS process. Consequently, the selected buses are most effective with capacity around this value, rather than an optimal capacity, which is determined in the capacity optimization stage. Besides, in practice, empirical or arbitrary capacity used for these CBS methods might lead to inaccurate and inconsistent zoning results (for instance, different selected candidate buses for different capacity values used). Although it is impossible to determine an optimal capacity before the candidate bus itself is determined, it is relatively easy to decide on a reasonable capacity range. In this chapter, the capacity sensitivity analysis (using Morris screening method [21]) is carried out to alleviate the inconsistency problem of conventional methods.

Morris screening method [21] was originally proposed to identify the most influential parameters by creating a multi-dimensional semi-global trajectory within its search space. Compared with local trajectory analysis, it is more accurate [22]. On the other hand, it is also more time-efficient than the time-consuming global sensitivity analysis. Although Latin hypercube sampling (LHS) might not be so advantageous in high-dimensional problems, it is still effective for the studied VAR planning problem since this thesis adopts the assumption of the area-based approach for VAR planning.

Firstly, LHS is used to generate random samples following the normal distribution. Compared with simple random sampling or uniform sampling, LHS is a better representative of the actual variability. Then, one variable at a time will be changed at a step of Δ_m (the step is determined based on the LHS results). So, an elementary effect of a Δ_m change is calculated as (16.7).

$$EE_{i,m} = \frac{f(\alpha_{i,m} + \Delta_m) - f(\alpha_{i,m})}{\Delta_m}, \forall i \in \mathscr{B}_l, m = 1, ..., N^m \tag{16.7}$$

$f(\alpha_{i,m})$ is an evaluation result when a STATCOM with a specific capacity ($\alpha_{i,m}$, determined by LHS) is installed at a candidate bus. Then, the mean (μ^*) of the elementary effects is calculated as (16.8). It represents the sensitivity strength between the input (different buses with different STATCOM/capacitor capacities) and the output (RI^{RVSI} or RI^{VCPI}). The standard deviation (σ_i^*), which is calculated as (16.9), represents the interaction between different variables.

$$\mu_i^* = \left(\sum_{m=1}^{N^m} |EE_{i,m}| \right) / N^c \tag{16.8}$$

$$\sigma_i^* = \sqrt{\sum_{m=1}^{N^m} (|EE_{i,m}| - \mu_i^*)^2 / N^c} \tag{16.9}$$

A smaller μ_i^* indicates that the deployment of reactive power compensation devices at ith bus has a relatively lower effect on the power system. On the other hand, a larger σ_i^* indicates a nonlinear impact on the corresponding performance index, and it has higher interactions with other buses.

16.3.3 Fuzzy C-mean (FCM) Clustering

Because of the proximity of voltage dip, the conventional methods adopting a direct index ranking method will inevitably select the most critical bus and its geographically adjacent buses. Obviously, this selection will adversely impact the efficiency of the compensation in a later stage (capacity optimization stage), since concentrated compensation devices in a relatively small area will have a decreasing marginal contribution to the system when the installed capacity increases. An effective way to address this issue is to categorize the buses according to their similarity of dynamic voltage responses after a contingency into different zones. In this chapter, an iterative clustering method called FCM clustering algorithm [23] is used to categorize the buses. It is formulated to minimize the objective function with respect to the membership function and the center vector.

An $N^c \times M$ matrix (N^c buses and M indices), is formed for a specific capacity as (16.10) and (16.11).

$$X_c^o = \left[x_{1,c}^o, x_{2,c}^o, ..., x_{M,c}^o \right] \tag{16.10}$$

$$x_{j,c}^o = \left[x_{1,j,c}^o, x_{2,j,c}^o, ..., x_{N^c,j,c}^o \right]^T, \, j = 1, 2, ..., M \tag{16.11}$$

The structure of the initial matrices is illustrated in Figure 16.3, based on the result of LHS. Specifically, as part of the capacity sensitivity analysis, $\left[x_{1,j,c}^o, x_{2,j,c}^o, ..., x_{N^c,j,c}^o \right]^T$ in the initial matrix for FCM is obtained from corresponding Morris screening results μ^* for different indices. Then, they are unified to make all the items in the matrices within [0,1], as calculated in (16.12).

$$x_{i,j,c} = \frac{x_{i,j,c}^o - x_{i,c}^{o,min}}{x_{i,c}^{o,max} - x_{i,c}^{o,min}} \tag{16.12}$$

FCM algorithm is based on the minimization of an objective function defined as (16.13).

$$Min \, J(U, \mathscr{V}) = \sum_{r=1}^{N^{center}} \sum_{i=1}^{N^e} (u_{r,i})^F D_{r,i}(x_i, y_r) \tag{16.13}$$

$$u_{r,i} = \frac{D_{r,i}(x_i, y_r)^{\frac{1}{1-F}}}{\sum_{j=1}^{N^{center}} D_{j,i}\left(x_i, y_j\right)^{\frac{1}{1-F}}} \tag{16.14}$$

$$y_r = \frac{\sum_{j=1}^{N^e} \left(u_{r,j}\right)^F x_j}{\sum_{j=1}^{N^e} \left(u_{r,j}\right)^F} \tag{16.15}$$

$$D_{r,i}^2 = \|x_i - y_r\|^2 \tag{16.16}$$

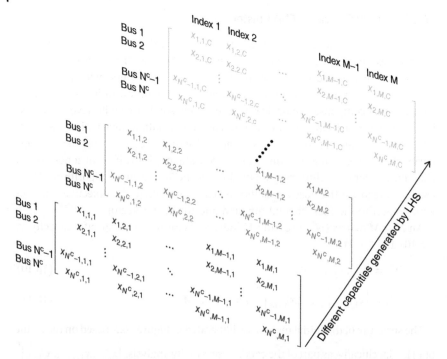

Figure 16.3 Structure of initial matrices group. *Source:* Chi and Xu (2020)/With permission of John Wiley & Sons.

The corresponding constraints are (16.17) and (16.18).

$$\sum_{r=1}^{N^{center}} u_{r,i} = 1 \tag{16.17}$$

$$0 < \sum_{i=1}^{N^e} u_{r,i} < 1 \tag{16.18}$$

The following four steps will be iteratively processed until the termination criterion is satisfied or the max iteration number is reached.

1) Choose the number for the clusters. Determine the weighting exponent and a proper termination tolerance ϵ. Then, conduct the initialization of the membership matrix $U_{(0)}$.
2) Calculate the center vector $\mathcal{Y}$ according to (16.15).
3) Update the membership matrix based on (16.14).
4) Go back to Step 2 until: (i) $\|U_z - U_{z-1}\| < \epsilon$ is satisfied (where U_z is z^{th} membership matrix) or (ii) maximum iteration number is reached.

16.3.4 Computation Steps of the Zoning-based Candidate Bus Selection

Figure 16.4 illustrates the computational flowchart of the proposed CBS scheme. The details of each computation step are described as follows:

1) **Contingency selection**: Contingency severity analysis is based on $TVSI^r$ and $ISGA$. Specifically, these two indicators are calculated according to voltage trajectories and rotor angles obtained from the time-domain simulation results. Then, the critical contingencies that represent large voltage disturbance will be selected according to the principles of the Pareto front [24].

2) **Performance evaluation**: Random capacity samples following the normal distribution are generated based on LHS. Optimal power flow is conducted according to these samples for the calculation of $VCPIs$. Then, $TVSI^r$, $TVSI^s$, and $ISGA$ are calculated to assess the voltage performance for each critical contingency, with all capacity samples.

3) **Categorizing buses**: Initial matrices are formed with evaluation results obtained from Step 2 for different capacity samples. A capacity analysis is conducted by the Morris screening method. FCM is initialized with cluster numbers, weighting exponent, and termination tolerance. Center set $\mathcal{Y}$ is calculated according to (16.15) with a randomly generated initial $U_{(0)}$. Then, the membership matrix is updated with a new $\mathcal{Y}$ until the variations of membership matrix reach tolerance or iterations reach the maximum number. Finally, FCM results are obtained and saved for the categorization of buses.

4) **Candidate bus selection**: Bus zones with low impact on the system (not influential on the assessment results) are excluded. For cases without predetermined candidate buses, the highest-ranking bus in each zone is selected iteratively. As for systems with m predetermined buses (e.g. the bus with wind turbines), m buses are selected firstly from zones with no predetermined buses. Then, other $(n - 2m)$ buses are chosen based on the same principle as the case without predetermined candidate buses.

5) **Termination**: The max number of candidate buses is reached.

16.4 Correlated Candidate Bus Selection Method

16.4.1 Dependent Operational Variable Sensitivity Analysis

Copula theory is an effective approach to representing the stochastic correlation between random variables and many copula families, including but not limited to Clayton copula, Frank copula, and Gaussian copula, can be utilized to represent the dependence between two or among multiple random variables. The key to an accurate representation is the selection of copula family.

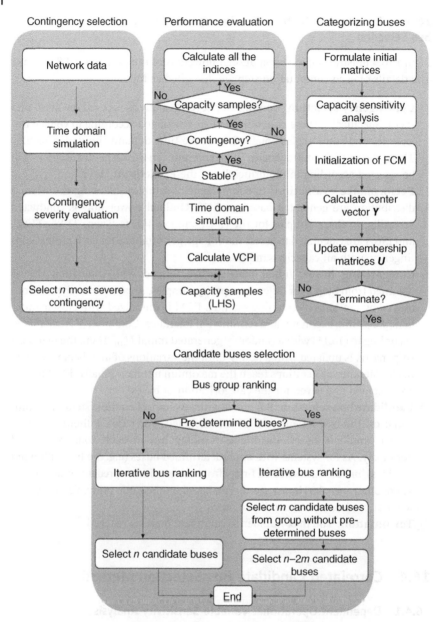

Figure 16.4 Computation flowcharts of the proposed candidate bus selection method. *Source:* Chi and Xu (2020)/With permission of John Wiley & Sons.

For the CBS of a VAR problem, it is difficult to determine the best suitable pair copulas, considering the complex patterns of loads and renewable generations in different locations and across different time periods. Therefore, instead of choosing copula family based on the case-by-case trial and error method, D-vine copula [25] is used in this chapter to select the suitable parameters of the pair-copula in a hierarchical and optimal way. Generally, D-vine copula is a two-step *statistical inference* process: (i) fit a number of probabilistic models to the available data, and (ii) select the best-fitting one through comparison. The advantages of D-vine copula in the application of the CBS problem can be summarized as follows: (i) it selects the copula family and the corresponding parameters in a systematic approach, instead of using a case-by-case trial and error method, and (ii) it is straightforward and easy to expand the structure to incorporate more correlated variables.

Let us take an M-dimensional copula as an example. It is defined as a multivariate distribution over $[0,1]^M$ with uniform marginals in the unit interval. *Sklar's theorem* [15] states that, for any M-variate distribution F_X of an M-dimensional input random vector $X = \{X_1, ..., X_M\}$ with marginals $F_{X_1}, ..., F_{X_M}$, there is a copula C linking F_{X_i} of a random vector to its joint cumulative distribution function F_X:

$$F_X(x) = C(F_{X_1}(x_1), F_{X_2}(x_2), \cdots, F_{X_M}(x_M)) \tag{16.19}$$

Based on (16.19), the joint probability distribution function (PDF) of X can be obtained by differentiation as (16.20), where the copula density function $c(\cdot)$ is expressed as (16.21). The conditional densities can be obtained as (16.22).

$$f_X(x) = c(F_{X_1}(x_1), F_{X_2}(x_2), ..., F_{X_M}(x_M)) \prod_{i=1}^{M} f_{X_i}(x_i) \tag{16.20}$$

$$c(x) = c(u_1, ..., u_M) = \frac{\partial^M C(u_1, ..., u_M)}{\partial u_1 \cdots \partial u_M} \tag{16.21}$$

$$f_{1|2,...,M}(x_1 \mid x_1, ..., x_M) = f_X(x) / \prod_{i=2}^{M} f_{X_i}(x_i) = c(F_{X_1}(x_1), ..., F_{X_M}(x_M)) f_1(x_1) \tag{16.22}$$

For high-dimensional problems, the total number of possible pair copulas is high. For instance, there are 240 different pair copulas constructed for a 5-dimensional X. In order to organize the pair copulas in an efficient way, D-vine copula construction [25] expressed as the product of pair copulas (possibly conditional) is used to represent multivariate copulas ($M > 2$) and decompose the copula density function in a specific way. According to the chain rule of probability, $f_X(x)$ can be expressed as (16.23).

$$f_X(x) = \prod_{j=1}^{M} f_{j|j+1,...,M}(x_j \mid x_{j+1}, ..., x_M) \tag{16.23}$$

Based on (16.20) and (16.23), it is straightforward to generate (16.24) as follows:

$$c\left(F_{X_1}(x_1), F_{X_2}(x_2), \ldots, F_{X_M}(x_M)\right) \prod_{i=1}^{M} f_{X_i}(x_i) = \prod_{j=1}^{M} f_{j|j+1,\ldots,M}\left(x_j \mid x_{j+1}, \ldots, x_M\right)$$

(16.24)

Each conditional PDF (on the right-hand side of (16.24) can be obtained in the form of a pair copula density (a differentiated pair copula). Eventually, a D-vine copula can be obtained as a tensor product of $M(M-1)/2$ pair copulas as follows:

- $M-1$ pair copulas (unconditioned) between variable X_i and X_{i+1}, $i = 1, \ldots,$ $M-1$
- $M-2$ pair copulas between variable X_i and X_{i+2} conditioned on X_{i+1}, $i = 1, \ldots,$ $M-2$
- one pair copula between variable X_1 and X_M conditioned on $X_1, \ldots, X_{M-1}$

Then, an M-dimensional D-vine copula can be defined as (16.25).

$$c(u) = \prod_{j=1}^{M-1} \prod_{i=1}^{M-j} c_{i,i+j|\{i+1,\cdots,i+j-1\}} \left(u_{i|\{i+1,\cdots,i+j-1\}}, u_{i+j|\{i+1,\cdots,i+j-1\}}\right)$$

(16.25)

A 5-dimensional D-vine with 4 trees and 10 edges is illustrated in Figure 16.5 as an example. Every node in the trees is connected to less than three edges. Each edge is associated with a specific pair copula. For example, the edge in T_4 is associated with the pair copula $C_{15|234}$. The iterative construction of pair copulas makes

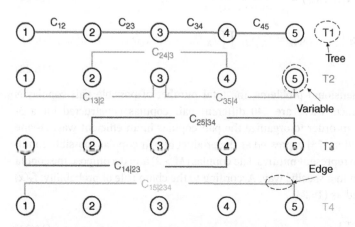

Figure 16.5 Illustration of a 5-dimensional D-vine copula. *Source:* Chi et al. [26]/ John Wiley & Sons/CC BY-4.0.

Table 16.1 Pair Copula families.

Name	Cumulative distribution function C(u, v; θ)	Parameter range
Frank	$-\dfrac{1}{\theta}\log\cdot\left(\dfrac{1-e^{-\theta}-\left(1-e^{-\theta u}\right)\left(1-e^{-\theta u}\right)}{1-e^{-\theta}}\right)$	$\theta \in \mathbb{R}\backslash\{0\}$
Gaussian	$\Phi_{2;\theta}(\Phi^{-1}(u),\ \Phi^{-1}(v))^{(a)}$	$\theta \in (-1, 1)$
Clayton	$(u^{-\theta}+v^{-\theta}-1)-1/\theta$	$\theta > 0$
Gumbel	$\exp\cdot(-((-\log\cdot u)^{\theta}+(-\log\cdot v)^{\theta})^{1/\theta})$	$\theta \in [1, inf)$
Student-t	$t_{2;v,\theta}\left(t_v^{-1}(u), t_v^{-1}(v)\right)^{(b)}$	$v > 1,$ $\theta \in (-1, 1)$

the D-vine copula effective and accurate in representing complex correlation patterns in a multivariate dataset.

For a specific dataset, a set of pair copulas that best fits the dataset should be identified. For example, the five most common pair copula families are listed in Table 16.1. Φ is the univariate standard normal distribution, and $\Phi_{2;\theta}$ is the bivariate normal distribution with "0" means, unit variance, and correlation parameter θ. t_v and $t_{v,\theta}$ are the univariate and bivariate t distributions (both are with v degrees of freedom and a correlation parameter θ).

Theoretically, the pair copulas that best fit the dataset can be selected by iterating maximum likelihood fitting over all vine structures and all copula families. However, for problems with many independent variables, the computation burden is unacceptable due to the large total number of possible vine structures and the different pair copulas that construct D-vine structure. To address this issue, an estimation approach with a light computation burden is used in this chapter, which takes advantage of the specific structure of D-vine copulas to find a compromised but acceptable solution. The structure of D-vine copula is constructed heuristically so as to first select the pairs with the highest correlation and then put these pairs in the upper trees of the D-vine before moving to the lower tree. Specifically, the correlation of the pairs (e.g. X_i and X_j) is assessed using *Kendall's τ* [27]. *Kendall's τ* is defined by (16.26), where $(\tilde{X}_i, \tilde{X}_j)$ is a pair of the independent copy of (X_i, X_j).

$$\tau_{ij} = \mathbb{P}\left((X_i - \tilde{X}_i)(X_j - \tilde{X}_j) > 0\right) - \mathbb{P}\left((X_i - \tilde{X}_i)(X_j - \tilde{X}_j) < 0\right) \tag{16.26}$$

Let the C_{ij} denotes the copula of (X_i, X_j), then we have:

$$\tau_K(X_i, X_j) = 4\iint_{[0\ 1]^2} C_{ij}(u,v)dC_{ij}(u,v) - 1 \tag{16.27}$$

where $K \in [1, ..., M]$. The variables $X_{i_1}, X_{i_1},..., X_{i_M}$ should be optimally arranged to get the maximum $\sum_{k=1}^{M-1} \in \tau_{i_k i_{k+1}}$. For a large M, it can be solved by a genetic algorithm or by a more general method proposed in [28].

For the sensitivity analysis, *Kucherenko indices*, which were first proposed in [29], are used in this chapter as a global sensitivity analysis technique for the dependent variables (the load and wind power uncertainties). They can be regarded as a generalization of *Sobol'* sensitivity indices [30], which are designed for models with independent variables. First, let us divide the input variables X into two complementary subsets X_v and $X_w = X_{\sim v}$. So, the model's total variance, $Y = \mathcal{M}(X)$, can be expressed as (16.28).

$$Var[Y] = Var[\mathbb{E}[Y \mid X_v]] + \mathbb{E}[Var[Y \mid X_v]] \tag{16.28}$$

After the normalization by $Var[Y]$, the first summand on the right-hand side of (16.27) becomes the closed indices of *Sobol's* index with respect to X_v [30], which means the sum of all univariate and interaction effects between these variables.

$$S_v = \frac{Var[\mathbb{E}[Y \mid X_v = x_v]]}{Var[Y]} \tag{16.29}$$

The second summand is written as (16.30), which indicates the total effect of the variables X_w, including all univariate and interaction effects between X_w, and the interaction effects between X_v and X_w.

$$S_w^T = \frac{\mathbb{E}[Var[Y \mid X_v - x_v]]}{Var[Y]}, w = \sim v \tag{16.30}$$

Equation (16.29) can also be expressed in terms of sample covariance as (16.31). Thus, the indices can be estimated as (16.31).

$$S_v = Cov[Y, Y_v]/Var[Y] \tag{16.31}$$

However, for the correlated variables, (16.31) cannot be used for the estimation. To address this limitation, the sample-based *Kucherenko indices* are employed: approximate conditioning on intervals $a_v^l < x_v < a_v^u$ replaces the exact conditioning on specific values $X_v = x_v$, where a_v^l and a_v^u are the multi-dimensional lower bound and upper bound, respectively. The approximate conditioning is realized by the selection of subsets of the sample $\mathcal{X}$ belonging to the interval. For example, for an input sample set $\mathcal{X}$ and a corresponding output $\mathcal{Y}$ (of size N), the total effect can be estimated as the weighted mean of conditional variance as (16.32).

$$S_v^T = \frac{1}{D} \sum_{j=1}^{B} \frac{N_{b,j}}{N} \widehat{Var}\left[\mathcal{Y} \mid \mathcal{X}_{w,j}\right] \tag{16.32}$$

$$D = \frac{1}{N} \sum_{l=1}^{N} \mathcal{M}^2(x_l) - \mathcal{M}_0^2 \tag{16.33}$$

$$\mathcal{M}_0 = \frac{1}{N} \sum_{l=1}^{N} \mathcal{M}(x_l) \tag{16.34}$$

where B and $N_{b,j}$ are the total number of histogram bins and the j-th bin, respectively. $\widehat{Var}\left[\mathcal{Y}|\mathcal{X}_{w,j}\right]$ is the sample variance of $\mathcal{Y}_j$ that correspond to $\mathcal{X}_{w,j}$ (in j-th bin). To calculate S_v^T in (16.32), $\mathcal{X}$ should be firstly divided into B hyperrectangles (bins) in the space.

Accordingly, the first-order index is the weighted variance of conditional mean estimates, which can be written as (16.35).

$$S_v = \frac{1}{D}\sum_{j=1}^{B}\frac{N_{b,j}}{N}\left(\widehat{\mathbb{E}}\left[\mathcal{Y}\mid\mathcal{X}_{v,j}\right]-\widehat{\mu}_{\widehat{\mathbb{E}}}\right)^2 \tag{16.35}$$

where $\widehat{\mathbb{E}}\left[\mathcal{Y}|\mathcal{X}_{v,j}\right]$ is the sample mean of $\mathcal{Y}_j$ with $\mathcal{X}_{w,j}$, and $\widehat{\mu}_{\widehat{\mathbb{E}}}$ is the weighted sample mean of $\widehat{\mathbb{E}}\left[\mathcal{Y}|\mathcal{X}_{v,j}\right]$ as (16.36).

$$\widehat{\mu}_{\widehat{\mathbb{E}}} = \sum_{j=1}^{B}\frac{N_{b,j}}{N}\,\widehat{\mathbb{E}}\left[\mathcal{Y}\mid\mathcal{X}_{v,j}\right] \tag{16.36}$$

16.4.2 Independent Capacity Sensitivity Analysis

Apart from the dependent uncertainties analyzed in the previous section, the capacity sensitivity analysis of STATCOM should also be analyzed. Unlike the operational uncertainties, the capacities of STATCOM are independent of each other. On the other hand, there are still interactions of STATCOM installed at different buses. So, an LHS-based Morris screening method is used to analyze the independent capacity sensitivity, considering the interaction between different STATCOM buses. Morris screening method [21] can efficiently identify the influential buses by creating a multi-dimensional semi-global trajectory within its searching space. Instead of using a single global perturbation, Morris screening method utilizes multiple different perturbations (in this chapter, it is the STATCOM capacities) through the generated grid. Then, the sensitivity measure is built based on the mean and standard deviation of the input variables. Furthermore, LHS is used to generate random samples following the normal distribution to improve representativeness. The uncertainty space is modeled as the range of STATCOM capacity. Mathematical details can be found in Section 16.3 "Capacity sensitivity analysis."

16.4.3 Candidate Bus Selection

Generally, there are four typical cases, as illustrated in Figure 16.6, based on sensitivity analysis results of dependent uncertainties and the independent capacity at the ith bus:

1) Both analysis results are of large values (marked red in the figure). Buses in these areas are sensitive to load variations and renewable generation uncertainties,

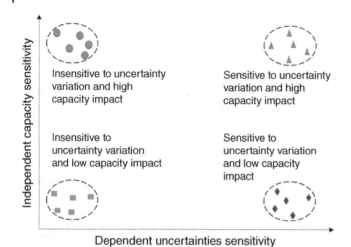

Figure 16.6 Illustration of two-dimension candidate bus selection criterion.
Source: Chi et al. [26]/John Wiley & Sons/CC BY-4.0.

and the placement of STATCOMs at these buses is expected to have a higher impact on the voltage stability.

2) Both analysis results are of small values (marked green in the figure). Buses in these areas are insensitive to load variations and renewable generation uncertainties, and the placement of STATCOMs at these buses is expected to have a lower impact on the voltage stability.

3) Dependent analysis result is of large value, and the independent analysis result is of small value (marked blue in the figure). Buses in these areas are insensitive to load variations and renewable generation uncertainties, but the placement of STATCOMs at these buses is expected to have a higher impact on the voltage stability.

4) Dependent analysis result is of small value, and the independent analysis result is of large value (marked purple in the figure). Buses in these areas are sensitive to load variations and renewable generation uncertainties, but the placement of STATCOMs at these buses is expected to have a lower impact on the voltage stability.

Based on the categorization, an ideal candidate bus should be one of the sensitive and high-impact buses (red ones in the figure). In engineering practice, if there is no such bus, or these buses are too few, the blue and purple ones in the figure can be selected as candidate buses according to the priorities and specific requirements of the decision-makers. Compared with a simple deterministic set of candidate buses, this two-dimensional selection approach (capacity sensitivity and uncertainty sensitivity) provides the decision-makers with more flexible solutions since realistic and practical needs can be considered.

16.4.4 Index for Dependent Uncertainty Sensitivity Analysis

Time-constrained short-term voltage stability index introduced in Section 14.3 is used as the index for dependent uncertainty sensitivity analysis to accurately reflect both the severity of a large disturbance and the impacts of varied uncertainties. Specifically, a risk-based index is used to quantify the short-term voltage performance based on probabilities of critical contingencies as (16.37).

$$TVSI^r = \sum_{k \in \mathcal{K}} \sqrt{\sum_{i \in \mathcal{B}} (TVSI_i^r)^2 / N^b} \cdot p_k \tag{16.37}$$

16.4.5 Index for Independent Capacity Sensitivity Analysis

In the CBS stage, the sensitivity of the indices to the capacity of the reactive power resource is very important. Conventionally, the indices designed for assessment of short-term voltage stability are usually with a set of constant priorities for buses or even without priorities at all. When a high-priority bus (with inferior voltage performance) has been compensated sufficiently, even though the marginal cost of further improvement is large, these buses are still playing important roles in the voltage performance evaluation due to their high and fixed priorities. Consequently, the evaluation results might be inaccurate, and the compensation efficiency might be compromised.

In this chapter, a new index based on $TVSI^r$, called $TVSI^a$, is proposed. It adaptively assigns a higher priority to the current worst bus (in terms of short-term voltage performance) as the capacity sensitivity analysis proceeds. Figure 16.7 illustrates the adaptive strategy of $TVSI^a$. For instance, in the capacity step 1, bus 3 is initially categorized into the high-priority group and then moved to the average priority group since it has been compensated. As the capacity step increases, bus 3 becomes adequately compensated and is moved to the lower priority group.

The overall voltage evaluation for the power system is calculated as (16.38), where β_i is the adaptive priority defined in (16.39) for each bus and σ is a threshold to determine the importance of a bus when compared with the average voltage performance of other buses in the power system.

$$TVSI^a = \sum_{k \in \mathcal{K}} \sqrt{\sum_{i \in \mathcal{B}} (TVSI_i^r \cdot \beta_i)^2 / N^b} \cdot p_k \tag{16.38}$$

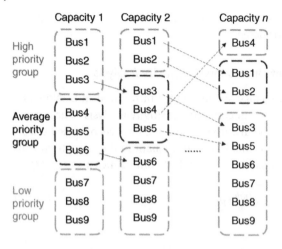

Figure 16.7 Evolution of bus groups with different priorities. *Source:* Chi et al. [26]/John Wiley & SonsCC BY-4.0.

$$\beta_i = \begin{cases} p^{\text{low}}, & TVSI_i^r \le (1 - \sigma) \cdot \sum_{i \in \mathscr{B}} TVSI_i^r / N^b \\ p^{\text{high}}, & TVSI_i^r \ge (1 + \sigma) \cdot \sum_{i \in \mathscr{B}} TVSI_i^r / N^b \\ p^{\text{ave}}, & \text{other} \end{cases} \qquad (16.39)$$

Compared with conventional indices with fixed priorities, $TVSI^a$ is more suitable for capacity sensitivity analysis in the CBS stage. At the initial stage, the relatively high priorities will reinforce the impacts of the buses with worse voltage performance. As the reactive power compensation for these buses increases, their detrimental impact on the system will decrease accordingly, and other buses, initially with lower priorities, might become more influential. So, the adaptive strategy can improve the accuracy of the sensitivity analysis and alleviate the issue of inefficient compensation.

16.4.6 Computation Steps of the Candidate Bus Selection Method

Figure 16.8 illustrates the flowchart of the CBS method, and the corresponding detailed computation steps are described as follows:

1) **Initialization**: Set up parameters of the algorithm and the probabilities of contingencies.
2) **Construction of D-vine structure**: The pair copulas for the D-vine structure are optimally selected based on Kendall's τ. Then, the D-vine structure is constructed by placing the pair with the strongest correlation in the upper tree level. The various correlations between different load buses and the buses with renewable generation are represented by the high-dimensional copula.

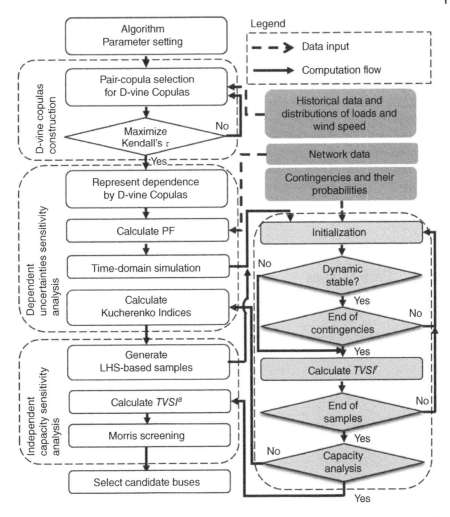

Figure 16.8 General computation flowchart. *Source:* Chi et al. [26]/John Wiley & Sons/CC BY-4.0.

3) **Dependent uncertainties sensitivity analysis**: Input samples (loads and renewable generation outputs) are generated by the D-vine copulas for both steady-state power flow (PF) calculations and time-domain simulations. $TVSI^r$ is calculated and utilized as "Y" in Kucherenko indices. Then, the first-order estimates of Kucherenko indices (Eq. (16.35)) are used for the selection of the candidate buses.

4) **Independent capacity sensitivity analysis**: LHS-based random capacity samples are generated for time-domain simulations with a set of contingencies.

Then, $TVSI^a$ is calculated to evaluate the dynamic voltage performance of the system for each contingency with all capacity samples. Rotor angle constraint [30] is used to check the dynamic stability of the system. Samples failed to satisfy the constraint are penalized to ensure their $TVSI^a$ are inferior to other stable cases. The mean and the standard deviation of the elementary effects are calculated, and the candidate buses are selected according to the principle of Morris screening method as described in Section "Capacity sensitivity analysis."

5) **Candidate bus selection**: After the dependent and independent sensitivity analysis results are obtained, the final candidate buses can be selected (as illustrated in Figure 16.4): an ideal candidate bus is a sensitive bus with high-capacity impact (located among the red triangles). At the same time, buses located among blue circles and purple diamonds can also be selected according to the practical needs of the decision-makers if there is no bus near the red area.

16.5 Case Studies

16.5.1 Case Study for Zoning-based Candidate Bus Selection

The New England 39-bus test system, as illustrated in Figure 16.9, is used to validate the effectiveness of the proposed method. The steady-state analysis is carried

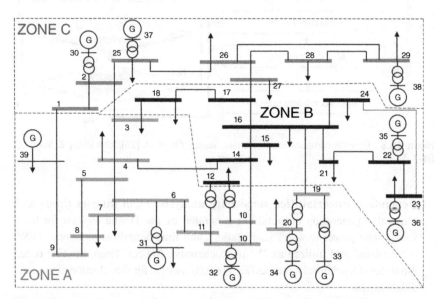

Figure 16.9 New England 39-bus test system. *Source:* Chi and Xu (2020)/With permission of John Wiley & Sons.

out in Matlab 2018b with MATPOWER 6.0 [31], and PSS®E 33.4 [32] is employed for time-domain simulation.

16.5.1.1 Case Study 1 (Without Wind Power Generation)

For case study 1, the original New England 39-bus system is used. For the conventional synchronous generators, they are modeled by GENROU as generators, IEEEVC as compensators, and ESDC1A as exciters. The SVSMO3 model introduced in Chapter 15 is used here. The industry-standard complex load model "CLOD" [33] is used here to simulate the load dynamics. Specifically, this load model can be defined by an 8-dimensional vector [LM, SM, DL, TS, CP, K_p, R_b, X_b]. The first five items stand for the proportions of different loads, respectively. The values of these parameters are shown in Table 16.2.

Figure 16.10 shows the results of contingency severity analysis, where the analysis results are normalized. The most severe contingencies, which have significant impact on the dynamic response of the system, are selected: faults on lines 2–25, line 16–17, line 16–21, and line 21–22.

Based on the membership matrix results, 18 load buses are categorized into three zones, as illustrated in Figure 16.9.

- Zone A: bus # 3, 4, 7, 8, 20;
- Zone B: bus # 12, 15, 16, 18, 21, 23, 24;
- Zone C: bus # 1, 25, 26, 27, 28, 29.

Table 16.2 Parameters for simulation setting.

Parameters	Value
N^{center}	3
$t_i^{s,th}$	4.2 s
t^{dl}	0.6 s
t^{du}	0.6 s
v^{dl}	0.9 pu
v^{du}	1.1 pu
α^l	2.0
α^u	2.0
Unit capacity range	0 ~ 100 Mvar
LHS level (w/o wind power)	80 (from 0 to 100 Mvar)
LHS level (with wind power)	20 (from 0 to 100 Mvar)
LM, SM, DL, TS, CP, K_p	25%, 15%, 10%, 10%, 10%, 2
Fault setup	3-phase short-circuit (cleared after 0.1 s)

Source: Chi and Xu [34]/With permission of John Wiley & Sons.

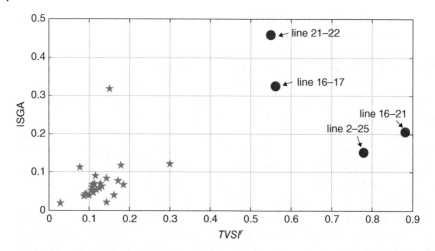

Figure 16.10 Contingency severity analysis results. *Source:* Chi and Xu (2020)/With permission of John Wiley & Sons.

Among them, as illustrated in Figure 16.11, zone C (marked as red) has a limited impact on the dynamic performance of the system (voltage response and rotor angle dynamics). Therefore, candidate buses are only selected from zone A and zone B. Specifically, bus 8 and bus 20 from zone A are selected because they are the most influential buses in terms of *ISGA* and $TVSI^s$; and bus 16 and bus 24 from zone B are selected because they are the highest-ranking buses for *VCPI* and $TVSI^r$ results, respectively.

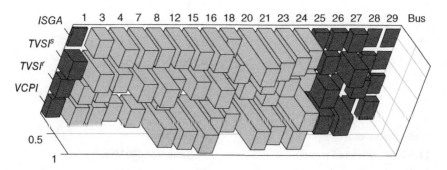

Figure 16.11 Evaluation results for each bus. *Source:* Chi and Xu (2020)/With permission of John Wiley & Sons.

Table 16.3 Simulation results for case study 1.

Method	Conventional (non-zoning)	Zone-based
Buses	8, 15, 12, 16, 20, 21, 24	8, 16, 20, 24
VCPI	**23.32**	23.55
TVSI^r	0.9590	**0.5282**
TVSI^s	1.6575	**1.6163**
ISGA	2586.3	**2515.9**

A comparison, as listed in Table 16.3, with a conventional non-zoning method is carried out to show the effectiveness of the proposed method. Both the proposed and conventional methods are with 210 MVAR STATCOM, and the capacity is the same for each selected bus. For instance, the installed capacity for each of the seven selected buses from the conventional method is 30 MVAR, and the installed capacity for each of the four selected buses from the proposed method is 52 MVAR. Since there are fewer buses selected in the proposed method (only 4 buses), the overall installation cost is lower than that of the conventional method, even if they are with the same total installed capacity. Furthermore, the performance evaluation of the proposed method is better than that of the conventional method, in terms of *TVSI^r*, *TVSI^s*, and *ISGA*. These three indicators are improved by 44.92%, 2.48%, and 2.74%, respectively. It should be noted that *TVSI^r* of the proposed method is significantly improved compared with the conventional method. This is because the zoning approach avoids the redundancy of CBS in a nearby area and improves the efficiency of the reactive power compensation (achieving a better evaluation result, *TVSI^r*, with the same total installed capacity).

16.5.1.2 Case Study 2 (With Wind Power Generation)

For case study 2, one wind farm replaces the synchronous generator at bus 30 and another two wind farms replace the synchronous generator at bus 32, with the same generation capacity as the original generators. These double-fed induction generators are modeled in PSS®E 33.4: the generator model is WT3G2, the electrical model is WT3E1, the mechanical model is WT3T1, and the pitch control model is WT3P1. The dynamic setup for these models can be found in [33]. The compliance of these wind power plants is checked according to the technical regulation of the Australian Energy Market Operator (AEMO) [35] with one of the most stringent LVRT/HVRT regulations compared with other international wind power integration regulations [36]. Two buses with wind farms are considered as the predetermined installation sites for STATCOMs. The wind power uncertainty is modeled through TOAT [37].

Like case study 1, four contingencies are selected as critical ones: faults on line 5–6, line 6–7, line 16–21, and line 21–22. Based on the membership matrix results, 18 load buses and 2 wind farms buses are categorized into three zones:

- Zone A: bus # 7, 23, 28;
- Zone B: bus # 1, 3, 4, 8, 20, 25, 26, 29, 30 (wind farm), 32 (wind farm);
- Zone C: bus # 12, 15, 16, 18, 20, 21, 24, 27.

For the conventional method, buses with the best evaluation results for different indicators are selected as candidate buses for STATCOM installation, as listed in Table 16.4. Specifically, bus 7 is for $TVSI^r$, bus 16 is for $VCPI$, bus 29 is for $TVSI^s$, and bus 24 is for $ISGA$. However, since wind farm buses 30 and 32 are predetermined candidate buses and they are all from zone B, there are three buses from zone B. This concentration is the cause of inefficient compensation. As for the proposed method, bus 7, bus 16, bus 23, and bus 24 are selected as candidate buses. Instead of selecting bus 29, which is the best bus for $TVSI^s$ improvement, the proposed method selects the second-best bus (bus 23) for $TVSI^s$ to avoid concentration in zone B. So, there are two buses from each zone, which is more balanced than the conventional method without consideration of the redundancy in the same zone.

The superiority of the proposed method is also shown in Table 16.4. Although both methods are with the same total installed STATCOM capacity (210 MVar), $TVSI^r$ and $TVSI^s$ of the proposed method is better than that of the conventional method, improved by 5.43% and 6.99%, respectively. The proposed method is also with better evaluation results for $VCPI$ and $ISGA$, although the improvements are relatively small. On the other hand, if the post-contingency performance is the top priority and the planner is more sensitive to the total investment, the evaluation results of the conventional method can be obtained by the proposed method with only three candidate buses, as listed in the last column of Table 16.4, which means a reduced installation cost.

Table 16.4 Simulation results for case study 2.

	Conventional (non-zoning)	Proposed (zone-based)	
		Four buses	Three buses
Buses	7,16,24,29	7,16,23,24	7,16,23
$VCPI$	23.48	**23.41**	23.45
$TVSI^r$	1.0842	**1.0241**	1.0946
$TVSI^s$	1.7205	**1.6002**	1.6568
$ISGA$	2728.1	**2700.1**	2710.4

One of the salient advantages of the proposed method is the consideration of capacity sensitivity analysis. Conventional zoning-based CBS methods, such as methods developed in reference, only consider one specific capacity in the CBS. This simplification might lead to inconsistent results in the CBS. A comparison is listed in Table 16.5, showing the zoning results based on three typical STATCOM capacities. It is worth noting that bus 18, bus 20, bus 23, and bus 32 are grouped into different zones when different capacities are used for the CBS. Among them, bus 23 (the second-best bus for $TVSI^s$ improvement) and bus 32 (one of the only two buses with wind farms) play critical roles in the overall effectiveness of the VAR compensation. Since it is impossible to find an optimal capacity in the CBS stage, the capacities used in references [10, 11] are arbitrary, and the inconsistency of categorization results might lead to a sub-optimal capacity optimization result since the candidate buses would be considered as predetermined inputs during the optimization of capacity.

Furthermore, to demonstrate the superiority of the proposed method, Figures 16.12–16.15 compare the evaluation results between the proposed method and other zoning-based methods without consideration of capacity sensitivity. All the methods presented in the figures are evaluated with the same indices proposed in this chapter. As listed in Table 16.4, bus 7, bus 16, bus 23, and bus 24 are candidate buses from the proposed method. Bus 7, bus 16, bus 29, and bus 24 are for the other two zoning-based methods. It is obvious that the proposed method has fewer unexpected sudden surges/drops. Besides, it improves the dynamic voltage performance of the system more efficiently, in terms of all the metrics, except for $TVSI^s$ in two small capacity scenarios (namely 7 and 21 MVar).

Table 16.5 Capacity impact on the result of other zoning methods.

Unit Capacity	Zones	Buses
20 MVar	A	3, 4, 7, 8, 12, **18**, 26, 27
	B	1, 25, 28, 29, 30, **32**
	C	15, 16, **20**, 21, **23**, 24
60 MVar	A	3, 4, 7, 8, 12, **23**, 26, 27, **32**
	B	1, 25, 28, 29, 30
	C	15, 16, **18**, **20**, 21, 24
100 MVar	A	3, 4, 7, 8, 12, **20**, **23**, 26, 27, **32**
	B	1, 25, 28, 29, 30
	C	15, 16, **18**, 21, 24

Source: Chi and Xu [34]/John Wiley & Sons/CC BY-4.0

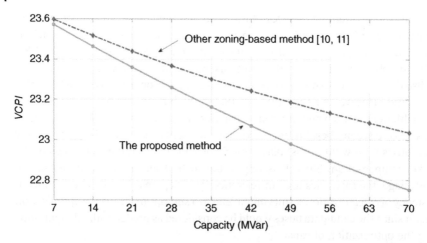

Figure 16.12 Capacity sensitivity comparison of *VCPI*. *Source:* Chi and Xu [34]/With permission of John Wiley & Sons.

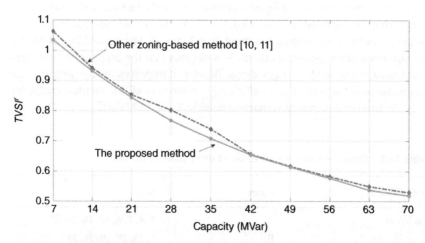

Figure 16.13 Capacity sensitivity comparison of *TVSI*. *Source:* Chi and Xu [34]/With permission of John Wiley & Sons.

Table 16.6 shows the advantage of the root-mean-square method used in this chapter over the direct averaging method (used in the original TVSI [12]).

Let us assume the evaluation results for each bus are as follows: bus 1 is 0.4, and buses 2–39 are all 0.1. After a reactive power compensation, the compensated results are changed: bus 1 is improved to 0.2, and buses 2-39 are all still 0.1. Although the improvement of TVSI for bus 1 is significant (50%), the improvement

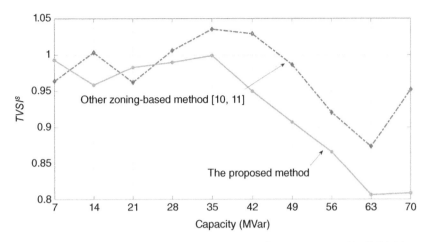

Figure 16.14 Capacity sensitivity comparison of *TVSI*^s. *Source:* Chi and Xu [34]/With permission of John Wiley & Sons.

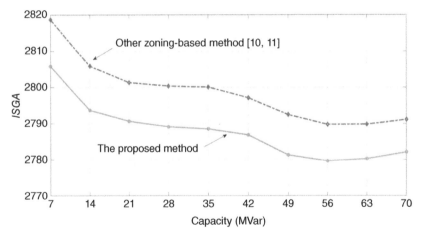

Figure 16.15 Capacity sensitivity comparison of *ISGA*. *Source:* Chi and Xu [34]/With permission of John Wiley & Sons.

of system-level TVSI is relatively small (4.76%) when the direct averaging approach is used, compared with the root-mean-square case (11.81%). So, the root-mean-square approach can make TVSI results more sensitive to reactive power compensation. In practice, there are usually only a few buses experiencing a large reactive power deficiency, while most of the buses in the system are with relatively minor voltage dips after a disturbance. But these few buses play much more important

Table 16.6 Comparison between direct averaging and RMS.

	System-level TVSI (Direct Average)		System-level TVSI (RMS)	
TVSI results for each bus	Value	Improvement	Value	Improvement
Bus 1: 0.4; Bus 2-bus 39: 0.1	0.10769	0.00513 (4.76%)	0.11767	0.01390 (11.81%)
Bus 1: 0.2; Bus 2-bus 39: 0.1	0.10256		0.10377	

Source: Adapted from Xu et al. [12].

roles than other buses, in terms of voltage stability. Therefore, compared with the direct averaging approach that only focuses on the overall voltage performance of the system, the root-mean-square approach, which is more sensitive to the voltage response of the worst buses, can identify the most influential buses more efficiently in the capacity sensitivity analysis.

16.5.2 Case Study for Correlated Candidate Bus Selection

The proposed method is tested on a modified Nordic 74-bus system [38], as illustrated in Figure 16.16. Generators g11, g14, and g17 are replaced with double-fed induction generators modeled in PSS®E [32]. The industry-standard complex load model "CLOD" [33] is used here to simulate the load dynamics. Specifically, this load model can be defined by an 8-dimensional vector [LM, SM, DL, TS, CP, K_p, R_b, X_b]. The first five items stand for the proportions of different loads, respectively. The values of these parameters are shown in Table 16.7. Contingency selection (lines 4031–4041, 4032–4044, and 4042–4044) is based on the method proposed in our previous work [12]. The steady-state analysis is carried out in Matlab with MATPOWER 6.0 [31], and PSS®E 33.4 [32] is employed for time-domain simulation.

The load variations and wind speed are obtained from [39, 40] over one year (8760 hours). Pearson correlation matrix [41] is used to form a $25 * 25$ matrix to show the correlation patterns between these 25 uncertainties, as presented in Figure 16.17. Columns 1–22 and rows 1–22 are the load buses, and columns 23–25 and rows 23–25 represent the wind speed data. The correlation between different uncertainties, including not only load-load and wind-wind correlation, but also load-wind correlation) can be identified in Figure 16.17. Three load zones can be clearly identified because of their relatively high correlation. Therefore, instead of considering load and wind uncertainties independently, it is critical to model the correlated uncertainties correctly for the evaluation of dynamic voltage performance.

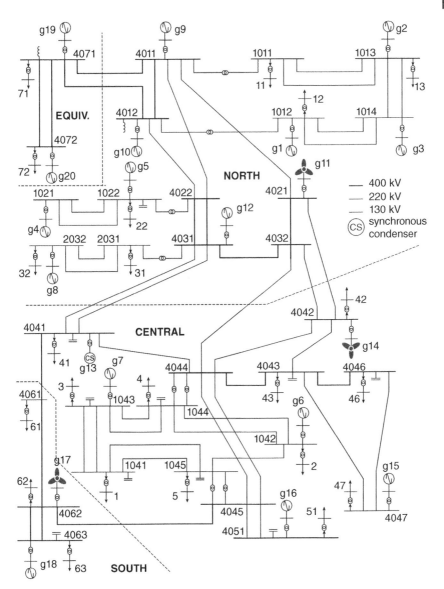

Figure 16.16 Single line diagram of the modified Nordic 74-bus test system. *Source:* Chi et al. [26]/John Wiley & Sons/CC BY-4.0.

The overall computation time is 50 168 seconds. Specifically, dependent uncertainty sensitivity analysis consumes 28 337 seconds while independent capacity sensitivity analysis costs 21 831 seconds. As for the alternative multivariate Gaussian copula method, the overall computation time is 27 901 seconds

Table 16.7 Parameters for simulation analysis.

Parameters	Value
ω	π
δ	0.1
t^{dl}	0.6 s
t^{du}	0.6 s
v^{dl}	0.9 pu
v^{du}	1.1 pu
α^{l}	2.0
α^{u}	2.0
$p^{low}, p^{avg}, p^{high}$	0.8, 1.0, 1.2
LHS levels	20 (0~100 MVar)
LM, SM, DL, TS, CP, K_p	25%, 15%, 10%, 10%, 10%, 2
Fault setup	3-phase short-circuit (cleared after 0.1 second)

Source: Chi et al. [26]/John Wiley & Sons/CC BY-4.0

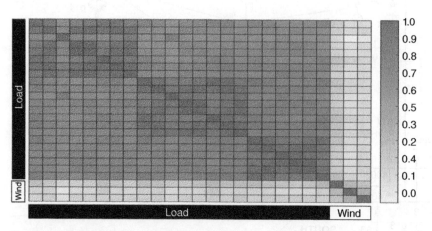

Figure 16.17 The correlation between different uncertainties over a year. *Source:* Chi et al. [26]/John Wiley & Sons/CC BY-4.0.

(only the dependent uncertainty sensitivity analysis is considered). The additional computation time of the proposed method is acceptable since the copula pair and the structure of D-vine are optimally determined, instead of a case-by-case trial and error method employed by the multivariate Gaussian copula approach. Figure 16.18 shows the optimized 25-dimensional D-vine structure, consisting

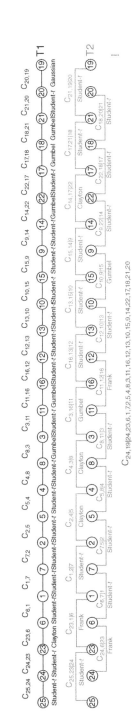

Figure 16.18 Optimized D-vine copulas. *Source:* Chi et al. [26]/John Wiley & Sons/CC BY-4.0.

Table 16.8 Rankings of different candidate selection methods.

| Bus | No correlation | | D-vine | | Gaussian | |
	Value	Rank	Value	Rank	Value	Rank
4072	0.0028	21st	0.0032	20th	0.0095	19th
4071	0.0025	22nd	0.0030	21st	0.0180	12th
2032	0.0068	18th	0.0039	15th	0.0206	10th
2031	0.0237	13th	0.0030	22nd	0.0109	18th
1045	0.0267	11th	0.0036	17th	0.0224	9th
1044	0.0334	8th	0.0039	14th	0.0121	17th
1043	0.0800	5th	0.0051	11th	0.0255	7th
1042	0.1009	**1st**	0.0042	**13th**	0.0165	15th
1041	0.0359	7th	0.0032	19th	0.0389	1st
1022	0.0210	14th	0.0047	12th	0.0177	13th
1013	0.0167	15th	0.0051	10th	0.0321	4th
1012	0.0037	**20th**	0.0067	**9th**	0.0034	20th
1011	0.0302	9th	0.0035	18th	0.0337	2nd
63	0.0111	16th	0.0037	16th	0.0284	5th
62	0.0867	3rd	0.0139	7th	−0.0093	21st
61	0.0301	10th	0.0171	3rd	0.0149	16th
51	0.0241	**12th**	0.0203	**1st**	0.0321	3rd
47	0.0067	**19th**	0.0149	**5th**	0.0225	8th
46	0.0973	2nd	0.0144	6th	0.0177	14th
43	0.0532	6th	0.0086	8th	−0.0110	22nd
42	0.0109	**17th**	0.0152	**4th**	0.0276	6th
41	0.0842	4th	0.0190	2nd	0.0191	11th

of $300(= 25 * 24/2)$ edges and $24(= 25 - 1)$ trees. In the figure, the numbers 1–22 represent the load buses (from bus 41 to bus 4072), and 23–25 denote the three buses with wind farms. Instead of applying one copula family to all the pairs in the D-vine copulas, the proposed method flexibly assigns different copula families to each pair and thus, the correlation between the variables of loads and wind speed can be better revealed.

Table 16.8 lists the sensitivity analysis results for three methods:

1) no correlation considered
2) statistical inference based on D-vine copula (the proposed method)
3) statistical inference based on multivariate Gaussian copula.

Figure 16.19 shows the sensitivity analysis results when the correlated uncertainties (load demand and wind power generation) are modeled as uncorrelated ones using the conventional sensitivity analysis method. In this case, the correlation between uncertainties, as illustrated in Figure 16.17, is not reflected in the sensitivity analysis results. Figure 16.20 shows the sensitivity analysis result based

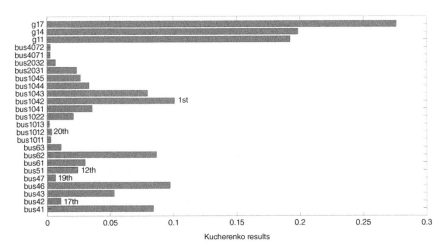

Figure 16.19 Uncertainties sensitivity analysis results (correlation is not considered, conventional method). *Source:* Chi et al. [26]/John Wiley & Sons/CC BY-4.0.

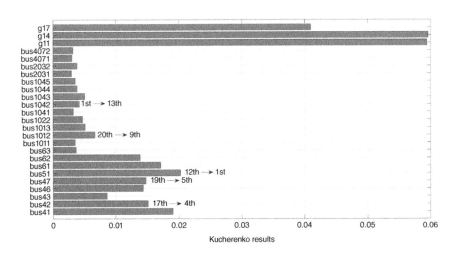

Figure 16.20 Dependent uncertainties sensitivity analysis results (D-vine copulas). *Source:* Chi et al. [26]/John Wiley & Sons/CC BY-4.0.

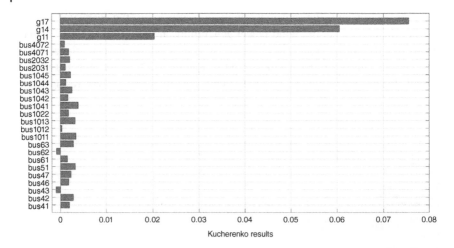

Figure 16.21 Dependent uncertainties sensitivity analysis results (Gaussian copulas). *Source:* Chi et al. [26]/John Wiley & Sons/CC BY-4.0.

on the proposed method that considers the correlation between the uncertainties. Figure 16.21 demonstrates that multivariate Gaussian copula also fails to reveal the dependencies between the uncertainties correctly. For instance, as listed in Table 16.8, the results of bus 62 and bus 43 are even negative.

Based on the comparison between the three methods, it can be concluded that some variables in sensitivity analysis without consideration of the correlation might be regarded as insignificant, but their importance will be revealed when their correlation with other variables is considered. For example, as listed in Table 16.8, the impact of load variations at buses 42 and 51 is relatively small (rank 17th and 12th among all the load buses as shown in Figure 16.19) when their correlation is ignored (considered as independent variables). Nonetheless, their impacts become the 4th and 1st, respectively, when their correlations are considered. On the other hand, the impact of load variations at bus 1042 is significant (ranked 1st) when their correlation is ignored. It decreases to the 13th once the correlations between the uncertainties are taken into consideration.

Figure 16.22 shows the independent capacity sensitivity analysis result based on Morris screening method. The advantage of these two-dimensional results is that, although μ^* is prioritized in decision making of the CBS, the σ^* can provide valuable information about the interaction between these buses (in terms of their voltage response with different capacities of STATCOMs) to the decision-makers. For example, the values of μ^* bus at bus 4042 and bus 2031 are very close, but bus 2031 is more suitable as a candidate bus because its σ^* is smaller.

With sensitivity analysis results for the dependent uncertainties and the independent capacity, candidate buses can be selected according to Figure 16.23.

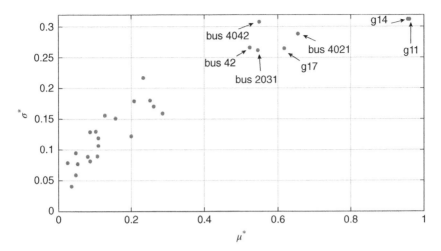

Figure 16.22 Independent capacity sensitivity analysis results. *Source:* Chi et al. [26]/John Wiley & Sons/CC BY-4.0.

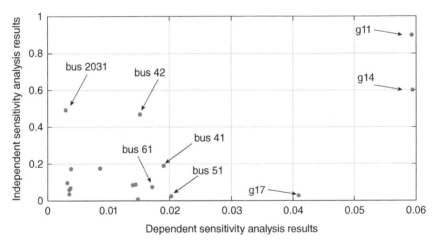

Figure 16.23 Two-dimensional candidate bus selection. *Source:* Chi et al. [26]/John Wiley & Sons/CC BY-4.0.

As explained in Figure 16.6, buses with wind farms (g11 and g14) are ideal candidate buses because their impacts on the system are larger than other buses. Instead of providing a set of deterministic candidate buses, this 2-dimensional selection strategy can provide decision-makers with more flexible options. For instance, for the planners who are more concerned about the voltage stability performance of the power system, the buses with high sensitivity to the STATCOM capacity (large value at the *y*-axis, like buses 42 and 2031) are preferred. For the planners who prioritize the robustness against load variation and uncertainties of wind generation, the buses with large values at the *x*-axis (like buses 51 and g17) can be selected. Alternatively, fuzzy membership function [42] can be adopted for the selection of the candidates.

As illustrated in Figure 16.24, for a conventional trajectory-based voltage stability index with a constant priority, the variation is insignificant even though the capacity changes significantly (from 10–120 MVar). This indifference might lead to inaccurate sensitivity analysis results because, as the capacity increases, more and more buses are no longer in deficiency of the reactive power, and the voltage performance of the power system mainly depends on other buses that were originally not prioritized. In contrast, the change rate of $TVSI^a$ is adaptive: at the initial stage, $TVSI^a$ is very sensitive to the capacity change and is becoming less sensitive as the installed capacity increases.

To show the advantage of the proposed method, the selected candidate buses are used for the VAR planning optimization problem (the solving method can be found in Chapter 18 in detail) in a Nordic 74-bus test system and a comparison is carried out with the Gaussian copula method and the conventional method (no correlation is considered). The optimized results (capacity of STATCOM) for the three methods are listed in Table 16.9. To further demonstrate the superiority of the proposed method, a *Robust Index* [26], which is designed to be the smaller, the better, is employed to quantitatively evaluate the robustness of the optimized results, in terms of their sensitivity to operational uncertainties (load demand variations and wind power generation uncertainty are considered in the case study). As listed in Table 16.10, the results from the proposed method are the most robust among

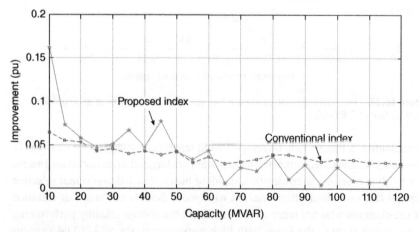

Figure 16.24 Comparison between $TVSI^a$ and conventional index. *Source:* Chi et al. [26]/ John Wiley & Sons/CC BY-4.0.

Table 16.9 Optimized capacity of STATCOM (in MVar).

Proposed method	Bus no.	41	42	51	2031	g11	g14	g17
	Capacity	39.92	101.9	119.9	40.67	41.23	62.94	40.56
Gaussian copula	Bus no.	51	1011	1013	1041	g11	g14	g17
	Capacity	96.03	83.46	119.6	40.00	52.11	47.12	74.99
No correlation	Bus no.	41	42	46	2031	g11	g14	g17
	Capacity	49.25	93.43	43.68	62.37	48.71	82.34	46.16

Source: Chi et al. [26]/John Wiley & Sons/CC BY-4.0

Table 16.10 Results comparison (in million US dollar).

Methods	Total Cost	$TVSI^a$	Robust Index
Proposed method	32.856	0.3024	1.3598
Gaussian copula	36.165	0.3209	1.4587
No correlation	31.797	0.3044	1.4998

Source: Chi et al. [26]/John Wiley & Sons/CC BY-4.0

the three methods. At the same time, a better $TVSI^a$ is also obtained compared with the Gaussian copula method and the conventional method.

Nomenclature

Indices

g index for generators
i index of buses
j index for voltage resilience indices
l index for lines
m index for LHS capacity
o index for an original state
r index for cluster centers
t index for time

Sets

$\mathcal{B}$ set of all buses
$\mathcal{B}_l$ set of all load buses
$\mathcal{G}_s$ set of synchronous generators

$\mathcal{L}$ set of lines
$\mathcal{K}$ set of contingencies
LHS   set of LHS results
$\mathcal{Y}$ set of clustering centers

Variables

D_{ri} square distance between the ith example and the rth clustering center
$EE_{i,m}$ elementary effect of Δ_m change
M_i inertia of machine i
M^{TOT} total inertia of all machines
P_l^r active power transferred to the receiving end through the line l
Q_l^r reactive power transferred to the receiving end through the line l
t_i^s time when voltage of bus i enters an acceptable fluctuation boundary
$TVSI^r$ time-constrained TVSI
$TVSI^s$ steady-state TVSI
U membership matrices
U_z zth membership matrix
$u_{r,i}$ membership value of x_i in cluster r
v_i voltage at bus i
v_i^d voltage deviation at bus i
v_i^o pre-contingency voltage at bus i
v_i^s post-contingency steady-state voltage at bus i
$x_{i,j,c}^o$ evaluation result of bus i when index j is used with capacity c
y_r rth clustering center
$\alpha_{i,m}$ a specific capacity determined by LHS for bus i, step m
μ^* mean of elementary effect
σ^* standard deviation of elementary effect
φ phase angle of load
θ phase angle of line impedance
δ_i rotor angle of machine i
δ^{COA} center of rotor angles
Δ_m offset in Morris Screening method

Parameters

F fuzziness of the clustering process
N^b number of buses
N^c number of all buses
N^{center} number of clustering centers
N^e number of examples

N^m	number of capacity levels
p^{avg}	priority of average priority group
p^{high}	priority of high priority group
p^{low}	priority of low priority group
p_k	probability of contingency k
t^c	fault clearing time
t^{dl}	maximum voltage deviation time (lower)
t^{du}	maximum voltage deviation time (upper)
t^{end}	end of time-domain simulation
t^f	fault time
t^{l1}	end of voltage deviation time within t^{dl} (lower)
t^{l2}	end of voltage deviation time beyond t^{dl} (lower)
t^o	start of time-domain simulation
$t_i^{s,th}$	time threshold for t_i^s
t^{u1}	end of voltage deviation time within t^{du} (upper)
t^{u2}	end of voltage deviation time beyond t^{du} (upper)
v^{dl}	lower voltage threshold
v^{du}	upper voltage threshold
α^l	penalty parameter for voltage violation (lower)
α^u	penalty parameter for voltage violation (upper)
ϵ	termination tolerance

References

1 Sode-Yome, A., Mithulananthan, N., and Lee, K.Y. (2006). A maximum loading margin method for static voltage stability in power systems. *IEEE Transactions on Power Systems* 21 (2): 799–808.

2 Ghahremani, E. and Kamwa, I. (2012). Optimal placement of multiple-type FACTS devices to maximize power system loadability using a generic graphical user interface. *IEEE Transactions on Power Systems* 28 (2): 764–778.

3 Aolaritei, L., Bolognani, S., and Dörfler, F. (2018). Hierarchical and distributed monitoring of voltage stability in distribution networks. *IEEE Transactions on Power Systems* 33 (6): 6705–6714.

4 Paramasivam, M., Salloum, A., Ajjarapu, V. et al. (2013). Dynamic optimization based reactive power planning to mitigate slow voltage recovery and short term voltage instability. *IEEE Transactions on Power Systems* 28 (4): 3865–3873.

5 Liu, H., Krishnan, V., McCalley, J.D., and Chowdhury, A. (2014). Optimal planning of static and dynamic reactive power resources. *IET Generation Transmission and Distribution* 8 (12): 1916–1927.

6 Rather, Z.H., Chen, Z., Thøgersen, P., and Lund, P. (2014). Dynamic reactive power compensation of large-scale wind integrated power system. *IEEE Transactions on Power Systems* 30 (5): 2516–2526.

7 State Grid Corporation of China (2010). *Security and Stability of the National Grid Computing Specification, Q/GDW 404-2010*. Beijing: State Grid Corporation of China.

8 Shoup, D.J., Paserba, J.J., and Taylor, C.W. (2004). A survey of current practices for transient voltage dip/sag criteria related to power system stability. *IEEE PES Power Systems Conference and Exposition*, New York, NY, USA (10–13 October 2004), pp. 1140–1147: IEEE.

9 NERC (2010). *Transmission System Planning Performance Requirements, TPL-001-4*. Atlanta: NERC.

10 Wang, Y., Li, F., Wan, Q., and Chen, H. (2011). Reactive power planning based on fuzzy clustering, gray code, and simulated annealing. *IEEE Transactions on Power Systems* 26 (4): 2246–2255.

11 Guan, L., Wu, L., Li, F., and Zhao, Q. (2017). Heuristic planning for dynamic VAR compensation using zoning approach. *IET Generation Transmission and Distribution* 11 (11): 2852–2861.

12 Xu, Y., Dong, Z.Y., Xiao, C. et al. (2015). Optimal placement of static compensators for multi-objective voltage stability enhancement of power systems. *IET Generation Transmission and Distribution* 9 (15): 2144–2151.

13 Liu, J., Xu, Y., Dong, Z.Y., and Wong, K.P. (2017). Retirement-driven dynamic VAR planning for voltage stability enhancement of power systems with high-level wind power. *IEEE Transactions on Power Systems* 33 (2): 2282–2291.

14 Hasan, K.N. and Preece, R. (2017). Influence of stochastic dependence on small-disturbance stability and ranking uncertainties. *IEEE Transactions on Power Systems* 33 (3): 3227–3235.

15 Li, P., Guan, X., Wu, J., and Zhou, X. (2015). Modeling dynamic spatial correlations of geographically distributed wind farms and constructing ellipsoidal uncertainty sets for optimization-based generation scheduling. *IEEE Transactions on Sustainable Energy* 6 (4): 1594–1605.

16 Joe, H. (2014). *Dependence Modeling with Copulas*. CRC press.

17 Park, H., Baldick, R., and Morton, D.P. (2015). A stochastic transmission planning model with dependent load and wind forecasts. *IEEE Transactions on Power Systems* 30 (6): 3003–3011.

18 Xu, X., Yan, Z., Shahidehpour, M. et al. (2017). Power system voltage stability evaluation considering renewable energy with correlated variabilities. *IEEE Transactions on Power Systems* 33 (3): 3236–3245.

19 Li, G. and Rovnyak, S.M. (2005). Integral square generator angle index for stability ranking and control. *IEEE Transactions on Power Systems* 20 (2): 926–934.

20 Zhang, Y., Xu, Y., Dong, Z.Y., and Zhang, R. (2018). A hierarchical self-adaptive data-analytics method for real-time power system short-term voltage stability assessment. *IEEE Transactions on Industrial Informatics* 15 (1): 74–84.

21 Iooss, B. and Lemaître, P. (2015). A review on global sensitivity analysis methods. In: *Uncertainty Management in Simulation-Optimization of Complex Systems* (ed. G. Dellino and C. Meloni), 101–122. Boston, MA: Springer.

22 Qi, B., Hasan, K.N., and Milanović, J.V. (2019). Identification of critical parameters affecting voltage and angular stability considering load-renewable generation correlations. *IEEE Transactions on Power Systems* 34 (4): 2859–2869.

23 Hathaway, R.J., Bezdek, J.C., and Hu, Y. (2000). Generalized fuzzy c-means clustering strategies using L_p distances. *IEEE Transactions on Fuzzy Systems* 8 (5): 576–582.

24 Miettinen, K. (2012). *Nonlinear Multiobjective Optimization*. Springer Science & Business Media.

25 Bedford, T. and Cooke, R.M. (2002). Vines – a new graphical model for dependent random variables. *Ann. Statist.* 30 (4): 1031–1068.

26 Chi, Y., Xu, Y., and Zhang, R. (2020). Many-objective robust optimization for dynamic var planning to enhance voltage stability of a wind-energy power system. *IEEE Transactions on Power Delivery* 36 (1): 30–42.

27 Annis, D.H. (2006). Kendall's Advanced Theory of Statistics, Vol. 1: Distribution Theory. Alan Stuart and J. Keith Ord; Kendall's Advanced Theory of Statistics, Vol. 2A: Classical Inference and the Linear Model. Alan Stuart and J. Keith Ord, and Steven F. Arnold. *Journal of the American Statistical Association*. 101: 1721.

28 Aas, K., Czado, C., Frigessi, A., and Bakken, H. (2009). Pair-copula constructions of multiple dependence. *Insurance: Mathematics & Economics* 44 (2): 182–198.

29 Kucherenko, S., Tarantola, S., and Annoni, P. (2012). Estimation of global sensitivity indices for models with dependent variables. *Computer Physics Communications* 183 (4): 937–946.

30 Saltelli, A. and Sobol', I.Y. (1995). Sensitivity analysis for nonlinear mathematical models: numerical experience. *Matematicheskoe Modelirovanie* 7 (11): 16–28.

31 Zimmerman, R.D., Murillo-Sanchez, C.E., and Thomas, R.J. (2011). MATPOWER: steady-state operations, planning, and analysis tools for power systems research and education. *IEEE Transactions on Power Systems* 26 (1): 12–19.

32 Siemens Power Technologies International (2013). *PSS®E 33.4 Program Application Guide*, vol. 2. New York: Siemens.

33 Siemens Power Technologies International (2013). *PSS®E 33.4 Model Library*. New York: Siemens.

34 Chi, Y. and Xu, Y. (2020). Zoning-based candidate bus selection for dynamic VAR planning in power system towards voltage resilience. *IET Generation, Transmission and Distribution* 14 (6): 1012–1020.

35 Australian Energy Market Commission (AEMC) (2018). *National Electricity Rules Version 110*. National Electricity.

36 Mokui, H.T., Masoum, M.A., and Mohseni, M. (2014). Review on Australian grid codes for wind power integration in comparison with international standards. *2014 Australasian Universities Power Engineering Conference (AUPEC)*, Perth, WA, Australia (28 September–01 October 2014), pp. 1–6: IEEE.

37 Stuart, P.G. (1993). *Taguchi Methods: A Hand on Approach*. Reading, MA: Addison-Wesley.

38 Cutsem, T.V., Glavic, M., Rosehart, W. et al. (2020). Test systems for voltage stability studies. *IEEE Transactions on Power Systems* 35 (5): 4078–4087.

39 ERCOT (2017). Hourly load data archives [Datafile]. Texas. [cited 2018 Nov 27]. http://www.ercot.org/gridinfo/load/load_hist (accessed 27 November 2018).

40 AgriMet (2017). Hourly dayfile data access [Datafile]. Texas. [cited 2018 Nov 9]. http://ww.usbr.gov/pn/agrimet/webaghrread.html/ (accessed 9 November 2018).

41 Dellino, G. and Meloni, C. (2015). *Uncertainty Management in Simulation-Optimization of Complex Systems*. Boston MA: Springer.

42 Agrawal, S., Panigrahi, B.K., and Tiwari, M.K. (2008). Multiobjective particle swarm algorithm with fuzzy clustering for electrical power dispatch. *IEEE Transactions on Evolutionary Computation* 12 (5): 529–541.

17

Multi-objective Dynamic VAR Resource Planning

17.1 Introduction

Voltage instability is a major threat in modern power systems. The phenomenon can manifest in many forms, such as progressive voltage drop or rapid voltage collapse after a disturbance. Generally, voltage instability originates from the attempt to restore power consumption beyond the capability of the combined transmission and generation system [1].

In the power system planning stage, deploying reactive power (VAR) resources is an effective method to enhance voltage stability. Conventional voltage stability assessment and associated VAR planning are mostly based on power flow methods. Although they are effective for steady-state voltage stability, they usually ignore the short-term voltage instability phenomenon. After a significant disturbance, the induction motors draw a very high reactive current, which impedes voltage recovery. Without adequate and fast VAR support, prolonged voltage depression, slow voltage recovery, and other unacceptable transient voltage performances are expected. Conventional capacitor banks alone are inadequate to mitigate the short-term voltage instability due to their slow response and that their VAR output decreases in a quadratic order with the voltage drop.

As promising candidates, the flexible AC transmission system (FACTS) devices such as static VAR compensator (SVC) and static compensator (STATCOM) are capable of providing rapid and dynamic VAR support during and after a significant voltage disturbance. This chapter focuses on dynamic VAR resource planning for enhancing short-term voltage stability.

Most previous studies on the optimal placement of VAR resources focused on enhancing the steady-state voltage stability level [2–6] or mitigating static voltage sags [7], leaving the short-term voltage instability issues unattended. Recently, some researchers have been working on the optimization of VAR resource

Stability-Constrained Optimization for Modern Power System Operation and Planning,
First Edition. Yan Xu, Yuan Chi, and Heling Yuan.
© 2023 The Institute of Electrical and Electronics Engineers, Inc.
Published 2023 by John Wiley & Sons, Inc.

planning considering the short-term voltage stability. A two-step optimization model is proposed in Ref. [8]. Firstly, an optimal power flow (OPF) model was used to optimize the VAR placement considering steady-state voltage stability. Secondly, the optimized VAR placement is adjusted through time-domain simulation to verify short-term voltage stability. To allocate the static and dynamic VAR sources against both steady-state and dynamic voltage instability problems, a mixed-integer programming (MIP) algorithm was developed in Ref. [9]. Based on linear sensitivities of performance measures with respect to the VAR injection, the MIP model can be solved efficiently with consideration of many contingencies simultaneously. However, due to the adoption of approximate linear sensitivities, it can only provide near-optimal solutions. Based on trajectory sensitivity analysis, the authors of reference [10] presented a heuristic approach for the STATCOM placement against the short-term voltage instability. The capacity of STATCOM is determined according to the existing capacitor banks at a bus. As a heuristic method, it cannot guarantee the optimality of the cost-effectiveness of the project. A mixed-integer dynamic optimization (MIDO) model was introduced in Ref. [11] to allocate SVCs optimally. The proposed model is converted to a mixed-integer nonlinear programming (MINLP) model, which is solved by standard solvers. However, the load model implemented in the model is the static ZIP (constant impedance, constant current, and constant power) model, which is unable to accurately represent the dynamic nature of various load components (in particular, induction motors). The MIDO model was improved later in Ref. [12] using the control vector parameterization (CVP), trajectory sensitivity, singular value decomposition, and linear programming techniques. The works [10–12] only considered a single contingency while allocating the dynamic VAR resources.

To address the limitations of existing works on this topic, a multi-objective optimization problem (MOP) with two conflicting objectives is proposed in this chapter for the optimal cost-effectiveness of the dynamic VAR placement. Two objectives are: (i) the total investment cost (including purchasing and installation cost) of the STATCOM device, and (ii) the expected unacceptable short-term voltage performance subject to a set of probable contingencies. System dynamic performance is realistically analyzed using full-time-domain simulations, and a complex load model comprising induction motors and other typical components is used to reflect the load dynamics.

17.2 Multi-objective Optimization Model

A VAR planning problem relates to many planning and operating criteria, such as device investment, steady-state and short-term voltage stabilities, and operational costs. These nonlinear and sometimes conflicting (such as high stability versus low cost) objectives should be optimized simultaneously instead of formulated as a single-objective model. Combining multiple objectives as a single objective by the weighting-sum method has some limitations: (i) difficult to subjectively determine the weighting factors (for a weighted-sum single-objective model) for each sub-objective; (ii) the units of each sub-objective might be inherently different; (iii) sensitivities of each sub-objective are different.

For optimal cost-effectiveness of the dynamic VAR placement, a MOP is formulated, minimizing two conflicting objectives: (i) the total investment cost (consisting of capacity cost and installation cost) of the STATCOM device, and (ii) the expected unacceptable short-term voltage performance subject to a set of probable contingencies. The dynamic performance of the power system is realistically analyzed through full-time-domain simulations and a complex load model comprising induction motors and other typical components is used to reflect the load dynamics. TVSI introduced in Chapter 14 is employed to quantitatively assess the short-term voltage stability. Critical contingencies are considered in the model to reflect the probabilistic nature of the contingencies. A multi-objective evolutionary algorithm called MOEA/D is introduced here and used to find Pareto optimal solutions of the MOP model.

17.2.1 Optimization Objectives

The objective function comprises two individual objectives:

$$\min_{u} F[f_1(\boldsymbol{x}, \boldsymbol{u}), f_2(\boldsymbol{x}, \boldsymbol{u})] \tag{17.1}$$

The decision variables are the STATCOM capacity and the installation location (candidate buses).

In previous research, such as [8, 10–12], only a single representative or the worst contingency was considered for the VAR planning problem. The contingency evaluation is based on a deterministic criterion. In engineering practice, more contingencies should be considered with various probabilities. In this chapter, a risk index (RI) is used to quantify the expected unacceptable transient voltage performance under a set of probable contingencies. So, the first objective is the risk level (RL) of unacceptable transient voltage performance under a set of possible contingencies, which is calculated as (17.2).

$$f_1 = RI = \sum_{k \in \mathscr{K}} TVSI_k \cdot p_k \tag{17.2}$$

TVSI is calculated as (13.16). It cannot be explicitly calculated and is based on the post-contingency voltage response obtained from time-domain simulation.

The second objective is the total investment cost of the STATCOM placement, calculated as (17.3).

$$f_2 = \sum_{i=1}^{H} I_i \cdot c_i^{install} + \sum_{i=1}^{H} I_i \cdot B_i \cdot c^{cap} \tag{17.3}$$

17.2.2 Steady-State Constraints

The steady-state constraints include the power flow balance (17.4) and the steady-state operational constraints (17.5) for pre-contingency state.

$$\begin{cases} P_g - V_i \sum_{j \in \mathcal{B}} V_j \left(G_{ij} \cos \theta_{ij} + B_{ij} \sin \theta_{ij} \right) = 0 \\ Q_g + Q^{sat} - V_i \sum_{j \in \mathcal{B}} V_j \left(G_{ij} \sin \theta_{ij} - B_{ij} \cos \theta_{ij} \right) = 0 \end{cases} \tag{17.4}$$

$$\begin{cases} V_i^{min} \leq V_i \leq V_i^{max} \\ L_l^{min} \leq L_l \leq L_l^{max} \\ P_g^{min} \leq P_g \leq P_g^{max} \\ Q_g^{min} \leq Q_g \leq Q_g^{max} \end{cases}, \forall i \in \mathcal{B}, \forall l \in \mathcal{L}, \forall g \in \mathcal{G} \tag{17.5}$$

17.2.3 Dynamic Constraints

It is possible that in some cases short-term voltage instability or collapse occurs along with the loss of synchronism of generators. However, it is difficult to identify the major driving force, voltage, or generator angle, for the short-term instability events [1]. Therefore, the rotor angle stability constraint is employed in the optimization model to avoid the voltage instability caused by the rotor angle instability.

Conventionally, the rotor angle stability is checked by examining the rotor angle deviation against a certain threshold [13]. In this chapter, the constraint (17.6) is used to ensure the rotor angle stability for each contingency:

$$\left[\max \cdot \left(\Delta \delta_{gg'}^T \right) \right]_k \leq \omega \; \forall g, g' \in \mathcal{G}_s, \forall k \in \mathcal{K} \tag{17.6}$$

ω can be set to for the extreme case [13]. The constraint (17.6) is consistent with electrical engineering practice since a real-world electric utility grid is always operated such that any generator rotor angle difference will not exceed a threshold [13]. Any violation will lead to tripping of a generator by an out-of-step relay to prevent potential damage. It should be noted that other rotor angle stability criteria can also be employed when necessary.

On the other hand, since the transient voltage performance is evaluated by the (17.2) and is also one of the optimization objectives, there is no need to constrain the post-contingency voltage magnitudes in the MOP model.

17.3 Decomposition-based Solution Method

17.3.1 Pareto Optimality

Contrary to an optimization problem with only one objective, a MOP has to deal with conflicting objectives. For instance, in this chapter, it is the lower cost versus higher voltage stability. There is no solution that can achieve a simultaneous minimization of all the objectives. Instead, a group of trade-off solutions is generated by an evolutionary algorithm. Pareto optimality theory [14] is a predominant concept for defining trade-off solutions. Given the feasible decision space $\mathcal{X}$, a solution $x^* \in \mathcal{X}$ is called Pareto optimal (or non-dominated), if there is no other solution, $x \in \mathcal{X}$, such that $\mathbf{f}(x) \leq \mathbf{f}(x^*)$ and $f_i(x) \leq f_i(x^*)$ for at least one function. In other words, no further improvement in the value of objective functions can be achieved without impairment in some other objective values. The Pareto optimal objective vector is defined as the vector of objective function values given by a Pareto optimal solution. And the set of all the Pareto optimal solutions is called the Pareto set (PS), and the set of all the Pareto optimal objective vectors is the Pareto front (PF). Figure 17.1 shows the non-dominated solutions (formulating the Pareto Front together) and the dominated solutions. For instance, point C is not on the PF because it is dominated by both point A and point B, while points A and B are Pareto optimal because they are not dominated by any other solutions.

To approximate the PF, aggregation (e.g. through weight factors) of the individual objectives into a composite one such that the MOP can be decomposed into a number of scalar objective optimization subproblems is a common approach. However, this approach has some limitations: (i) difficult to find an appropriate weight

Figure 17.1 Illustration of Pareto Front.

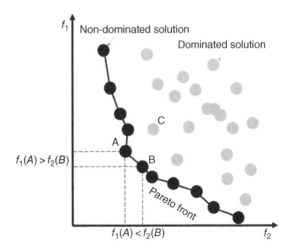

factor, (ii) the convergence performance largely depends on the initial starting point, and (iii) only one single Pareto optimal solution for each simulation run.

Alternatively, multi-objective evolutionary algorithm (MOEA) [15] can be used to approximate the PF. Based on the emulation of the mechanism of natural selection, the MOEA treats the MOP as a whole and can find multiple optimal solutions for just one simulation run. The PF can be systematically approximated without any subjectivity. Besides, the convexity and/or differentiability of the mathematical model of the problem do not affect the application of the MOEA.

17.3.2 Decomposition-based MOEA

State-of-the-art MOEAs are non-dominated sorting genetic algorithm II (NSGA-II) and multi-objective genetic local search (MOGLS) [14, 15]. This chapter introduces a decomposition-based algorithm called MOEA/D [16] and utilizes it to approximate the PF of the MOP model. This algorithm can be considered for solving other MOP problems in power engineering.

MOEA/D adopts an aggregation strategy based on the principles of the original MOEA [16]. Firstly, it explicitly decomposes the MOP into a set of scalar optimization subproblems. Then, these subproblems are solved by evolving a population of solutions. Compared with other prevalent algorithms, such as MOGLS, MOEA/D has been demonstrated to be able to find higher-quality solutions for many benchmark problems [16] with less computation burden.

For a MOP with m objectives and a set of evenly spread weighting vectors λ^1, ..., λ^j, ..., λ^S, where $\lambda^j = \left(\lambda_1^j, ..., \lambda_m^j\right)$, the MOP can be decomposed into S scalar optimization subproblems by *Tchebycheff* method and the objective function of the *j*th subproblem can be written as (17.7).

$$\min\left[\vartheta^{te}\left(x \mid \lambda^j, z^*\right) = \max_{1 \le i \le m}\left\{\lambda_i^i \mid f_i(x) - z^*\right\}\right. \tag{17.7}$$

The population of each generation consists of the best solutions so far for each subproblem. The neighborhood relations among these subproblems are defined based on the distance between their aggregation coefficient vectors, and the optimal solutions to two neighboring subproblems should be very similar. The subproblem is optimized only based on the information obtained from its neighboring subproblems. The major procedures of MOEA/D are listed as follows. More details can be found in [16].

1) **Initialization**: Generate a set of evenly spread weighting vectors λ^1, ..., λ^j, ..., λ^S. Set the external population (EP) = Ø. Compute the *Euclidean* distance between any two weighting vectors and find the R closest weighting vectors to each

weighting vector. For each $i = 1, ..., S$, set $E(i) = \{ i_1, ..., i_R \}$, where $\lambda^{i_1}, ..., \lambda^{i_R}$ are the R closest neighboring weighting vector to λ^i. Generate an initial population $x^1, ..., x^S$ randomly. Set the objective function value $FV^i = f(x^i)$. Initialize $z = (z_1, ..., z_m)$ by a problem-specific approach.

2) **Updating**: For $i = 1, ..., S$, perform the following steps:
   a) Reproduction: randomly select two indices k, l from $E(i)$ and generate a new solution y from x^k and x^l by a specific operator.
   b) Improvement: apply a specific repair/improvement heuristic on y to produce y'.
      Update z: for each $j = 1, ..., m$, if $z_j < f_j(y')$, set $z_j = f_j(y')$.
      Update neighboring solutions: for each index, $j \in E(i)$, if $\vartheta^{le}(y' \mid \lambda^j, z) \leq \vartheta^{le}(x^j \mid \lambda^j, z)$, set $x^j = y'$ and $FV^i = F(y')$.
   c) Update EP: remove from EP all the vectors dominated by $F(y')$, and add $F(y')$ to EP otherwise.

3) **Termination**: If any stopping criterion (maximum iteration number or no change of EP in a pre-defined number of successive iterations) is satisfied, stop and generate the final EP. Otherwise, go to Step 2.

17.3.3 Coding Rule

The decision variables consist of binary installation decisions and the STATCOM size decisions (MVAR). For H candidate buses, the length of an individual is $2 * H$.

17.3.4 Computation Steps

The details of computation steps are elaborated as follows:

1) **Initialization**: The system model, contingencies with the corresponding probabilities, and other parameters are prepared as inputs.
2) **Candidate bus selection**: The candidate buses for STATCOM installation can be selected based on the methods introduced in Chapter 16. In the subsequent optimization process, only the candidate buses will be considered.
3) **Generation of the initial population**: Generate the initial population based on the corresponding data from Step 1.
4) **Calculation of objectives**: Time-domain simulation is carried out for the evaluation of the short-term voltage stability, and the objectives are calculated based on (17.2) and (17.3) for each individual.
5) **Assessment**: The individuals are assessed according to the values of the objectives and the constraints. Any individual who fails to satisfy all of the

constraints is considered "infeasible." On the other hand, the "feasible" individuals are saved. The EP is updated according to Step 2.

6) **Stopping criterion**: If the maximum iteration number is reached or there is no change of EP in a pre-defined number of successive iterations, the whole optimization is terminated with the current EP as the final output; otherwise, new individuals are generated, and the computation goes to Step 4.

17.4 Case Studies

17.4.1 Test System and Parameters

The proposed method is verified on the New England 39-bus test system, as illustrated in Figure 17.2. The time-domain simulation is performed using the PSS®E package [17], and the MOEA/D algorithm is implemented in the MATLAB platform. The interface between MATLAB and PSS®E is developed according to reference [18]. Major parameters for the simulation setting are listed in Table 17.1.

To study the load dynamics, the complex load model called "CLOD" [18] is used in this chapter. It should be noted that other load models capable of reflecting the load dynamics can also be employed when necessary. As illustrated in Figure 17.3, this load model consists of small induction motors, large induction motors, lighting, and other typical equipment that are fed from a common substation.

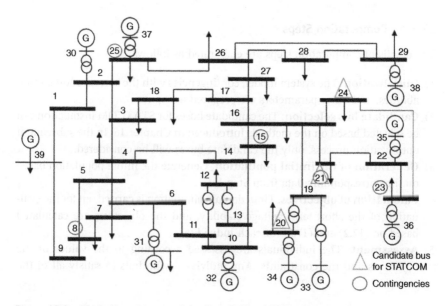

Figure 17.2 Single line diagram of the New England 39-bus test system.

Table 17.1 Parameters for simulation setting.

Parameters	Value settings
$c^{install}$	1.5 million/unit
c^{cap}	0.05 million/MVAR
ω	π
δ	0.2
LM, *SM*, *DL*, *CP*, K_p	20%, 30%, 5%, 30%, 2
Fault setup	3-phase short-circuit (cleared after 0.11 s)
Maximum iteration number	100
Number of successive iterations without change	10
Maximum RI (to select the final planning decision)	1.0
Simulation time	2 s

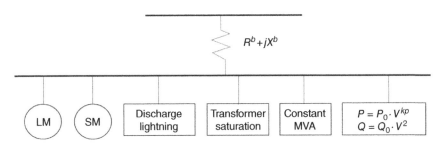

Figure 17.3 Complex load model "CLOD".

Furthermore, "CLOD" is defined by an 8-dimensional vector [*LM*, *SM*, *DL*, *TS*, *CM*, R^b, X^b, K_p]. The former five parameters indicate the percentage value of large induction motor, small induction motor, discharge lighting, transformer exciting current, and constant power load, respectively. It should be noted that the flux linkage dynamics of the machine are not considered in this complex load model.

17.4.2 Numerical Simulation Results

To demonstrate the impact of load dynamics, a comparison study is carried out between two cases. The first one is with a constant load model, and the load dynamics are not considered during the transient period, while the second case is with the complex load model "CLOD," and the load dynamics are considered. The corresponding short-term voltage stability evaluation results for the two cases

Table 17.2 Short-term voltage stability evaluation results with different load models and without STATCOMs.

Fault bus no.	8	21	15	23	25
TVSI with constant power load model	0.018	0.061	0.043	0.067	0.041
TVSI with CLOD	0.56	3.70	6.87	2.30	2.30

are listed in Table 17.2. It is clear that the load dynamics significantly worsen the short-term voltage stability of the system. Based on (17.2), the *RI* values (the second objective of the optimization model) are 0.046 and 3.15 for the cases without and with consideration of load dynamics, respectively.

The proposed two-objective optimization model (17.1)–(17.6) is solved using the MOEA/D. A PF with nine Pareto optimal solutions is obtained and illustrated in Figure 17.4, providing decision-makers with a series of trade-off solutions. A smaller RI indicates a larger investment cost. Based on the PF, a final planning decision can be selected according to electrical engineering practices and other specific requirements. In this chapter, the following criterion is utilized: given the maximum acceptable RL, select the solution with the lowest investment cost on the PF. It should be noted that other decision-making criteria can also be considered,

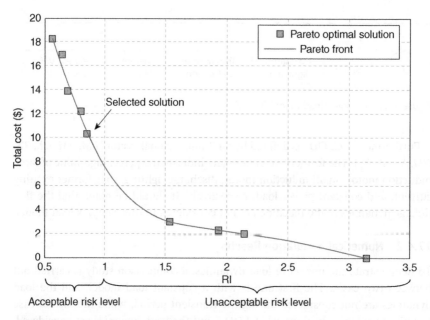

Figure 17.4 Pareto optimal solutions and the Pareto Front.

Table 17.3 STATCOM installation decision.

Bus no.	16	20	21	23	24	Total
Size (MVAR)	0	33	42	0	41	116

Table 17.4 Short-term voltage stability results with STATCOMs.

Fault bus no.	8	21	15	23	25	RI
TVSI	0.42	0.85	1.14	0.58	1.28	0.85

such as fuzzy logic, and Nash equilibrium [19, 20]. In the case study, the selected solution is marked in Figure 17.4, and its details are listed in Table 17.3. The corresponding installation locations are also marked as red triangles in Figure 17.2. The total capacity is 116 MVAR, the total investment cost is US$ 10.3 million, and the short-term voltage stability evaluation result (RI) is 0.85. Specifically, the TVSI for each contingency is given in Table 17.4. It can be concluded that, with STATCOMs, the short-term voltage stability of the studied power system is significantly improved for all contingencies, and the overall RL is reduced by 73%.

Nomenclature

Indices

i, j indices for buses
l indices for lines
k index for contingencies
g index for generators
m index for the objective

Sets

$\mathcal{B}$ set of all buses
$\mathcal{G}_s$ set of synchronous generators
$\mathcal{L}$ set of all the lines
$\mathcal{K}$ set of contingencies
u set of control variables
x set of state variables

Variables

B_i	integer capacity decision variables for bus i
B_{ij}	susceptance between i and j
G_{ij}	conductance between i and j
I_i	binary installation decision variables for bus i
L_l	power flow on line l
P_g	max and min active power of generator g
Q_g	max and min reactive power of generator g
Q^{stat}	output of STATCOM
V_i, V_j	voltage magnitude of bus i (or bus j)
z^*	reference point for MOEA/D
λ^j	weighting vector for MOEA/D
$\Delta\delta_{gg'}^T$	rotor angle deviation between generator g and g'
θ	phase angle of line impedance

Parameters

c^{cap}	unit capacity cost of STATCOM
$c_i^{install}$	installation cost of STATCOM at bus i
CP	percentage of constant power loads in CLOD model
DL	percentage of discharging lighting loads in CLOD model
H	number of candidate buses
K_p	parameter for voltage-dependent real power load in CLOD model
LM	percentage of large motor loads in CLOD model
p_k	probability of contingency k
SM	percentage of small motor loads in CLOD model
TS	percentage of transformer saturation loads in CLOD model
ω	rotor deviation in extreme conditions

References

1 Cutsem, T.V. and Vournas, C. (2007). *Voltage Stability of Electric Power Systems*. Springer Science & Business Media.

2 Gerbex, S., Cherkaoui, R., and Germond, A.J. (2001). Optimal location of multi-type FACTS devices in a power system by means of genetic algorithms. *IEEE Transactions on Power Systems* 16 (3): 537–544.

3 Yorino, N., El-Araby, E.E., Sasaki, H., and Harada, S. (2003). A new formulation for FACTS allocation for security enhancement against voltage collapse. *IEEE Transactions on Power Systems* 18 (1): 3–10.

4 Sode-Yome, A., Mithulananthan, N., and Lee, K.Y. (2006). A maximizing loading margin method for static voltage stability in power systema. *IEEE Transactions on Power Systems* 21 (2): 799–808.

5 Chang, Y.C. (2012). Multi-objective optimal SVC installation for power system loading margin improvement. *IEEE Transactions on Power Systems* 27 (2): 984–992.

6 Ghahremani, E. and Kamwa, I. (2013). Optimal placement of multiple-type FACTS devices to maximize pow system loadability using a generic graphical user interface. *IEEE Transactions on Power Systems* 28 (2): 764–778.

7 Milanovic, J.V. and Zhang, Y. (2010). Modelling of FACTS devices for voltage sag mitigation studies in large power systems. *IEEE Transactions on Power Delivery* 25 (4): 3044–3052.

8 Pourbeik, P., Koessler, R.J., Quaintance, W., and Wong, W. (2006). Performing comprehensive voltage stability studies for the determination of optimal location, size and type of reactive compensation. *2006 IEEE Power Engineering Society General Meeting*, Atlanta, GA, USA (29 October–01 November 2006). p. 6: IEEE.

9 Krishnan, V., Liu, H., and McCalley, J.D. (2009). Coordinated reactive power planning against power system voltage instability. *2009 IEEE/PES Power Systems Conference and Exposition (PSCE '09)*, Seattle, WA, USA (15–18 March 2009): IEEE.

10 Sapkota, B. and Vittal, V. (2009). Dynamic VAr planning in a large power system using trajectory sensitivities. *IEEE Transactions on Power Systems* 25 (1): 461–469.

11 Tiwari, A. and Ajjarapu, V. (2010). Optimal allocation of dynamic VAR support using mixed integer dynamic optimization. *IEEE Transactions on Power Systems* 26 (1): 305–314.

12 Paramasivam, M., Salloum, A., Ajjarapu, V. et al. (2013). Dynamic optimization based reactive power planning to mitigate slow voltage recovery and short term voltage instability. *IEEE Transactions on Power Systems* 28 (4): 3865–3873.

13 Gan, D., Thomas, R.J., and Zimmerman, R.D. (2000). Stability-constrained optimal power flow. *IEEE Transactions on Power Systems* 15 (2): 535–540.

14 Miettinen, K. (2012). *Nonlinear Multiobjective Optimization*. Springer Science & Business Media.

15 Coello, C.C. (2006). Evolutionary multi-objective optimization: a historical view of the field. *IEEE Computational Intelligence Magazine* 1 (1): 28–36.

16 Zhang, Q. and Li, H. (2007). MOEA/D: a multiobjective evolutionary algorithm based on decomposition. *IEEE Transactions on Evolutionary Computation* 11 (6): 712–731.

17 Siemens Power Technologies International (2013). *PSS®E 33.4 Program Application Guide*, vol. 2. New York: Siemens.

18 Meng, K., Dong, Z.Y., Wong, K.P. et al. (2010). Speed-up the computing efficiency of power system simulator for engineering-based power system transient stability simulations. *IET Generation Transmission and Distribution* 4 (5): 652–661.

19 Agrawal, S., Panigrahi, B.K., and Tiwari, M.K. (2008). Multiobjective particle swarm algorithm with fuzzy clustering for electrical power dispatch. *IEEE Transactions on Evolutionary Computation* 12 (5): 529–541.

20 Zhou, B., Chan, K.W., Yu, T., and Chung, C.Y. (2013). Equilibrium-inspired multiple group search optimization with synergistic learning for multiobjective electric power dispatch. *IEEE Transactions on Power Systems* 28 (4): 3534–3545.

18

Retirement-Driven Dynamic VAR Resource Planning

18.1 Introduction

Equipment aging is a critical issue in the planning and operation of power systems. Existing methods to quantify the uncertainty of failures are accurate enough for the estimation of potential losses [1]. The approximation of the retirement date requires an enormous amount of historical data [2]. Among the common methods, life cycle cost (LCC) is a prevalent approach to transforming a practical equipment aging problem into a quantitative economic assessment [3]. It is usually utilized as a tool to support decision-making for industrial investment and is also utilized for retirement decisions of equipment. For instance, in Victorian, Australia, although some overhauling and upgrade arrangements were scheduled for some aged VAR devices in five years [4], these plans had been deferred considering the absence of an appropriate planning strategy and the high cost [5].

For the electric power industry, equipment retirement plays a critical role in facility management. The reasons for retirement can vary from ages to irreparable damages to concerns about the carbon footprint. For instance, AEMO has arranged a series of thermal generator retirement plans [6] to improve the low-carbon energy structure in the future. However, there is still a lack of a systematic model for equipment retirement in VAR planning.

In this chapter, to overcome the inadequacies in the existing works, a practical economic planning model with both the installation and the retirement is elaborated. This model has five salient features: (i) LCC-based retirement and installation of dynamic VAR resources from a financial perspective, (ii) optimal retirement timing to balance stability requirements and capital flow, (iii) steady-state and short-term stability metrics to evaluate performance of wind-penetrated power system against voltage instability, (iv) LVRT and HVRT capabilities to maintain a secure

Stability-Constrained Optimization for Modern Power System Operation and Planning,
First Edition. Yan Xu, Yuan Chi, and Heling Yuan.
© 2023 The Institute of Electrical and Electronics Engineers, Inc.
Published 2023 by John Wiley & Sons, Inc.

operation condition and enhance their adaptability to power system transient disturbances, and (v) incorporation of dynamic load models represented by a representative set of scenarios.

18.2 Equipment Retirement Model

As the installation and the retirement are both evaluated from a financial perspective, it is reasonable to describe them through their combination as (18.1):

$$TC = IC + LCC \tag{18.1}$$

where TC is the total cost and IC is the installation cost of the new device, which will be extended in detail later in the next chapter, and LCC is the retirement equipment evaluation.

In this chapter, LCC assessment is adopted to optimize the retirement timing of aged equipment. LCC is expressed as follows:

$$LCC = CI + CO + CM + CF + CD \tag{18.2}$$

Since the proposed model is for a long-term planning problem, net present value (NPV) is utilized as (18.3).

$$NPV(C, r) = C(1 + d)^{-r} \tag{18.3}$$

For a long-term planning problem, the horizon of the planning is usually divided into several stages for the cash flow and the decision-making. The planning horizon is described as follows:

$$n = SC \times l \tag{18.4}$$

The average annual installation cost will decrease as time moves forward due to inflation and performance deterioration. On the contrary, the annual maintenance cost and the failure cost are expected to increase because of the aging of the equipment.

Five different cost terms related to equipment retirement are considered in this chapter as follows.

18.2.1 Investment Cost

Investment cost is a one-off purchase and installation of all equipment. The amount of CI is huge, and in the long-term planning, the influence of inflation is essential.

$$CI^{n+h} = NPV\left(Cost^{Inv}, n + h\right) \tag{18.5}$$

18.2.2 Operation Cost

Operation cost is a sum of money spent during operation, including salaries of agents, resource purchase fees, and environment tax [3]. In a power system, line loss cost is often used to represent operation cost. The previous research work uses the cost of production or electricity price to transfer energy into money. However, the line loss cost has a far deeper meaning than that: on the one side, it stands for the economic loss of industries due to energy loss; on the other side, it should be considered from the consumers' perspective. In this case, value of customer reliability [7] is used to evaluate the financial loss. VCR is the value that electricity consumers place on avoiding service interruptions; on the other hand, it could be a key valuation component valuing the benefit of expected reduction of profit from consumers' perspective. VCR is how much profit consumers will get from the unity electricity, which can efficiently represent the economic loss of consumers because of unserved energy.

$$CO = \sum_{i=1}^{n} LL_i \times 8760 \times VCR_i \tag{18.6}$$

18.2.3 Maintenance Cost

Maintenance cost is an annual expenditure for maintaining performance. Maintenance includes component replacement, annual preventive enhancement, and corrective maintenance. It has to be applied for the whole life of the equipment to maintain a healthy operating condition and extend the mean life [3]. Because of long-term continuously serving, the performance of the equipment will get worse, and it is also harder and harder to restore it to its best performance. So, the maintenance cost will increase annually. In this case, there is an increasing model to estimate the real-world situation.

$$CM = \sum_{i=1}^{n} (1+\delta)^{h+i-1} \times M \tag{18.7}$$

18.2.4 Failure Cost

Failure cost stands for those costs associated with instability. Some industries that are sensitive to the quality of power supplies would suffer an economic loss if blackouts happened. If electricity quality cannot be guaranteed, they will lose their trust in their providers. In previous research [3, 8–10], the failure cost estimation is not well-developed enough because they cannot explain why it does not increase linearly. When dividing it into fault probabilities and economic loss per fault, it is revealed that failure cost should increase quadratically. The probabilities increase annually because of aging issues, and consequences also become more severe because load capacity, electricity price, and system topology change as time goes by.

$$CF = \sum_{i=1}^{n} \left((1 + \xi)^{h+i-1} \times F \right) \times \left((1 + \lambda)^{h+i-1} \times p \right) \tag{18.8}$$

18.2.5 Disposal Cost

Disposal cost is the expenditure to deal with the retired devices. The major components of it are: (i) manpower and other resources spent in uninstallation (ii) income from recycling:

$$CD = DC - RB \tag{18.9}$$

18.3 Retirement-Driven Dynamic VAR Planning Model

The retirement-driven VAR planning described in this chapter consists of three objectives, including upgrade cost with retirement, steady-state voltage stability index, and short-term voltage stability index. Steady state and dynamic constraints are used to maintain a stable operation of power systems. Apart from these constraints, LVRT and HVRT also serve as major selection criteria for the planning solutions, by which the candidate solution, in any scenario and any operation stage, will be abandoned immediately if violated.

18.3.1 Optimization Objectives

$$\min_{u} F[f_1(x,u), f_2(x,u), f_3(x,u)] \tag{18.10}$$

The control variables include STATCOM location, STATCOM size, and retirement time of aged capacitor bank.

The first objective f_1 is for the economic cost, consisting of LCC cost of capacitors and investment cost of STATCOM. Since the goal is to minimize the overall cost, they are combined as one objective.

$$f_1 = TC_1 = IC_1 + LCC_1 \tag{18.11}$$

$$IC_1 = \sum_{i \in \mathscr{B}} NPV \left(I_i \times c^{install}, J_i \times l \right) + \sum_{i \in \mathscr{B}} NPV(I_i \times B_i \times c^{cap}, J_i \times l) \tag{18.12}$$

$$LCC_1 = \sum_{r \in \mathcal{R}} LCC(h + S_r \times l) \tag{18.13}$$

It should be noted that the decision variable of STATCOM has an initial value of zero, while the decision variable of retirement has an initial value of the maximum stage.

$$I_i = \begin{cases} 1, \text{if } J_i > 0 \\ 0, \text{otherwise} \end{cases} \tag{18.14}$$

The second and third objective function, f_2 and f_3, respectively, stand for the risk level for short-term voltage stability and steady-state stability. The risk-based criteria, as introduced in Chapters 16 and 17, are used in these objective functions to quantitatively evaluate the contingency impact on stability. The definitions and mathematical details of TVSI and VCPI can be found in Chapter 14.

$$f_2 = \sum_{k \in \mathcal{K}} TVSI_k \times p_k \tag{18.15}$$

$$f_3 = \sum_{k \in \mathcal{K}} VCPI_k \times p_k \tag{18.16}$$

18.3.2 Constraints

The steady state includes power flow balance and steady-state operational limits. The constraint for the active and reactive power equilibrium of each bus is written as (18.17). Operational limits are represented by constraint (18.18), including voltage magnitude, line capacity, and generation output capacity.

$$\begin{cases} P_g + \tilde{P}_w - P_d - V_i \sum_{j \in \mathcal{B}} V_j \left(G_{ij} \cos \theta_{ij} + B_{ij} \sin \theta_{ij} \right) = 0 \\ Q_g + \tilde{Q}_w + Q_s^{stat} - Q_d - V_i \sum_{j \in \mathcal{B}} V_j \left(G_{ij} \sin \theta_{ij} - B_{ij} \cos \theta_{ij} \right) = 0 \end{cases} \tag{18.17}$$

$$\begin{cases} V_i^{min} \leq V_i \leq V_i^{max} \\ L_l^{min} \leq L_l \leq L_l^{max} \\ P_g^{min} \leq P_g \leq P_g^{max} \\ Q_g^{min} \leq Q_g \leq Q_g^{max} \end{cases} \begin{array}{l} \forall i \in \mathcal{B} \\ \forall l \in \mathcal{L} \\ \forall g \in \mathcal{G} \end{array} \tag{18.18}$$

Rotor angle is utilized as a transient constraint as in (18.19). For any two generators (g and g') within a time period of T during a contingency k, the maximum allowed rotor angle difference between them is ω. Generally, for an extreme condition, the parameter ω can be set to 180° [11].

$$\left[\max\left(\Delta\,\delta^{T}_{gg'}\right)\right]_{k} \leq \omega \quad \forall g,g' \in \mathscr{G}_{s}, \forall k \in \mathscr{K} \tag{18.19}$$

The LVRT and HVRT profiles are converted into a set of voltage magnitude values. They are used to compare with the obtained post-contingency trajectories. Any candidate solution with violations of LVRT or HVRT as expressed in (18.20) will be abandoned.

$$\begin{cases} V_{i}^{post}(t) - LVRT(t) \geq 0, & \forall t \in T, \forall i \in B \\ V_{i}^{post}(t) - HVRT(t) \leq 0, & \forall t \in T, \forall i \in B \end{cases} \tag{18.20}$$

The LVRT and HVRT criteria are designed to be binary and strict, since the wind farm that fails to satisfy LVRT and HVRT criteria will definitely be cut off from the grid according to the industry requirements and the stringent grid codes [12].

18.4 Solution Method

As one of the most widely-used multi-objective evolutionary algorithms, NSGA-II [13] is utilized in this chapter to solve the proposed VAR planning model. The mathematical details can be found in [13] and will also be elaborated on in Chapter 19. NSGA-II consists of a repaid non-dominated searching scheme and a simple but effective bionics method, in which every solution is called a gene and works under natural selection and evolution.

In the proposed model, there are two sectors of decision variables. The first sector includes three planning stages with an installation decision variable and a capacity decision variable for each stage. The second and third stages refer to upgrades. For instance, if a certain STATCOM has been constructed in the first stage, a positive result of the second or third stage represents a capacity upgrade decision. Otherwise, it will be a construction decision at the corresponding stage. For the second sector, it is a one-off decision for the retirement of capacitor banks. The values of the decisions indicate the stage of retirement (from one to four). The initial value is the maximum stage of planning, and an unchanged value indicates corresponding capacitor bank has not been retired at the end of the planning year.

18.4.1 Computation Steps

The computation steps of the proposed method are elaborated as follows:

1) **Initialization**: After the initialization of algorithms and the inputs (network data, loads, and generations), candidate buses can be selected based on methods introduced in Chapter 16 or using conventional sensitivity analysis.
2) **Main procedure**: Loop the main optimization procedure.

3) **Voltage stability analysis**: Evaluate every candidate planning decision, in terms of steady-state voltage stability (VCPI) and short-term voltage stability (TVSI), under various load scenarios.

4) **Candidate evaluation**: Return results of voltage stability indices (VCPI and TVSI) evaluation for each contingency under each dynamic load condition to the main procedure. The averaged stability results for different scenarios and contingencies will be integrated with the total cost as objectives.

5) **Break criterion**: If LVRT and HVRT constraints are violated at any time, the assessment of the corresponding planning decision will be terminated, and the next planning decision will be evaluated.

6) **Termination**: If any stopping criterion is satisfied, the main procedure will stop, and the final optimization result will be presented as a Pareto front. Two stopping criteria are adopted in this chapter: (i) change in fitness function is smaller than a specific tolerance, and (ii) the maximum iteration number is reached.

18.5 Case Studies

18.5.1 Test System and Parameters

The effectiveness and efficiency of the proposed method are validated on the New England 39-bus test system with several modifications, as illustrated in Figure 18.1. Compared with the original test system, the synchronous generator installed at bus 30 is replaced by a wind farm. All the wind turbines are modeled as double-fed induction generators in PSS®E [15]. The generator model, the mechanical model, the electrical model, and the pitch control model are the same ones used in previous chapters. The STATCOM model is SVSMO3, as illustrated in Figure 15.4. An industry-standard complex load model "CLOD" [15] is also used to represent the time-varying load dynamics. Major parameters for case studies are listed in Table 18.1. In a real-world application, the parameters for the CLOD model are determined by TOAT method [16]. Among the parameters in Table 18.1, the parameters of LCC are the most difficult to be determined. Some are based on previous studies and common sense, while others are based on assumptions and engineering practices [17].

18.5.2 Numerical Simulation Results

Figure 18.2 illustrates the planning results as a 3-dimensional Pareto front, which is a series of trade-off cost-benefit choices of solutions. For the decision-makers, it

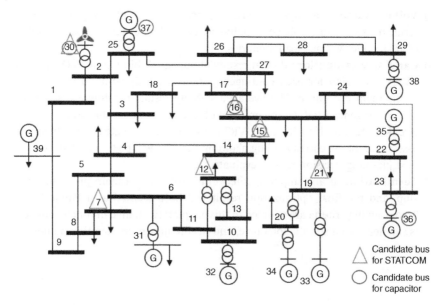

Figure 18.1 One-line diagram of the New England 39-bus system. *Source:* Liu et al. [14]/ With permission of John Wiley & Sons.

Table 18.1 Parameters for simulation setting.

Parameters	Value settings
Lifetime of capacitor banks	45 years
Maintenance cost	5% of purchase cost per year
Maintenance cost increase rate	3% per year
Residual benefit of capacitor banks	5% of purchase cost per year
Disposal cost	US$ 5000
Failure rate	8%
Increase rate of failure rate	5% per year
Penalty of failure	US$ 100 000 every time
Discount rate	0.05
Planning length	15 years
Planning stage	3
VCR	US$ 2 000/MWh
Installation cost of STATCOM	1.5 million
Capacity cost of STATCOM	US$ 50 000/MVAR

Source: Liu et al. [14]/With permission of John Wiley & Sons.

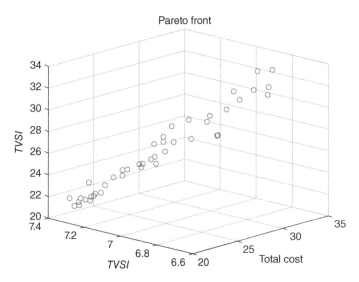

Figure 18.2 Illustration of Pareto front. *Source:* Liu et al. [14]/With permission of IEEE.

is practically valuable to have trade-off candidate planning results instead of a single decisive one since they can provide a margin for the industrial profit.

Based on the results from Figure 18.2, it can be concluded that planning with high stability comes with the high cost of installation and aged equipment overhauls. To avoid unfixable failures, it is necessary to retire the aged devices. Putting off the retirement plans may be favored, in terms of the short-term profit, but will lead to an overall loss of benefit. Meanwhile, it is also flexible for the decision-makers to select candidate plans based on engineering practices and specific industrial requirements.

In this chapter, the lowest acceptable stability evaluation value is determined, and the candidate solution with the lowest overall investment is selected. As a result, the 26th solution in the Pareto front is selected, and the details of this solution are presented in Tables 18.2–18.4.

The details of a comparison between the proposed method and the state-of-the-art in the literature are listed in Table 18.5. None of the existing methods considers equipment retirement, and very few of them consider the long-term multi-stage planning, dynamic load scenarios, and wind energy simultaneously. However, these considerations are critical in the engineering practice, without which the planning decisions might be impractical or even require too much re-adjustment to satisfy the industrial requirements before application. Compared with other methods, the proposed method is more comprehensive. As for the computation burden, the proposed method requires much less time than conventional methods based on Monte Carlo simulation. Due to the implementation of NSGA-II, the overall

Table 18.2 Installation and upgrade scheme.

Bus	Stage 1 (MVAR)	Stage 2 (MVAR)	Stage 3 (MVAR)
30	79	33	37
7	55	42	38
12	66	0	0
15	0	0	25
16	46	33	43
21	73	34	40
20	58	0	38

Source: Liu et al. [14]/With permission of IEEE.

Table 18.3 Capacitor banks retirement scheme.

Bus	30	36	37	15	16
Stage	N/A	N/A	3	2	2

Source: Liu et al. [14]/With permission of IEEE.

Table 18.4 Objective values.

Objective	Total cost	VCPI	TVSI
Value	27.31 million	6.972	0.6562

Source: Liu et al. [14]/With permission of IEEE.

Table 18.5 The comparison of the proposed method and previous research.

Bus	Proposed method	Ref. [5]	Ref. [18]	Ref. [19]	Ref. [20]
Wind power penetration	√	X	√	X	X
Dynamic load modeling	√	X	√	X	X
Multi-stage horizon planning	√	X	√	X	X
Equipment retirement	√	X	X	X	X
Static voltage stability	√	X	√	√	X
Short-term voltage stability	√	√	X	√	√
Computation burden	Medium	Low	Low	High	High

computational burden is heavier than classic programming algorithms. It should be noted that, for a long-term planning problem, the overall computation time of the proposed method is still acceptable.

Nomenclature

Indices

i,j	indices for buses
k	index for contingencies
m	index for the objective
r	indices for years
t	index for the time

Sets

$\mathcal{B}$	set of all buses
$\mathcal{G}$	set of all the generators
$\mathcal{G}_s$	set of synchronous generators
$\mathcal{G}_w$	set of wind turbines
$\mathcal{L}$	set of all the lines
$\mathcal{R}$	set of years
$\mathcal{K}$	set of contingencies
$\mathcal{T}$	set of time
$\boldsymbol{u}$	set of control variables
$\boldsymbol{x}$	set of state variables

Variables

B_i	integer capacity decision variables for bus i
B_{ij}	susceptance between i and j
C	cost influenced by inflation
CD	disposal cost (including residual benefit and disposal fee)
CF	failure cost
CI	investment cost of aged devices
CM	maintenance cost
CO	operation cost
$Cost^{Inv}$	cost of capacitor banks
DC	disposal cost
d	discount rate

G_{ij}	conductance between i and j
h	age of the device
$HVRT(t)$	time-varying Constraints for HVRT
I_i	binary installation decision variables for bus i
IC	installation cost of STATCOM
J_i	stage decision variable of STATCOM
LL_i	line loss in year i
$LVRT(t)$	time-varying Constraints for LVRT
l	time length per stage
n	total planning period
P_d	active power of load
P_g	active output of generator g
$\widetilde{P_w}$	uncertain active power output of wind turbine w
Q_d	reactive power of load at bus d
Q_g	reactive output of generator g
$\widetilde{Q_w}$	uncertain reactive power output of wind turbine w
Q_s^{stat}	output of STATCOM
RB	residual benefit
r	total time
S_r	decision variable of retirement
SC	selected stage when a capacitor is retired
TC	total cost
VCR_i	value of Costumer Reliability in year i
V_i, V_j	voltage magnitude of bus i (or bus j)
$V_i^{post}(t)$	post-contingency voltage at bus i
θ	phase angle of line impedance
$\Delta\delta_{gg'}^T$	rotor angle deviation between g and g'

Parameters

c^{cap}	unit capacity cost of STATCOM
$c^{install}$	installation cost of STATCOM
F	penalty factor
H	number of candidate buses
M	maintenance rate
p_k	probability of contingency k
δ	increase rate of maintenance
ω	rotor deviation in extreme conditions
ξ	increase rate of load
λ	increase rate of p because of aging

References

1 Cutsem, T.V. and Vournas, C. (2007). *Voltage Stability of Electric Power Systems.* Springer Science & Business Media.

2 Kundur, P., Paserba, J., Ajjarapu, V. et al. (2004). Definition and classification of power system stability IEEE/CIGRE joint task force on stability terms and definitions. *IEEE Transactions on Power Systems* 19 (3): 1387–1401.

3 Baran, M. and Wu, F.F. (1989). Optimal sizing of capacitors placed on a radial distribution system. *IEEE Transactions on Power Delivery* 4 (1): 735–743.

4 Senjyu, T., Miyazato, Y., Yona, A. et al. (2008). Optimal distribution voltage control and coordination with distributed generation. *IEEE Transactions on Power Delivery* 23 (2): 1236–1242.

5 Huang, W., Sun, K., Qi, J., and Ning, J. (2017). Optimal allocation of dynamic var sources using the Voronoi diagram method integrating linear programing. *IEEE Transactions on Power Systems* 32 (6): 4644–4655.

6 Operator, A.E.M. (2011). National transmission network development plan. Australian Energy Market Operator.

7 Operator, A.E.M. (2013). *Value of Customer Reliability Issues Paper.* Australian Energy Market Operator.

8 Li, W. (2002). Incorporating aging failures in power system reliability evaluation. *IEEE Transactions on Power Systems* 17 (3): 918–923.

9 Morea, F., Viciguerra, G., Cucchi, D., and Valencia, C. (2007). Life cycle cost evaluation of off-grid PV-wind hybrid power systems. *INTELEC 07-29th International Telecommunications Energy Conference*, Rome, Italy (30 September–04 October 2007), pp. 439–441: IEEE.

10 Nilsson, J. and Bertling, L. (2007). Maintenance management of wind power systems using condition monitoring systems – life cycle cost analysis for two case studies. *IEEE Transactions on Energy Conversion* 22 (1): 223–229.

11 Hatziargyriou, N., Milanović, J., Rahmann, C. et al. (2020). Stability Definitions and Characterization of Dynamic Behavior in Systems with High Penetration of Power Electronic Interfaced Technologies. *IEEE PES Technical Report PES-TR77*.

12 Australian Energy Market Commission (AEMC) (2018). *National Electricity Rules Version 110*. National Electricity.

13 Deb, K., Pratap, A., Agarwal, S., and Meyarivan, T.A. (2002). A fast and elitist multiobjective genetic algorithm: NSGA-II. *IEEE Transactions on Evolutionary Computation* 6 (2): 182–197.

14 Liu, J.W., Xu, Y., Dong, Z.Y., and Wong, K.P. (2018). Retirement-driven dynamic VAR planning for voltage stability enhancement of power systems with high-level wind power. *IEEE Transactions on Power Systems* 33 (2): 2282–2291.

15 Siemens Power Technologies International (2013). *PSS®E 33.4 Program Application Guide, vol. 2. New York: Siemens.*

16 Stuart, P.G. (1993). *Taguchi Methods: A Hand on Approach*. Reading, MA: Addison-Wesley.

17 Australian Energy Market Operator Limited (2011). Economic planning criteria in Queensland.

18 Fang, X., Li, F., Wei, Y. et al. (2015). Reactive power planning under high penetration of wind energy using Benders decomposition. *IET Generation Transmission and Distribution* 9 (14): 1835–1844.

19 Xu, Y., Dong, Z.Y., Xiao, C. et al. (2015). Optimal placement of static compensators for multi-objective voltage stability enhancement of power systems. *IET Generation Transmission and Distribution* 9 (15): 2144–2151.

20 Xu, Y., Dong, Z.Y., Meng, K. et al. (2014). Multi-objective dynamic VAR planning against short-term voltage instability using a decomposition-based evolutionary algorithm. *IEEE Transactions on Power Systems* 29 (6): 2813–2822.

19

Multi-stage Coordinated Dynamic VAR Resource Planning

19.1 Introduction

In 2016, South Australia experienced a state-wide electricity outage [1]. The blackout is triggered by the tripping of several wind farms, and the insufficient reactive power reserve is one of the major causes [2]. Considering the prosperous renewable energy industry and the inherent uncertainty of extreme weather conditions, the shortage of fast-response reactive power sources and the corresponding consequences will present themselves on a much larger scale in the future. Firstly, the post-contingency performance of wind turbines plays a vital role in the overall stability of the power system. South Australia's 2016 blackout [1] is a vivid example that VAR compensation devices are necessary but not sufficient to help the power system survive extreme weather conditions and a large voltage disturbance. Reduction in the wind power generation or a complete cut-off will not only break the generation–demand balance but also hold the potential of triggering cascading events. When wind power is considered in a VAR source placement problem, the focus is usually on the LVRT/HVRT capability of the wind turbines [3] or the general voltage stability of a wind-penetrated power system [4]. Although the adjustment of wind power generation is important (since 40% of large blackout events are caused by bad weather [5]), the curtailment of wind power, which is an effective preventive control for better stability performance before a coming emergency, is usually ignored. Furthermore, the inherent stochasticity of wind power is not addressed in these studies.

Furthermore, other effective mitigations, including but not limited to reactive current compensation from synchronous generators [6], on-load tap changers [7], and load shedding, are rarely incorporated in existing studies related to VAR planning with considerations of short-term voltage stability. As a dispatchable unit itself, STATCOM can also contribute to the system's steady-state voltage

Stability-Constrained Optimization for Modern Power System Operation and Planning,
First Edition. Yan Xu, Yuan Chi, and Heling Yuan.
© 2023 The Institute of Electrical and Electronics Engineers, Inc.
Published 2023 by John Wiley & Sons, Inc.

performance by participating in the generation dispatch. Since the cost of these operational mitigations is much less than the investment of dynamic VAR sources, the planning decisions from the models that do not consider the operational mitigations are conservative. Furthermore, the development of advanced smart devices enables the operators to schedule the demand more flexibly, and the impact of DR [8] on power system planning should also be considered. For instance, DR is usually implemented in generation planning problems [9] and transmission planning problems [10], respectively, to cut down the peak load.

Therefore, for a power system with a large portion of induction motor load and high penetration level of wind power, STATCOM alone is not sufficient for the power system to survive significant transient voltage disturbance. Prevalent single-stage multi-objective model [3, 11–13] for the VAR planning problem might generate arbitrary and conservative planning decisions due to the lack of coordination between the planning of the compensation devices and operational mitigation measures (in both pre-contingency stage and post-contingency stage).

To overcome the inadequacies of the previous works, this chapter proposes a multi-stage coordinated dynamic VAR resource placement method for both static and dynamic voltage stability improvement of the wind-penetrated power system.

19.2 Coordinated Planning and Operation Model

19.2.1 Multi-stage Coordination Framework for Voltage Stability Enhancement

Conventionally, VAR planning is considered an independent process. However, since different voltage stability mitigations have different features in terms of deployment time and cost, as illustrated in Figure 19.1, the conventional single-stage approach, which neglects the operational control measures, usually generates conservative planning decisions with a relatively high investment cost.

In this chapter, different measures are employed in a multi-stage planning framework, as illustrated in Figure 19.2, aiming to coordinate the planning and operation stages to enhance voltage stability optimally, particularly after a large disturbance. Specifically, it includes the planning stage and operation stage. For the operation stage, it has been divided into two substages: pre-contingency stage and post contingency stage. The multi-stage VAR planning framework, as illustrated in Figure 19.2, is further explained in the following sub-chapters.

19.2.1.1 Planning Stage

In the planning stage, STATCOMs are deployed as a network hardening measure to improve both the static and short-term voltage stability of the wind-penetrated

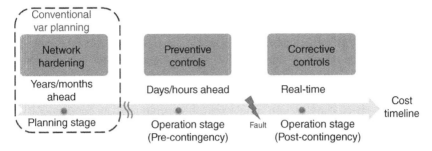

Figure 19.1 Different voltage stability mitigations comparison. *Source:* Chi and Xu [14]/With permission of John Wiley & Sons.

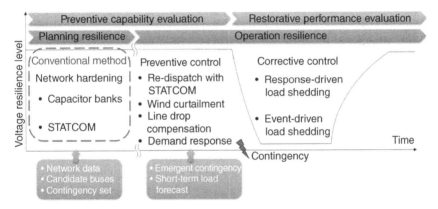

Figure 19.2 Multi-stage coordinated VAR planning framework. *Source:* Chi et al. [15]/With permission of IEEE.

power system, which is similar to the conventional approach. Generally, given the heavy computational burden, candidate buses for VAR installations should be determined as inputs before capacity optimization to reduce the complexity of the problem. The critical contingencies should be the most influential ones on the voltage stability.

In this stage, the inputs consist of: (i) network data (including network structure, generation information, and load information), (ii) candidate buses for the installation of VAR sources, (iii) selected critical contingencies, (iv) specific grid codes for wind power generation (if any), and (v) forecasted peak load. It should be noted that, apart from STATCOM and other FACTS devices, conventional VAR compensation devices such as capacitor banks can also be considered in this stage since they are relatively cost-effective in improving the steady-state voltage stability of the power system.

19.2.1.2 Operation Stage

For a power system with a large portion of induction motor load and high penetration level of electronic-interfaced renewable generation, STATCOM alone is insufficient when there is a significant voltage disturbance. Thus, it is necessary and economically favored to incorporate operational control measures to further alleviate the detrimental impacts of a large voltage disturbance.

In the proposed coordination framework, the operation stage consists of two substages: pre-contingency stage and post-contingency stage. In the pre-contingency stage, there are three preventive controls implemented to improve the capability of the system to withstand critical contingencies: (i) generation re-dispatch with VAR compensators, (ii) reactive power compensation (by line drop compensation), and (iii) wind power curtailment. Firstly, the generation re-dispatch will optimize the power-sharing among the generators with the participation of STATCOMs. Then, line drop compensation will enhance the dynamic voltage response of the conventional generator in a contingency. Finally, with credible information about the weather condition (as an emergency contingency), wind power can be curtailed to improve the system's capability to withstand voltage disturbance.

In the post-contingency stage, event-driven load shedding (EDLS, can also be referred as under voltage load shedding, UVLS) and response-driven load shedding (RDLS) are utilized as corrective control (CC) to help the power system restore back to a stable state. An appropriate and timely EDLS can effectively alleviate the detrimental impacts of the contingency and accelerate the voltage recovery process. Specifically, under voltage load shedding is triggered as an RDLS once the voltage drops below the lower threshold. On the other hand, since the RDLS is relatively slow, EDLS, which is activated instantly when a specific contingency occurs, is employed for a critical contingency with a severe detrimental impact on voltage stability. The application of EDLS can compensate for the slow response of RDLS and improve the capability of the system to stop the potential propagation of instability or cascading events.

Economically speaking, the instantaneous costs for PCs and CCs (US$ 8/MWh for load shedding) are neglectable compared with the investment of VAR source deployment (US$ 1.5 million/unit for site installation and US$ 0.05 million/MVAR for unit capacity), but their aggregate cost over a long time is still considerable and cannot be ignored. For instance, the load shedding cost over 10 years is comparable to the investment of a single VAR source deployment. On the other hand, investment of STATCOM is much higher [16], but it can provide immediate and direct reactive power support before and after a voltage disturbance, improving both the steady-state and short-term voltage stability of the wind-penetrated power system. However, South Australia's 2016 blackout is a vivid example that dynamic VAR resources (network enhancement) are necessary but insufficient. In practice, operational control measures will be used in certain scenarios (for instance, if there is a forecast of extreme weather conditions, appropriate wind power curtailment is reasonable), regardless of whether they are considered in the planning stage. Therefore, coordination of the very effective but expensive network enhancement measure and the less effective but more flexible and economical operational control measures will not only improve the voltage stability of the power system but also keep the total cost at an acceptable level and avoid unnecessary conservative investment.

Different planning and operation measures have their advantages and disadvantages. These measures are compared in the five aspects (in degrees from one to five, the larger area indicates better performance in a specific aspect), as illustrated in Figure 19.3: investment cost, deployment time, response time, steady-state, and short-term voltage stability improvement. Different characteristics of these measures bring challenges to the decision-makers: a balance between the performance and the total capital investment. For instance, although STATCOM outperforms others in improving short-term voltage stability, its relatively long deployment time and high investment cost are not preferable. It will be very difficult to comprehensively evaluate the effectiveness of STATCOM installations, PCs, and CCs, due to their different characteristics. On the other hand, conventional metrics, such as TVSI and VCPI introduced in Chapter 16, are usually designed for a generic voltage stability assessment and are no longer accurate enough if these different planning and operation measures are all implemented.

19.2.2 Voltage Resilience Indices

The different characteristics of these mitigations bring a new challenge to the decision-makers: how to quantitatively evaluate the effectiveness of different mitigations and the overall performance of the wind-penetrated power system. So, in this chapter, a set of new voltage resilience indices is proposed in this chapter to assess the situation more accurately. Generally, considering the different priorities in the planning and operation stages, the voltage resilience level is assessed from two perspectives: the preventive capability before a contingency and restorative performance after a contingency. All the indices are grouped based on the two perspectives and designed to be the smaller the better unless otherwise specified.

19.2.2.1 Reactive Power Margin of DFIG
The maximum reactive power output of DFIG is usually modeled as a fixed value subject to the dynamic reactive power limits of the stator and the inverters of DFIG, as shown in Figure 19.4.

For the stator, its reactive power regulation potential is mainly limited by the rotor current and the active power output [17], as illustrated in Figure 19.5.

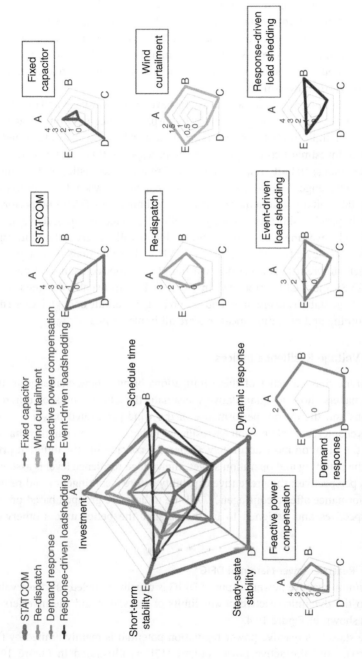

Figure 19.3 Comparison of characteristics of different measures. *Source*: Adapted from Chi et al. [15].

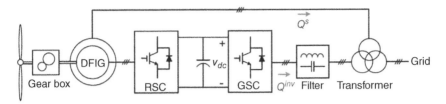

Figure 19.4 Illustration of a general DFIG.

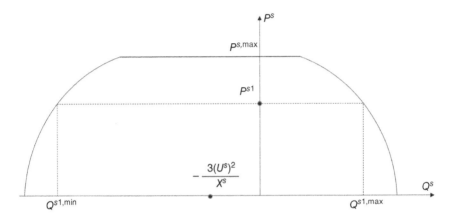

Figure 19.5 Active and reactive power limits of stator (DFIG).

Q_s^{max} is changing according to the active power output. For instance, the real-time reactive power capability of the stator, $Q^{s1,max}$, will decrease as the actual active power output (P^{s1}) approaches the maximum active power ($P^{s,max}$). The maximum and minimum stator reactive powers can be calculated as in (19.1) and (19.2), respectively.

$$Q^{s,max} = \frac{-3(U^s)^2}{X^s} + \sqrt{\left(3U^s \cdot X^m \cdot \frac{I^{r,max}}{X^s}\right)^2 - (P^s)^2} \tag{19.1}$$

$$Q^{s,min} = -\frac{3(U^s)^2}{X^s} + \sqrt{\left(3U^s \cdot X^m \cdot \frac{I^{r,max}}{X^s}\right)^2 - (P^s)^2} \tag{19.2}$$

As for the inverter, $Q^{inv,max}$ is mainly limited by the inverter capacity. It can be calculated as (19.3). As shown in Figure 19.5, the maximum reactive power output of DFIG is determined by the stator side and grid side together. So, the maximum output can be calculated as the sum of $Q^{s,max}$ and $Q^{inv,max}$, as shown in (19.4). A larger margin of reactive power is favorable to the wind-penetrated power system when subjected to a voltage disturbance, as well as the wind turbine itself. Accordingly, an index RI^{margin} is proposed in this chapter as (19.5) to evaluate the reactive power margin of wind turbines quantitatively. It is calculated as the ratio of the difference between the maximum possible reactive power output and the reactive power output before the contingency, and the maximum possible reactive power output. A larger margin to the maximum possible reactive power indicates a larger reactive power regulation capability, while a smaller one means the reactive power regulation capability of a wind turbine is close to its maximum upper limit.

$$Q^{inv,\,max} = \sqrt{\left(P^{inv,\,max}\right)^2 - \left(s \cdot P^s\right)^2} \tag{19.3}$$

$$Q^{w,\,max} = Q^{s,\,max} + Q^{inv,\,max} \tag{19.4}$$

$$RI^{margin} = \left(\sum_{w \in \mathscr{I}} \frac{Q_w^{w,\,max}}{Q_w^{w,\,max} - Q_w^w}\right) / N^w \tag{19.5}$$

19.2.2.2 Wind Power Curtailment

Figure 19.6 illustrates the impacts of an increasing wind power penetration level (installed wind power capacity as a percentage of total installed generation) on the voltage response after a contingency.

When there is no wind power generation in the power system, the bus voltage can restore to a stable state quickly after the contingency. As the penetration level increases, the voltage response becomes inferior and the recovery slows down. When the penetration level is increased to 45%, the bus voltage is unable to recover to a stable state after the contingency. So, an appropriate preventive wind

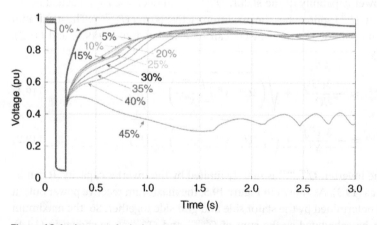

Figure 19.6 Impact of wind power penetration level on the voltage profile. *Source:* Adapted from Chi and Xu [14].

curtailment is effective in enhancing voltage stability during a critical contingency. For instance, it can be implemented hours before the storms or other extreme weather conditions, since the forecast about these extreme weather conditions is usually of high credibility.

To quantitatively evaluate the effectiveness of the proactive wind power curtailment in helping the power system withstand a disturbance, a straightforward wind power curtailment index is proposed as (19.6).

$$RI^{curtail} = \frac{\sum_{w \in \mathscr{G}_w} \left(\tilde{P}_w - P_w^{curt} \right)}{\sum_{w \in \mathscr{G}_w} \tilde{P}_w} \tag{19.6}$$

19.2.2.3 Dynamic VAR Resource Margin

In this chapter, STATCOMs are also participating in the generation dispatch as a reactive power source. So, the actual output of STATCOM is usually different from its rated capacity. Obviously, this capacity margin has its own impact on the voltage performance of the system during the voltage recovery process. RI^{dev}, calculated as the ratio of rated capacity to the capacity margin in (19.7), is proposed to assess this capacity margin. A smaller RI^{dev} indicates a higher potential for reactive power output after a contingency.

$$RI^{dev} = \sum_{i \in \mathscr{B}_s} \sum_{i \in \mathscr{B}} Q_i^{statrated} / \left(Q_i^{statrated} - Q_i^{stat} \right) \tag{19.7}$$

19.2.2.4 Load Shifting (Due to DR Program)

After a critical contingency, an accurate forecast of the peak load is important since the power system is generally in a state of deficiency of active power and reactive power. On the other hand, the impact of load shifting after a demand response is rarely addressed. Since the responded loads will be restored later, there is a chance that the restoration will lead to a new peak load (larger than the original one), as illustrated in Figure 19.7.

$RI^{DRshift}$ is proposed to quantitatively assess the potential peak load shifting in the recovery process after a contingency. It can be calculated as (19.8). A smaller $RI^{DRshift}$ is favorable since it indicates a lower possibility of creating a higher peak load after the responded load is restored.

$$RI^{DRshift} = \sum_{k \in \mathscr{K}} \sum_{m \in \mathscr{A}} DR_{m,k}^{shift} \cdot p_k \tag{19.8}$$

$$DR_{m,k}^{shift} = \begin{cases} \left(DR_k^{res} + P_m^{pred} \right) / P^{peak}, & DR_k^{res} + P_m^{pred} > P^{peak} \\ 0, & \text{otherwise} \end{cases} \tag{19.9}$$

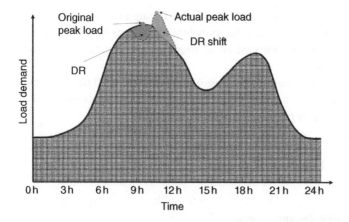

Figure 19.7 Impact of DR shifting on the peak load after a contingency.

$$DR_k^{res} = \sum_{i \in \mathscr{B}_{DR}} P_{i,k}^{DR} \cdot \varepsilon \tag{19.10}$$

19.2.2.5 Steady-State Voltage Stability

A line-based VCPI used in Section 19.1 is also employed here to quantify the steady-state voltage stability of a power system. So, RI^{VCPI} is calculated as a probability-weighted sum of VCPIs of all the lines as (19.11).

$$RI^{VCPI} = \sum_{k \in \mathscr{K}} \sum_{l \in \mathscr{L}} VCPI(power)_l \cdot p_k \tag{19.11}$$

19.2.2.6 Wind Turbine Voltage Violation

As part of the protection mechanism, wind turbines will experience forced cut-off temporarily if there are several moderate low voltage dip incidents, even if these voltage dips are actually above the LVRT threshold, within a short transient period after a voltage disturbance [1]. So, RI^{WV} is proposed in this chapter to assess this cut-off potential when the wind turbines are subjected to minor but more frequent voltage dips. It can be calculated as (19.12). WTVI is a transient voltage deviation index for buses with wind turbines (grid codes of these buses might be different from the grid codes of buses without wind turbines), evaluating the extent of voltage violation after a contingency, based on voltage trajectories from time-domain simulation.

$$RI^{WV} = \sum_{k \in \mathscr{K}} \frac{\sum_{w \in \mathscr{G}_w} \sum_{t \in [T^c, T^{end}]} W \, TVI_{w,t}}{N^w} \cdot p_k \tag{19.12}$$

$$W\,TVI_{w,t} = \begin{cases} \dfrac{|V_{w,t} - V_{w,t0}|}{V_{w,t}}, \dfrac{|V_{w,t} - V_{w,t0}|}{V_{w,t}} \geq \delta_w \\ 0, \text{otherwise} \end{cases}, \forall t \in \left[T^c, T^{end}\right], \forall w \in \mathscr{B}_w$$

(19.13)

19.2.2.7 Short-term Voltage Stability

TVSI is successfully used in Chapter 16 to assess the short-term voltage performance. However, since TVSI adopts a direct averaging method, it cannot fully reveal a slow voltage recovery of a specific bus if the recovery speed of others is relatively fast. To improve this shortcoming, A root-mean-square method [18] is used in this chapter to evaluate the voltage performance in a more accurate way as (19.14), with the consideration of probabilities of different contingencies. TVDI is defined as (19.15). Like TVSI, a smaller RI^{RVSI} is favorable since it indicates less voltage violation and faster voltage recovery.

$$R\,I^{RVSI} = \sum_{k \in \mathscr{K}} \sqrt{\left(\sum_{i \in \mathscr{B}_t} \sum_{t \in \left[T^c, T^{end}\right]} TVDI_{i,t} \right)^2 / N^b \cdot p_k}$$

(19.14)

$$TVDI_{i,t} = \begin{cases} \dfrac{|V_{i,t} - V_{i,t0}|}{V_{i,t}}, \dfrac{|V_{i,t} - V_{i,t0}|}{V_{i,t}} \geq \delta \\ 0, \text{otherwise} \end{cases} \forall t \in \left[T^c, T^{end}\right], \forall i \in \mathscr{B}$$

(19.15)

19.2.2.8 Rotor Angle Deviation

Conventionally, the rotor angle is considered as a constraint. The constraint can only provide binary information about the transient stable state, while there is a huge difference between a very stable state and a marginal one. RI^{angle}, the ratio of maximum rotor angle deviation to the rotor angle threshold, is proposed to quantitatively evaluate the rotor angle deviation during a restoration, as calculated in (19.16). A smaller RI^{angle} is favorable because it indicates a larger distance between the actual rotor angle deviation and the constraint.

$$R\,I^{angle} = \sum_{k \in \mathscr{K}} \max\left(\Delta\,\delta_{gg'}^T \right) \cdot p_k / \omega$$

(19.16)

19.2.2.9 Load Loss

$RI^{loadloss}$ is used to indicate the total load loss, including RDLS, EDLS, and demand response. Considering the probabilities of different contingencies, it can be calculated as (19.17). $RI^{loadloss}$ is helpful in coordinating DR, load shedding, and other voltage stability countermeasures.

$$RI^{loadloss} = \sum_{k \in \mathscr{K}} \left(\sum_{i \in \mathscr{B}_{EDLS}} x_{i,k} + \sum_{i \in \mathscr{B}_{EDLS}} x_{j,k} + \sum_{i \in \mathscr{B}_{DR}} P_{i,k}^{DR} \cdot \varepsilon \right) \cdot p_k$$

(19.17)

19.2.3 Compact Mathematical Model

The compact mathematical model for this multi-objective VAR planning problem is as follows.

$$\min_u F[f_1(x,u,\tilde{w},\mathscr{K}), f_2(x,u,\tilde{w},\mathscr{K}), f_3(x,u,\tilde{w},\mathscr{K})]$$

(19.18)

$$\text{s.t. } G_0(y_0, u, \tilde{w}) = 0$$

(19.19)

$$H_0^S(x_0, y_0, u, \tilde{w}) \leq 0$$

(19.20)

$$G_l(y_l, u, \tilde{w}, \mathscr{K}) = 0$$

(19.21)

$$H_t^D(x_t, y_t, u, \tilde{w}, \mathscr{K}) \leq 0$$

(19.22)

where F is the overall objective function, consisting of three sub-objectives: (i) total planning cost (f_1), including the investment cost and the operating cost, (ii) short-term voltage stability index (f_2), and (iii) static voltage stability index (f_3).

x denotes the power system's state, and y denotes the algebraic variables during the transient period. u denotes the control variables, and $\tilde{w}$ indicates the uncertain variables. $\mathcal{K}$ denotes the critical contingencies. Constraints (19.19) are the steady-state power balance constraints, and constraints (19.20) represent the steady-state operational limits. Constraints (19.21) are the post-contingency power balance constraint, and constraints (19.22) represent the post-contingency rotor angle constraint, LVRT/HVRT constraint, voltage recovery constraint, and load shedding constraint. This compact mathematical model will be expanded equation by equation in the following subchapters.

19.2.4 Optimization Objectives

19.2.4.1 Total Planning Cost

Total planning cost f_1 is calculated as (19.23). It includes investment cost and operating cost. The operating cost consists of preventive control cost and CC cost.

$$f_1 = C^{invest} + C^{re-dis} + C_w^{curtail} + C^{DR} + C^{EDLS} + C^{RDLS}$$

(19.23)

The investment cost is calculated as (19.24), consisting of installation costs for fixed capacitor banks and STATCOMs. The installation decision at a specific candidate bus is implicitly determined by the installed capacity as (19.26).

$$C^{invest} = \sum_{i \in \mathscr{B}_s} C_i^{invest\_s} + \sum_{j \in \mathscr{B}_c} C_j^{invest\_c} \tag{19.24}$$

$$C_j^{invest\_c} = c^{cap\_c} \cdot cap_j \tag{19.25}$$

$$C_i^{invest\_s} = \begin{cases} c_i^{install} + c^{cap\_s} \cdot cap_i, \text{ if } cap_i \neq 0 \\ 0, \text{ if } cap_i = 0 \end{cases} \tag{19.26}$$

Four preventive controls are considered in this chapter. First, wind power curtailment before a critical contingency is optimized to avoid aggressive curtailment or a conservative one. The curtailment cost is calculated as (19.27).

$$C^{curt} = \sum_{w \in \mathscr{C}_w} C_w^{curt} = \sum_{w \in \mathscr{C}_w, k \in \mathscr{K}} c^{curt} \cdot P_w^{curt} \cdot t^c \cdot p_k \tag{19.27}$$

With curtailment decisions for each wind farm, synchronous generators will be re-dispatched along with the dispatchable STATCOMs. The corresponding cost is calculated as (19.28), which is the difference between the new operating state (optimized) and the original state.

$$C^{re-dis} = C^{gen}\left(x^{ori}, u^{ori}\right) - C^{gen}\left(x^{opt}, u^{opt}\right) \tag{19.28}$$

$$C^{gen} = \sum_{g \in \mathscr{C}_s} \left(a_s + b_g \cdot P_g + c_g \cdot P_g^2\right) \cdot t^a \tag{19.29}$$

The 3rd preventive control is line drop compensation. It is available in most excitation systems of a synchronous machine as an effective compensation method [19]. If reactive output power increases, a rising voltage profile will be produced by LDC at generator terminals. LDC could be interpreted as the equivalent inner reactance of generators, as (19.30), where V^T and I^T are terminal voltage and current, R^c and X^c are the resistive and reactive components of the compensation. The compensated voltage is generated by processing the feedback signal V^{c1} through a first-order system. As a cost-effective way to boost the utilization of generators' reactive capabilities, the cost of LDC is negligible compared to other preventive and corrective controls.

$$V^{c1} = \left|V^T + (R^c + jX^c)I^T\right| \tag{19.30}$$

However, existing studies on dynamic VAR source planning did not incorporate LDC, and the default parameters are usually used. Given that the detailed parameters of the LDC are typically based on specific simulation results, a stability verification of the system is necessary. In this thesis, different LDC levels will be verified under various contingencies and wind power scenarios to improve the flexibility of the setting of LDC.

The last preventive control is demand response. In this chapter, incentive-based DR [10] is employed because it has a relatively higher response rate compared with the price-based DR. This characteristic is advantageous when the operators of the system are expecting an extreme condition. The cost of DR is calculated as (19.31).

$$C^{DR} = \sum_{n \in \mathscr{P}_{DR}, k \in \mathscr{K}} cap_{n,k}^{DR} \cdot t_n^{DR} \cdot c_{n,t}^{DR} \cdot \varepsilon \cdot p_k \tag{19.31}$$

The total preventive control cost can be calculated as (19.32) since the cost of LDC is negligible compared with DR and generation re-dispatch.

$$C^{PC} = C^{curt} + C^{re-dis} + C^{DR} \tag{19.32}$$

Given the potential for quick propagation of short-term voltage instability in a wind-penetrated power system with a high portion of induction motor loads, event-driven load shedding is used together with UVLS to help the system restore to a stable state. Therefore, CC cost can be calculated as (19.33), where $C_{i,j}$ is the cost function for different load types. For a practical description of differentiated impacts of loads on the voltage response after a contingency, three load shedding cost functions (linear, quartic, and sigmoid) are used.

$$C^{EDLS} + C^{RDLS} = \sum_{i \in \mathscr{P}_{EDLS}, k \in \mathscr{K}} C_i(x_{i,k}) \cdot p_k + \sum_{i \in \mathscr{P}_{EDLS}, k \in \mathscr{K}} C_j(x_{j,k}) \cdot p_k \tag{19.33}$$

Conventionally, a linear function, like (19.34), is used as the load shedding model, where the priority factors are assigned to different load buses accordingly. A basic assumption for this approach is that the least amount of load shedding leads to the least load shedding cost [20]. This might be true for the residential loads. Still, the application on the industrial loads is an oversimplification because most of the loads in the industrial plant are actually dependent on each other. Therefore, considering the priorities and characteristics of different loads, three different load shedding cost functions are used as $C_i(x_{i,k})$ in (19.33) to reflect the actual practice and the distinctive impacts of loads on the dynamics of the system [20].

C^{linear} aims to describe the loads, usually residential loads, with a linear correlation between the amount and the cost of load shedding. $C^{quartic}$ is used to describe the loads, usually common industrial loads, with a quartically increasing load shedding cost. As for $C^{sigmoid}$, it is for important loads with a load threshold for fundamental electricity demand, such as an industrial plant requiring different production lines for a working procedure. For these critical loads, the loss of any production lines (or specific equipment) will bring an immediate shutdown of the main production. A general illustration of these three load shedding functions is illustrated in Figure 19.8. The sudden surge in (19.36) is a realistic description of the post-contingency scenario because the load shedding cost will increase significantly after a certain threshold due to the potential chain effect of load loss.

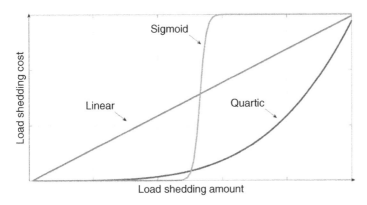

Figure 19.8 Characteristics of load shedding cost functions.

$$C^{linear}(x_{i,k}) = d_i \cdot x_{i,k} \cdot c^{ls}, \quad \forall i \in \mathscr{B}_l, k \in \mathscr{K} \tag{19.34}$$

$$C^{quartic}(x_{i,k}) = d_i \cdot x^4_{i,k} \cdot c^{ls}, \quad \forall i \in \mathscr{B}_l, k \in \mathscr{K} \tag{19.35}$$

$$C^{sigmoid}(x_{i,k}) = \frac{z \cdot c^{ls}}{1 + e^{-s \cdot \left(x_{i,k} - x_i^{threshold}\right)}} \quad \forall i \in \mathscr{B}_l, k \in \mathscr{K} \tag{19.36}$$

19.2.4.2 Resilience Indices of Preventive Capability

The 2nd objective function is a weighted sum of the indices RI^{margin}, $RI^{curtail}$, RI^{dev}, $RI^{DRshift}$ and RI^{VCPI} as (19.37). The determination of weighting factors will be explained in the following sub-chapter.

$$f_2 = RI^{margin} \cdot \alpha_1 + RI^{curtail} \cdot \alpha_2 + RI^{dev} \cdot \alpha_3 + RI^{DRshif\,t} \cdot \alpha_4 + RI^{VCPI} \cdot \alpha_5 \tag{19.37}$$

19.2.4.3 Resilience Indices of Restorative Performance

The third objective function is a weighted sum of the indices RI^{WV}, RI^{RVSI}, RI^{angle}, and $RI^{loadloss}$ as (19.38). The determination of weighting factors will be explained in the following subchapter.

$$f_3 = RI^{WV} \cdot \beta_1 + RI^{RVSI} \cdot \beta_2 + RI^{angle} \cdot \beta_3 + RI^{loadloss} \cdot \beta_4 \tag{19.38}$$

19.2.4.4 Determination of Weighting Factors

In a practical application, the weighting factors α and β in (19.37) and (19.38) should be determined based on the following two aspects:

a) Empirical priorities of the indices in a practical planning problem. These priorities are usually determined according to the experts' judgment of their importance and the corresponding operational requirements.
b) Sensitivity analysis of the indices. It is used to figure out the range of potential index variations when subjected to different scenarios and uncertainties.

The empirical priorities adopted in this chapter are illustrated in Figure 19.9 (a larger area of the pie means a higher priority). Then, all-at-a-time (AAT) method is used to carry out the sensitivity analysis. The output variations (numerical variation of indices) are obtained by changing all the inputs (the capacities of STATCOMs) simultaneously. By using AAT, both the direct impact of increasing capacity and joint impact of interactions among STATCOMs installed at different buses. Finally, with empirical priorities and the results of sensitivity analysis, α and β can be determined. A typical example is presented as follows:

a) Determine the priorities based on the empirical estimations of their importance and the corresponding operational requirements. RI^{margin}, $RI^{curtail}$, RI^{dev}, $RI^{DRshift}$, and RI^{VCPI} are assigned with priorities α_{1a}, α_{2a}, α_{3a}, α_{4a}, and α_{5a}, while RI^{RVSI}, RI^{wv}, RI^{angle}, and $RI^{loadloss}$ are assigned with β_{1a}, β_{2a}, β_{3a}, and β_{4a}. For instance, $\alpha_{1a} = 20\%$, $\alpha_{2a} = 15\%$, $\alpha_{3a} = 10\%$, $\alpha_{4a} = 15\%$, and $\alpha_{5a} = 40\%$; $\beta_{1a} = 50\%$, $\beta_{2a} = 20\%$, $\beta_{3a} = 20\%$, and $\beta_{4a} = 10\%$.
b) Increase the capacities of STATCOMs (2 MVAR a step) from 0 to 100 MVAR simultaneously.

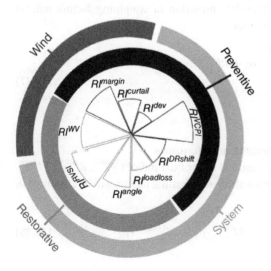

Figure 19.9 Illustration of priorities assigned to different indices. *Source:* Chi et al. [15]/With permission of IEEE.

c) Normalize the initial evaluation results. Then, compute the variation range of each index with the priorities. For example, the variation range of "$RI^{RVSI} \cdot \beta_{1a}$"

d) Adjust the results obtained from Step c by multiplying a group of parameters ($\alpha_{1b} \sim \alpha_{5b}$ and $\beta_{1b} \sim \beta_{5b}$). This is to avoid the situation that the values of f_2 and f_3 are largely determined by some very sensitive indices, even though these indices are actually with relatively low priorities in Step a.

e) Calculate the α and β. For instance, $\alpha_1 = \alpha_{1a} \cdot \alpha_{1b}$, $\beta_1 = \beta_{1a} \cdot \beta_{1b}$.

19.2.5 Constraints

Pre-contingency constraints are the power flow balance constraints as in (19.39) and steady-state operational constraints as in (19.40). Constraint (19.41) is the wind curtailment constraint determined by the operators with a modified curtailment ratio.

$$\begin{cases} P_g + \tilde{P}_w - P_w^{curt} - P_d - V_i \sum_{j \in \mathscr{B}} V_j \left(G_{ij} \cos \theta_{ij} + B_{ij} \sin \theta_{ij} \right) = 0 \\ Q_g + \tilde{Q}_w + Q^{capacitor} + Q^{stat} - Q_w^{curt} - \\ Q_d - V_i \sum_{i \in \mathscr{B}} V_j \left(G_{ij} \sin \theta_{ij} - B_{ij} \cos \theta_{ij} \right) = 0 \end{cases} \tag{19.39}$$

$$\begin{cases} V_i^{min} \le V_i \le V_i^{max} \\ L_l^{min} \le L_l \le L_l^{max} \\ P_g^{min} \le P_g \le P_g^{max} \\ Q_g^{min} \le Q_g \le Q_g^{max} \end{cases}, \forall i \in \mathscr{B}, \forall l \in \mathscr{L}, \forall g \in \mathscr{G} \tag{19.40}$$

$$\sum_{w \in \mathscr{G}_w} P_w^{curt} \le \sum_{w \in \mathscr{G}_w} \tilde{P}_w \cdot W^c, \forall W^c \in [0, 1] \tag{19.41}$$

Constraint (19.42) is the rotor angle stability constraint (transient stability). It is used to guarantee that the maximum rotor angle difference between any two generators during a transient period of T is no larger than ω. Constraints for LVRT/HVRT are modeled as (19.43). Constraint (19.44) is a constraint for voltage recovery speed, indicating that the time of a bus voltage under the voltage threshold (like 0.9 pu) should be no more than the maximum allowed time. Constraints (19.45) and (19.46) are constraints for event-driven load shedding and response-driven load shedding, respectively. Constraint (19.47) is the demand response constraint, indicating that the responded loads before a contingency should be equal to the shifted loads after the contingency. Furthermore, the responded loads are also limited by the maximum loads participated in DR program as (19.48).

$$\left[\max \left(\Delta \delta_{gg'}^T \right) \right]_k \le \omega \ \forall g, g' \in \mathscr{G}_s, \forall k \in \mathscr{K} \tag{19.42}$$

$$\begin{cases} V_{i,t} - LVRT(t) \ge 0, \forall t \in T, \forall i \in B \\ V_{i,t} - HVRT(t) \le 0, \forall t \in T, \forall i \in B \end{cases} \tag{19.43}$$

$$T_i^r - T^C \leq T^{\max}, \forall i \in \mathscr{B} \tag{19.44}$$

$$x_{i,k} + x_{j,k} \leq P_i^{LS,\max}, \ \forall i \in \mathscr{B}_{EDLS}, \ \forall j \in \mathscr{B}_{RDLS}, \ \forall k \in \mathscr{K} \tag{19.45}$$

$$x_{i,k} \leq P_j^{LS,\max}, \ \forall i \in \mathscr{B}_{RDLS}, \ \forall k \in \mathscr{K} \tag{19.46}$$

$$\sum_{n \in \mathscr{B}_{DR}} P_n^{DR} \cdot \varepsilon = \sum_{n \in \mathscr{B}_{DR}} P_n^{DRshif\ t} \tag{19.47}$$

$$P_n^{DR} \cdot \varepsilon \leq P_n^{DR,\max}, \ \forall n \in \mathscr{B}_{DR} \tag{19.48}$$

19.2.6 Candidate Bus Selection

To reduce the computation burden of a VAR planning problem, candidate buses should be selected before capacity optimization. In this chapter, an LHS-based Morris screening method is developed to choose the most influential buses for STATCOMs installation, aiming to alleviate the computation complexity of the planning optimization problem. Morris screening method [21] is originally proposed to identify the most influential parameters by creating a multi-dimensional semi-global trajectory within its search space. Compared with local trajectory analysis, it is more accurate [22]. On the other hand, it is also more time-efficient than the time-consuming global sensitivity analysis.

Firstly, LHS is used to generate random samples following the normal distribution. Compared with simple random sampling or uniform sampling, LHS is a better representative of the actual variability. Then, one variable at a time will be changed at a step of Δ_m (the step is determined based on the LHS results). So, an elementary effect of a Δ_m change is calculated as (15.7).

$f(\alpha_{i,m})$ is an evaluation result when a STATCOM with a specific capacity ($\alpha_{i,m}$, determined by LHS) is installed at a candidate bus. In this chapter, RI^{RVSI} is chosen as the criterion for STATCOM candidate bus selection since it is mainly responsible for the post-contingency voltage restoration, and RI^{VCPI} is employed to select candidate buses for capacitors because capacitors only improve the steady-state voltage stability of the power system. Then, the mean (μ^*) of the elementary effects is calculated as (15.8). It represents the sensitivity strength between the input (different buses with different STATCOM/capacitor capacities) and the output (RI^{RVSI} or RI^{VCPI}). The standard deviation (σ_i^*), which is calculated as (15.9), represents the interaction between different variables.

A smaller μ_i^* indicates that the deployment of reactive power compensation devices at i^{th} bus has a relatively lower effect on the power system, in terms of RI^{RVSI} or RI^{VCPI}. On the other hand, a larger σ_i^* indicates a nonlinear effect on RI^{RVSI} or RI^{VCPI}, and it has higher interactions with other buses. Therefore, the buses with high and linear impact are the best candidate buses. However, in practice, if the ideal candidate buses are insufficient or even no such bus at all, the high and nonlinear impact can also be selected as candidate buses.

19.2.7 Wind Power Uncertainty Modeling

The output of a wind turbine is stochastic by nature and difficult to predict. Although wind speed usually follows Weibull distribution, wind power output is a nonlinear function of the wind speed. Therefore, the probability density function (PDF) of wind power output is much more complicated than Weibull distribution, and thus not easy to determine. Besides, such a complicated PDF is even more difficult to obtain for a long-term VAR planning problem (e.g. 10 years).

In this chapter, instead of the PDF approach, we use the upper and lower bounds to model the stochastic wind power output. It is a more straightforward method and easier to implement for a long-term VAR planning problem. Specifically, TOAT [23], a robust parameter design technique, is used to decompose a large-scale stochastic programming problem into a series of deterministic problems. Generally, the decomposition is realized by approximating the whole uncertainty space with a smaller set of representative scenarios. For instance, if there are N stochastic variables, $\tilde{w} = [\tilde{w}_1, \tilde{w}_2, ..., \tilde{w}_N]$ in an uncertainty space and M levels of values for each variable, the uncertainty space will have M^N combinations. It is of huge computational burden when N is large. TOAT is employed to reduce the number of scenarios used to represent all the combinations. The relatively fewer scenarios are determined by orthogonal arrays, and their mathematical details can be found in [23]. In this chapter, there are two realistic assumptions for the application of TOAT on the wind power output uncertainty: (i) wind power prediction is given as a range with maximum value and minimum value, and (ii) deviation of the actual output from its prediction follows the normal distribution.

For example, as listed in Table 19.1, each of the three wind farms ($\tilde{w}_1, \tilde{w}_2, \tilde{w}_3$) has two levels (maximum and minimum values) of output predictions. So, the total number of scenarios is 2^3. With the implementation of TOAT, the whole uncertainty space (eight scenarios) is decomposed as a TOAT library consisting of only four scenarios as scenarios 1–4 in Table 19.1. The "+" indicates the maximum value of wind power outputs, and "−" stands for the minimum value of wind power outputs.

Table 19.1 Decomposition of wind power scenarios.

TOAT Library	Variable levels		
	$\tilde{w}_1$	$\tilde{w}_2$	$\tilde{w}_3$
Scenario 1	−	−	−
Scenario 2	−	+	+
Scenario 3	+	−	+
Scenario 4	+	+	−

19.3 Solution Method

In this chapter, NSGA-II is adopted as the solving algorithm for this multi-objective optimization problem. It can provide a group of trade-off solutions, which are a balance or compromise achieved between two or three desirable but incompatible objectives, for decision-makers to choose from based on their practical needs (for instance, the preference for low investment over high short-term voltage stability).

19.3.1 Basics of NSGA-II

Proposed by Deb in Ref. [24], NSGA-II has been proven to be an efficient algorithm for multi-objective benchmark optimization problems. One of the salient advantages of NSGA-II is Pareto optimal solutions from Pareto front. Figure 19.10 illustrates the general computation steps of NSGA-II, where P_t is the parent population, Q_t denotes the offspring population, and R_t is the combined population. Firstly, the population is grouped after initialization based on non-domination levels into fronts. The first front consists of a set of non-dominant individuals. The individuals from the second front are only dominated by the ones from the first front, and the following fronts follow the same pattern. Apart from the front ranking, the crowding distance is also computed for each individual to measure how close an individual is to its neighbors. A larger average crowding distance indicates a more diversified population. Based on *crowded-comparison operator*, parents are from the population, following two general rules [24]:

1) higher front ranking individual is always selected regardless of crowding distance;

2) for individuals in the same front, the one with a larger crowding distance is selected.

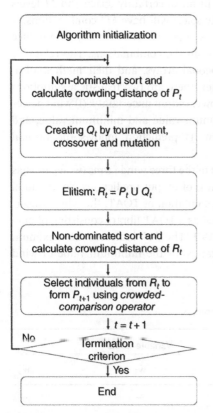

Figure 19.10 Computation flowchart of NSGA-II.

After the operation of crossover and mutation operators, the parent population generates offspring. Then, the population is grouped again, and the best individuals are selected. Since all previous and current population members are included in R_t, elitism is ensured in the procedures.

19.3.2 Coding Rules

The coding rules adopted in this chapter are similar to the ones used in Section 19.1. As illustrated in Figure 19.11, control variables include installation decisions, wind power curtailment, and DR scheme.

19.3.3 Computation Steps

Since "higher resilience performance" always conflicts with "lower cost," a group of trade-off solutions can be generated to satisfy the practical needs of the decision-makers. Figure 19.12 shows the computation flowchart of the proposed method. Detailed computation steps are described as follows:

1) **Initialization:** After the algorithm parameters setting, candidate buses are selected. Then, the first generation of the population is generated. Each individual consists of decision variables for STATCOMs, capacitor banks, DR, and wind power curtailment, as illustrated in Figure 19.11.
2) **Pre-contingency OPF and preventive capability evaluation:** Preventive controls are implemented in the pre-contingency stage. Two OPF calculations under different scenarios are carried out. The first one is without preventive controls, and the second one is with both preventive controls and the dispatchable STATCOM. With the OPF results, RI^{margin}, $RI^{curtail}$, RI^{dev}, $RI^{DRshift}$, and RI^{VCPI} are calculated. The preventive controls cost is also calculated for the total cost calculation.
3) **Time-domain simulation, and resilience performance evaluation:** Time-domain simulations are conducted in this stage. Specifically, load shedding (corrective controls) is activated if: pre-determined voltage thresholds are violated (as in UVLS), or one of the pre-determined critical contingencies

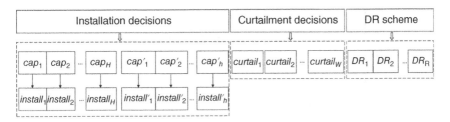

Figure 19.11 An individual code structure.

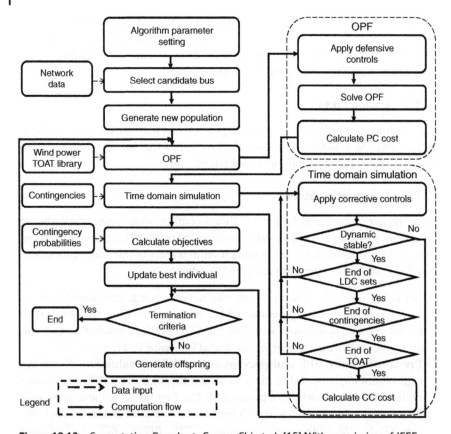

Figure 19.12 Computation flowchart. *Source:* Chi et al. [15]/With permission of IEEE.

occurs (as in event-driven load shedding). The individuals of the population generated by NSGA-II are verified under different LDC parameters, wind power generation scenarios, and critical contingencies. After all the scenarios are checked, RI^{RVSI}, RI^{wv}, RI^{angle}, and $RI^{loadloss}$ are calculated. CC cost is also calculated for the calculation of a total investment.

4) **Objective calculations:** The total investment and voltage resilience indices (including preventive capability indices and resilience performance indices) are calculated in a risk-based manner with different contingency probabilities.

5) **Individual update:** Find and update the best individual of the current generation.

6) **Termination:** If one of the following stopping criteria is satisfied, the main procedure will be terminated, and a Pareto front will be presented: (i) reached the maximum iteration number, or (ii) variations in the fitness function are smaller than or equal to a pre-determined tolerance.

7) **Offspring generation**: If the stopping criteria are not satisfied, the new off-spring of the current generation will be generated. Then, the computation procedure will go back to Step 2. Typical operations for the offspring generation include mutation, recombination, and crossover [25].

8) **Break criteria**: The evaluation of each individual will be terminated if any post-contingency constraint is violated, and the cost objectives will be calculated with high penalty parameters. Then, the new offspring of the current generation will be generated, and the computation procedure will go back to Step 2.

19.4 Case Studies

19.4.1 Test Systems and Parameter Settings

The proposed method is tested on a modified New England 39-bus system, as illustrated in Figure 19.13. The models for DFIG, STATCOM, and synchronous generators are also the same ones adopted in Chapter 16. The short-term load forecast data (used in (19.8)–(19.10) for $RI^{DRshift}$) are from reference [26]. In practice, load forecast data can be obtained from artificial neural network-based

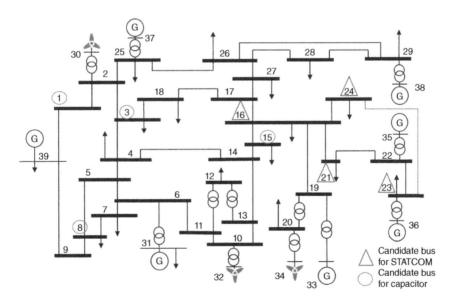

Figure 19.13 Single-line diagram of the New England 39-bus test system. *Source:* Chi et al. [15]/With permission of IEEE.

Table 19.2 Parameters for simulation setting.

Parameters	Value settings
$c^{install}$	1.5 million/unit
c^{cap_s}	0.05 million/MVAR
c^{cap_c}	US\$ 1400/MVAR
u^{1st}, u^{2nd}, u^{3rd}	5% each at 0.9, 0.85, and 0.8 pu
W^c	15% of original load
$c^{curtail}$	8/MWh
t^c, t^a	4 h, 4500 h
ω, δ	π, 0.1
LHS levels	20 (0 $\sim$ 100 MVAR)
LM, SM, DL, CP, K_p	25%, 15%, 10%, 10%, 2
NSGA II population and generation	100 and 15
NSGA II tolerance	1.0×10^{-3}
NSGA II mutation probability	0.06
NSGA II crossover rate	0.8
Fault setup	3-phase short-circuit (cleared after 0.1 s)

Source: Chi et al. [15]/With permission of IEEE.

methods, such as the one proposed in reference [27]. CLOD model is also used in this chapter. Other important parameters are listed in Table 19.2. Based on the reference [11], faults on line 2–3, line 6–7, and line 16–24 are selected as three critical contingencies. MATPOWER 6.0 [28] is used to solve OPF. The time-domain simulation is carried out using PSS®E 33.4 [29] (through an interface with MATLAB). NSGA-II is implemented in MATLAB 2018b. The case studies are carried out on a laptop workstation with Intel®Core™ i7-8750H CPU @ 4.6 GHz and 32 GB DDR4 2400 MHz RAM.

19.4.2 Numerical Simulation Results and Comparisons

The overall computation time is 62 046 seconds (17.235 hours). In Figure 19.13, buses with triangle marks are candidate buses for STATCOMs, and buses with circle marks are candidate buses for capacitor banks. As illustrated in Figure 19.13, the STATCOM candidate buses are selected because of their relatively high values of μ. As for the conventional sensitivity analysis method, it only considers the direct impact. In this case, if more candidate buses (for instance, five buses) are required, it would be quite difficult to select one from the buses in the dashed

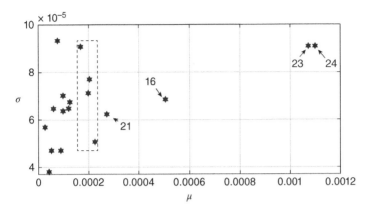

Figure 19.14 Candidate bus selection results for STATCOMs. *Source:* Chi et al. [15]/With permission of IEEE.

square, as illustrated in Figure 19.14, since these buses are very close to each other on the x-axis. By using Morris Screening method, the four buses are distinguishable at the y-axis (σ) since their nonlinearity is different. So, the 5th candidate bus (if needed) can be selected as the one at the bottom of the dashed square.

A Pareto front is obtained and shown in Figure 19.15. These are 42 trade-off solutions, and these solutions are more valuable and practical than a single optimized result because they can support a cost-effective balanced decision.

Fuzzy membership function [30] is used in this chapter to select a compromise solution. The compromise solution is marked in Figure 19.15, and the installation plan of this solution is listed in Table 19.3. For each Pareto optimal solution x_p, its

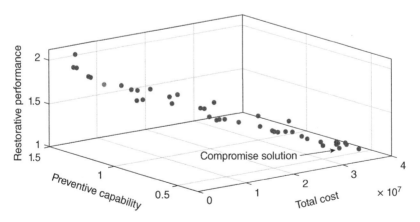

Figure 19.15 Pareto optimal solutions. *Source:* Chi et al. [15]/With permission of IEEE.

Table 19.3 Installation decisions (MVAR, 11 candidate buses).

	STATCOM							Capacitor bank			
Bus	16	21	23	24	30	32	34	1	3	8	15
MVAR	86.2	17.2	21.8	94.0	31.6	87.8	76.0	12.2	13.1	12.0	10.4

Source: Chi et al. [15]/With permission of IEEE.

satisfactory degree for fth objective function, and the satisfactory degrees for all objective functions can be expressed as (19.49) and (19.50). The solution with the largest μ^p (0.0159) is selected as the fuzzy membership result.

$$\mu_f^p = \frac{f_f^{\max} - f_f(x_p)}{f_f^{\max} - f_f^{\min}} \tag{19.49}$$

$$\mu^p = \frac{\sum_{f \in [1,2,3]} \mu_f^p}{\sum_{p \in \mathscr{P}} \sum_{f \in [1,2,3]} \mu_f^p} \tag{19.50}$$

The wind curtailment for wind farms at bus 30, bus 32, and bus 34 are 0.90%, 0.41%, and 5.16%, respectively. Table 19.4 shows the values of the three objectives. The preventive capability and restorative performance indices of the original system (base case) are 0.962 and 1.871. After the deployment of STTACOMs and the implementation of PCs and CCs, the two indices are improved by 51.6% and 43.2%, respectively.

To demonstrate the effectiveness of the candidate bus selection method proposed in this chapter, four more buses are selected (the buses in the dashed box in Figure 19.14) as additional candidate buses for comparison. The compromise solution for this 15-bus case (11 buses + 4 additional buses) is listed in Table 19.4, and the corresponding installation plan is listed in Table 19.5. It can be concluded from the two tables that the additional four buses are redundant, because: (i) the performance of 15-bus case is inferior to that of the 11-bus case, and (ii) the total cost of the 15-bus case is even higher than that of the 11-bus case. On the other hand, the optimized

Table 19.4 Compromise solutions.

Objectives	Total cost ($)	Preventive capability	Restorative performance
11-bus	32.35 m	0.4650	1.062
15-bus	37.93 m	0.4771	1.073

Source: Chi et al. [15]/With permission of IEEE.

Table 19.5 Installation decisions (MVAR, 15 candidate buses).

Bus	STATCOM										Capacitor bank		
	7	8	16	18	21	23	24	27	30	32	34	1	3	8	15
MVAR	9.4	4.2	85.3	4.8	27.7	33.1	72.3	8.3	39.3	72.2	70.5	17.0	10.0	11.9	12.6

capacities of the additional buses (buses 7, 8, 18, and 27) are much smaller than that of the buses in Table 19.3.

To demonstrate the advantages and the effectiveness of TOAT, the Monte Carlo method (100 wind power generation scenarios) is implemented as a comparison. The numerical results are in Table 19.6. These two methods have similar performance, in terms of preventive capability and restorative performance results. But the proposed method only uses four scenarios, which significantly alleviates the computation burden of the planning problem without any deterioration in the performance.

To further demonstrate the advantages of the proposed method, five methods and a base case are compared in Table 19.7.

Method A: The proposed method with the proposed resilience indices. The solution is a compromise solution.

Method B: The proposed method with conventional indices (*TVSI* and *VCPI*). The solution is also a compromise solution.

Table 19.6 Robust design comparison.

Uncertainty modeling	Preventive capability	Restorative performance	Total number of scenarios
TOAT	0.4650	1.062	4
Monte Carlo	0.4651	1.060	100

Source: Chi et al. [15]/With permission of IEEE.

Table 19.7 Comparison results (MVAR, MW, and Million).

	Total capacity						
Method	Capacitor banks (MVAR)	STATCOM (MVAR)	Total LS (MW)	Total DR (MW)	Total cost (million)	Preventive capability	Restorative performance
A	47.7	414.7	146.0	30.1	32.35	0.4650	1.062
B	88.2	445.6	141.7	31.0	33.93	0.5569	1.095
C	92.6	504.7	148.5	67.6	37.00	0.4741	1.094
D	45.7	411.5	146.5	30.0	31.34	0.5364	1.073
E	58.1	490.0	—	—	35.08	0.4922	1.099
Base	—	—	—	—	—	0.9615	1.871

Source: Chi et al. [15]/With permission of IEEE.

Method C: The proposed method with conventional indices (*TVSI* and *VCPI*). It is the same with Method B, but the solution is selected from the Pareto front based on resilience performance.

Method D: The proposed method with the proposed resilience indices (same as method A). But the solution is selected from the Pareto front based on investment cost.

Method E: It is the conventional method (considers only the deployment of STATCOMs and capacitor banks) based on the *TVSI* and *VCPI*. The solution is a compromise solution.

As illustrated in Figure 19.2, method E (conventional method) only considers the planning stage countermeasures (installation of STATCOMs). Its planning decisions are relatively conservative compared with the proposed method. As in Table 19.7, the total cost of method E is 8.44% higher than that of method A, and at the same time, both the preventive capability and restorative performance of method E are inferior to those of method A. On the other hand, even with the implementation of operational control measures, the conventional voltage metrics (TVSI and VCPI) are no longer adequate for a comprehensive evaluation of the voltage performance before and after a contingency. For instance, although PCs and CCs are implemented in method A and method B, method A still has better resilience results in terms of preventive capability and restorative performance, improved by 16.50% and 3.01%, respectively. At the same time, the total cost of method A is 1.58 million less than that of method B.

Another disadvantage of using the conventional TVSI and VCPI is that the load shedding and demand response are not directly evaluated, although they are both implemented as control measures to mitigate the reactive power deficiency after a voltage disturbance. This might make the load shedding and the demand response aggressive or conservative. For an aggressive scheme, although load shedding or demand response will be helpful in the voltage recovery process, the cost is too high. On the other hand, the installation capacity of STATCOMs will be large if the load shedding or demand response is too conservative. For instance, as listed in Table 3.6, the load shedding and DR of method C (the conventional metrics) are higher compared with those of method A (the proposed metrics), while the resilience evaluation results of method C are still inferior to that of method A.

Figure 19.16 shows the voltage trajectory of bus 24 after a critical line contingency. It is clear that method A has a faster voltage recovery than method B. Once the fault is cleared, method A experiences an acceleration of voltage recovery, while the recovery speed of method B is relatively slow. Furthermore, two cross points of the two trajectories indicate that method A is with a smaller fluctuation after restoring to the acceptable voltage range. This voltage trajectories comparison is consistent with the performance results listed in Table 19.7. Two more critical

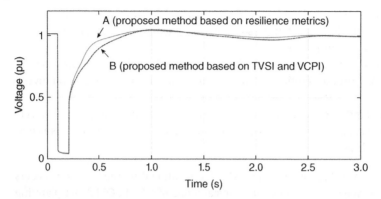

Figure 19.16 Post-contingency voltage response (fault on line 16–24) at bus 24.
Source: Chi et al. [15]/With permission of IEEE.

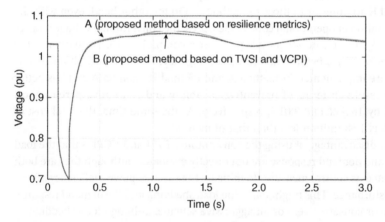

Figure 19.17 Post-contingency voltage response (fault on line 2–3) at bus 24.
Source: Chi et al. [15]/With permission of IEEE.

contingencies are also compared in Figure 19.17 (fault on line 2–3) and Figure 19.18 (fault on line 6–7). Similar to the case in Figure 19.16, method A outperforms method B under both contingencies.

Table 19.8 summarizes RI^{RVSI} results in different wind power generation scenarios for all critical contingencies. Compared with method B, method A has better results in all representative scenarios. On the other hand, the performance improvement of these three critical contingencies is relatively large (average improvement of 2.914%) compared with other contingencies (average improvement of 0.890%), which demonstrates that the contingencies selected are representative and adequate for the contingency analysis in this problem.

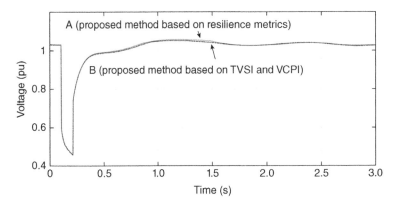

Figure 19.18 Post-contingency voltage response (fault on line 6–7) at bus 24. *Source:* Chi et al. [15]/With permission of IEEE.

Table 19.8 RI^{RVSI} results comparison.

Contingency/method		Scenario 1	Scenario 2	Scenario 3	Scenario 4
Line 16–24	Method A	0.343	0.388	0.371	0.375
	Method B	0.352	0.401	0.379	0.381
	Improvement	2.62%	3.35%	2.16%	1.60%
Line 2–3	Method A	0.344	0.389	0.372	0.375
	Method B	0.352	0.402	0.379	0.382
	Improvement	2.33%	3.34%	1.88%	1.87%
Line 6–7	Method A	0.334	0.377	0.361	0.363
	Method B	0.347	0.396	0.374	0.375
	Improvement	3.89%	5.04%	1.93%	3.31%

Source: Chi et al. [15]/With permission of IEEE.

Figure 19.19 further demonstrates why conventional TVSI and VCPI are insufficient, when there is a large voltage disturbance, for a comprehensive voltage resilience evaluation of the power system. When there is an $N-2$ contingency in the system (wind turbines at bus 30 and bus 32 are forced off-line), which is not included in the contingency analysis in both methods, the voltage of bus 8 becomes unstable at 1.3 second with method B, and it still remains stable with method A. This is because method A has a better evaluation result of RI^{angle} (as listed in Table 19.9), which means there is a larger margin in terms of the distance away from the edge of rotor angle instability. On the other hand, Method A also

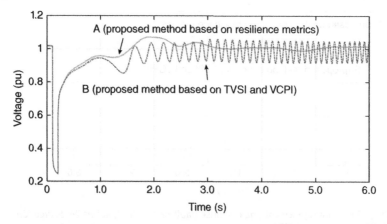

Figure 19.19 Post-contingency (N-2) voltage response at bus 8. *Source:* Chi et al. [15] / With permission of IEEE.

Table 19.9 Details of resilience evaluation results.

Method	$RI^{curtail}$	$RI^{DRshift}$	RI^{VCPI}	RI^{angle}	RI^{RVSI}
A	0.1136	0.070	0.1236	0.0628	0.6455
B	0.1870	0.072	0.1125	0.0757	0.6713

Source: Chi et al. [15]/With permission of IEEE.

outperforms method B in $RI^{curtail}$, which indicates a lower penetration level during the voltage recovery stage.

19.4.3 Scalability and Computation Efficiency Discussion

Due to the highly nonlinear and nonconvex features, as well as the time-consuming time-domain simulations, the VAR planning problem considering short-term voltage stability issues is naturally of high computational complexity. The total execution time T can be estimated based on the parameters listed in (19.51)

$$T \propto \left(N^c, N^{gen}, N^{pop}, N^{dec}, N^{TOAT}, T^{dts}, T^{OPF}\right) \qquad (19.51)$$

where N^c is the number of contingencies considered in time-domain simulations. N^{gen}, N^{pop}, and N^{dec} are the number of generations, population size, and the number of decision variables for NSGAII, respectively. N^{TOAT} is the number of

Table 19.10 Computation times for different test systems.

Test system	Candidate bus selection	Main optimization	Total time
New England 39	2 912 s	59 134 s	62 046 s (17.24 h)
Nordic 74	3 057 s	61 820 s	64 877 s (18.02 h)

Source: Chi et al. [15]/With permission of IEEE.

scenarios in TOAT library. T^{dts} and T^{OPF} represent the computation time for a single time-domain simulation and a single OPF run, respectively.

In this chapter, the computation complexity is alleviated through the contingency selection and the representative scenarios for wind power generation. When the size of the power system increases, the corresponding increase in computation time is relatively moderate since the number of candidate buses for a larger system does not necessarily increase proportionally. As listed in Table 19.10, when the proposed method is applied in a Nordic 74-bus test system [31], there is only a small increase in the total computation time, even though the number of buses in this test system nearly doubled (from 39 buses to 74 buses). On the other hand, long-term VAR planning is not a time-critical problem, and the total computation time is around 17–18 hours for both test systems, which is acceptable in practice.

In practice, VAR planning of a large-scale power system is usually studied in an area-based way rather than in a way that takes the whole system as one entity. By employing the area-based method, the self-sufficiency of each area can be maximized, and the reactive power transmission between different areas can be minimized [32]. Specifically, a large-scale power system can be divided into different areas, and then the VAR planning can be conducted within one area at a time while the other areas are modeled as equivalent subsystems. So, the proposed method in this chapter is applicable for VAR planning of a real-world transmission network as long as the network can be divided into areas, which is also the operational practice for most electric utility corporations.

Nomenclature

Indices

D	index for dynamic states
d	index for loads
f	index for objective functions
g, g'	indices for generators
i, j, n	indices for buses

k	index for contingencies
l	index for lines
m	index for post-contingency hours (or LHS levels)
o	index for an original state
S	index for static states
t	index for the time
w	index for wind turbines

Sets

$\mathcal{B}$	set of all the buses
$\mathcal{B}_c$	set of capacitor bank buses
$\mathcal{B}_l$	set of load buses
$\mathcal{B}_s$	set of STATCOM buses
$\mathcal{B}_{DR}$	set of buses for demand response
$\mathcal{B}_{RDLS}$	set of buses for response-driven load shedding
$\mathcal{B}_{EDLS}$	set of buses for event-driven load shedding
$\mathcal{G}$	set of all generators
$\mathcal{G}_s$	set of synchronous generators
$\mathcal{G}_w$	set of wind turbines
$\mathcal{K}$	set of contingencies
$\mathcal{L}$	set of lines
$\mathcal{M}$	set of shift hours after the demand response
$\mathcal{P}$	set of Pareto optimal solutions
$\mathcal{T}$	set of time
u	set of control variables
$\tilde{w}$	set of uncertainties
x	set of state variables

Variables

B_{ij}	susceptance between i and j
$c_{n,t}^{DR}$	demand response price at bus n at time t
$cap_{i \text{ or } j}$	reactive power source capacities at bus i or j
$cap_w^{curtail}$	curtailment of wind turbine w
DR^{res}	total amount of demand response
DR_m^{shift}	violation index at m^{th} hour after DR
G_{ij}	conductance between i and j
I^r	rotor current of DFIG

L_l	apparent power flow on line l
P_d	active power of load
P_g	active power output of generator g
$\widetilde{P_w}$	active output of wind turbine w
P_w^c	curtailed active power of wind turbine w
P_w^{curt}	active power curtailment of wind turbine w
P_i^{DR}, P_n^{DR}	demand response of load at bus i (or at bus n)
P^{inv}	inverter active power of DFIG
P^{peak}	original load peak within the shifted hours
P_m^{pred}	load peak at m^{th} hour after DR
P^s	stator active power of DFIG
P_d^{shed}	active power of load shedding at bus d
Q_d	reactive power of load
Q_g	reactive power output of generator g
$\widetilde{Q_w}$	reactive output of wind turbine w
$Q^{capacitor}$	reactive power output of a capacitor
Q_w^c	curtailed reactive power of wind turbine w
Q_w^{curt}	reactive power curtailment of wind turbine w
Q^{inv}	inverter reactive power of DFIG
Q^s	stator reactive power of DFIG
Q_d^{shed}	reactive power of load shedding
Q^{stat}	actual reactive power output of STATCOMs
$Q^{statrated}$	rated reactive power output of STATCOMs
Q^w	actual reactive power output of DFIG
s	slip ratio
T_i^r	time when V_i reaches a recovery threshold
U^s	stator voltage of DFIG
V_i, V_j	voltage magnitude of bus i, j
$V_i^{post}(t)$	post-contingency voltage at bus i
V^S	voltage magnitude for VCPI calculation
x^{opt}, u^{opt}	optimal steady-state operating state
x^{ori}, u^{ori}	original steady-state operating state
$x_{i,k}, x_{j,k}$	amount of EDLS and RDLS after contingency k at bus i and j
x_p	Pareto optimal solution
$\Delta\delta_{gg'}^T$	rotor angle deviation between g and g'
Δ_m	offset in Morris screening method
φ	phase angles of load
θ	phase angles of line impedance
μ_i^p	satisfactory degree for ith objective function

Parameters

a_g, b_g, c_g	generation cost coefficients
c^{curt}	unit wind power curtailment cost
c^{cap_c}	unit capacity cost of capacitor bank
c^{cap_s}	unit capacity cost of STATCOM
$c_i^{install}$	installation cost of STATCOM at bus i
c^{ls}	unit load shedding cost
$cap_{n,k}^{DR}$	load participated in Demand Response program at bus n for contingency k
d_i	priority assigned to loads at bus i
$HVRT(t)$	time-varying constraints for HVRT
$LVRT(t)$	time-varying constraints for LVRT
N^b	number of total buses
N^w	number of wind turbines
p_k	probability of contingency k
$P_n^{DR,\max}$	maximum DR allowed at bus n
$P_i^{LS,\max}$	maximum load shedding allowed at bus i
$q, z, x_i^{threshold}$	coefficients for Sigmoid cost function
t^a	annual maximum load equivalent hours
t^c	wind power curtailment hours
t_n^{DR}	demand response duration at bus n
T^c	fault clearing time
T^{end}	end time of set $\mathcal{T}$
$T^{\max}$	max time allowed under recovery voltage
$u^{1st}, u^{2nd}, u^{3rd}$	first, second and third load shedding stage
v^r	recovery voltage
W^c	wind power curtailment ratio
X^m	excitation reactance of DFIG
X^s	stator reactance of DFIG
Δcap	small perturbation in capacity
ϵ	demand response rate
δ	maximum acceptable voltage deviation
ω	rotor deviation in extreme conditions
$\alpha_{1,2,3,4,5}$	weighting factors for preventive capability indices
$\beta_{1,2,3,4}$	weighting factors for restorative performance indices

Functions

C_i, C_j	load shedding cost functions
$C^{curtail}$	total cost of wind power curtailment
C^{CC}	total cost of corrective controls
C^{PC}	total cost of preventive controls

C^{DR}	total cost of DR
C^{EDLS}	total cost of EDLS
C^{linear}	linear load shedding cost
C^{ls}	total load shedding cost
C^{RDLS}	total cost of RDLS
C^{gen}	generation cost of all generators
C^{invest}	total investment cost
C^{invest_c}	cost of capacitor bank
C^{invest_s}	cost of STATCOM
C^{PC}	total cost of preventive controls
$C^{quartic}$	quartic load shedding cost
C^{re-dis}	total cost of re-dispatch
$C^{sigmoid}$	sigmoid load shedding cost
$f(\cdot)$	sub-objective functions
F	overall objective function
g_i	equality constraint i
G	equality constraints
h_i	inequality constraint i
H	inequality constraints

References

1 AEMO (2016). South Australia Separation Event–Final Report. https://www.aemo. com.au/Gas/Gas-Bulletin-Board/-/media/Files/Electricity/NEM/ Market_Notices_and_Events/Power_System_Incident_Reports/2017/Final-report---SA-separation-event-1-December-2016.pdf.

2 Steinmetz, C.P. (1920). Power control and stability of electric generating stations. *Transactions of the American Institute of Electrical Engineers* 39 (2): 1215–1287.

3 Liu, J., Xu, Y., Dong, Z.Y., and Wong, K.P. (2017). Retirement-driven dynamic VAR planning for voltage stability enhancement of power systems with high-level wind power. *IEEE Transactions on Power Systems* 33 (2): 2282–2291.

4 Fang, X., Li, F., Wei, Y. et al. (2015). Reactive power planning under high penetration of wind energy using benders decomposition. *IET Generation Transmission and Distribution* 9 (14): 1835–1844.

5 Wang, Y., Chen, C., Wang, J., and Baldick, R. (2015). Research on resilience of power systems under natural disasters – a review. *IEEE Transactions on Power Systems* 31 (2): 1604–1613.

6 Kosterev, D. (2001). Design, installation, and initial operating experience with line drop compensation at John Day powerhouse. *IEEE Transactions on Power Systems* 16 (2): 261–265.

7 Daratha, N., Das, B., and Sharma, J. (2013). Coordination between OLTC and SVC for voltage regulation in unbalanced distribution system distributed generation. *IEEE Transactions on Power Systems* 29 (1): 289–299.

8 Qdr, Q.J. (2006). *Benefits of Demand Response in Electricity Markets and Recommendations for Achieving them*. Washington, DC, USA, Tech. Rep. 2006 Feb: US Dept. Energy.

9 Samadi, M., Javidi, M.H., and Ghazizadeh, M.S. (2013). Modeling the effects of demand response on generation expansion planning in restructured power systems. *Journal of Zhejiang University SCIENCE C.* 14 (12): 966–976.

10 Li, C., Dong, Z., Chen, G. et al. (2015). Flexible transmission expansion planning associated with large-scale wind farms integration considering demand response. *IET Generation Transmission and Distribution* 9 (15): 2276–2283.

11 Xu, Y., Dong, Z.Y., Xiao, C. et al. (2015). Optimal placement of static compensators for multi-objective voltage stability enhancement of power systems. *IET Generation Transmission and Distribution* 9 (15): 2144–2151.

12 Han, T., Chen, Y., Ma, J. et al. (2017). Surrogate modeling-based multi-objective dynamic VAR planning considering short-term voltage stability and transient stability. *IEEE Transactions on Power Systems* 33 (1): 622–633.

13 Tiwari, A. and Ajjarapu, V. (2010). Optimal allocation of dynamic VAR support using mixed integer dynamic optimization. *IEEE Transactions on Power Systems* 26 (1): 305–314.

14 Chi, Y. and Xu, Y. (2020). Multi-stage coordinated dynamic VAR source placement for voltage stability enhancement of wind-energy power system. *IET Generation, Transmission and Distribution* 14 (6): 1104–1113.

15 Chi, Y., Xu, Y., and Ding, T. (2020). Coordinated VAR planning for voltage stability enhancement of a wind-energy power system considering multiple resilience indices. *IEEE Transactions on Sustainable Energy* 11 (4): 2367–2379.

16 Paramasivam, M., Salloum, A., Ajjarapu, V. et al. (2013). Dynamic optimization based reactive power planning to mitigate slow voltage recovery and short term voltage instability. *IEEE Transactions on Power Systems* 28 (4): 3865–3873.

17 Wang, S., Li, G., and Zhou, M. (2014). The reactive power adjusting mechanism and control strategy of double fed induction generator. *Proceedings of the CSEE* 34 (16): 2714–2720.

18 Zhang, Y., Xu, Y., Dong, Z.Y., and Zhang, R. (2018). A hierarchical self-adaptive data-analytics method for real-time power system short-term voltage stability assessment. *IEEE Transactions on Industrial Informatics* 15 (1): 74–84.

19 IEEE Excitation System Model Working Group (1992). Excitation system models for power system stability studies. IEEE Standard p421, pp. 5–1992.

20 Ruppert, M., Bertsch, V., and Fichtner, W. (2015). Optimal load shedding in distributed networks with sigmoid cost functions. *2015 International Symposium on*

Smart Electric Distribution Systems and Technologies (EDST), Vienna, Austria (08–11 September 2015). pp. 159–164: IEEE.

21 Iooss, B. and Lemaître, P. (2015). A review on global sensitivity analysis methods. In: *Uncertainty Management in Simulation-Optimization of Complex Systems* (ed. D. Gabriella and C. Meloni), 101–122. Boston, MA: Springer.

22 Qi, B., Hasan, K.N., and Milanović, J.V. (2019). Identification of critical parameters affecting voltage and angular stability considering load-renewable generation correlations. *IEEE Transactions on Power Systems* 34 (4): 2859–2869.

23 Stuart, P.G. (1993). *Taguchi Methods: A Hand on Approach*. Reading, MA: Addison-Wesley.

24 Deb, K., Pratap, A., Agarwal, S., and Meyarivan, T.A. (2002). A fast and elitist multiobjective genetic algorithm: NSGA-II. *IEEE Transactions on Evolutionary Computation* 6 (2): 182–197.

25 Miettinen, K. (2012). *Nonlinear Multiobjective Optimization*. Springer Science & Business Media.

26 New York Independent System Operator (2018). *Open Access Same-Time Information System: Load Data/ISO Load Forecast Data [Datafile]*. New York: New York Independent System Operator http://mis.nyiso.com/public/csv/pal/20180601pal_csv.zip.

27 Zhang, R., Dong, Z.Y., Xu, Y. et al. (2013). Short-term load forecasting of Australian National Electricity Market by an ensemble model of extreme learning machine. *IET Generation Transmission and Distribution* 7 (4): 391–397.

28 Zimmerman, R.D., Murillo-Sánchez, C.E., and Thomas, R.J. (2010). MATPOWER: steady-state operations, planning, and analysis tools for power systems research and education. *IEEE Transactions on Power Systems* 26 (1): 12–19.

29 Siemens Power Technologies International (2013). *PSS®E 33.4 Program Application Guide, vol. 2. New York: Siemens*.

30 Agrawal, S., Panigrahi, B.K., and Tiwari, M.K. (2008). Multiobjective particle swarm algorithm with fuzzy clustering for electrical power dispatch. *IEEE Transactions on Evolutionary Computation* 12 (5): 529–541.

31 Cutsem, T.V., Glavic, M., Rosehart, W. et al. (2015). *Test Systems for Voltage Stability Analysis and Security Assessment*. IEEE.

32 Lee, Y. and Song, H. (2019). A reactive power compensation strategy for voltage stability challenges in the Korean power system with dynamic loads. *Sustainability* 11 (2): 326.

20

Many-objective Robust Optimization-based Dynamic VAR Resource Planning

20.1 Introduction

The VAR planning to enhance the voltage stability of the power system is usually formulated as a deterministic optimization problem. Different uncertainties, including but not limited to the wind power generation and the initial status of STATCOM before a contingency that has influential impacts on the dynamic voltage response of the power system, are usually neglected [1–3]. A significant change in the operating condition, compared with the condition used in the planning models, might decrease the robustness of the solutions. Recently, some researchers have explored different approaches to improve the robustness of planning decisions. In the earlier stage of Var planning research, the uncertainty of load level is usually considered [4, 5]. These works focus on the impacts of the load variations on their optimization objectives, such as the operational cost in Ref. [4] and the voltage stability in Ref. [5]. The robustness of the solutions is explicitly considered. In Ref. [6], the uncertainty of load models is considered. The impact of this operational uncertainty is studied in terms of voltage stability. A robust VAR planning model is proposed in Ref. [7, 8], considering the uncertainty of load level. However, these two studies are only focusing on steady-state voltage stability enhancement. A multi-objective VAR planning model is proposed in Ref. [9], considering several uncertainties simultaneously. Probability distribution function (PDF) of wind power output and the probability of load variations are needed in the model. The accuracy and effectiveness of the method largely depend on the quality of these probability data. However, it is difficult to obtain these data for a long-term planning problem accurately. The wind power uncertainty is formulated as a multi-scenario generation resource in Ref. [10], but the probability data for the "load factor" and "capacity factor" are still needed. A scenario-based method is another alternative to address the uncertainty issue [11, 12], and the main concern

Stability-Constrained Optimization for Modern Power System Operation and Planning,
First Edition. Yan Xu, Yuan Chi, and Heling Yuan.

with this method is the computation burden. In a word, previous VAR planning models failed to consider the operational uncertainties in a systematic way and were overdependent on the quality of probability data. Given these limitations, the planning solutions from these methods are not robust to operational uncertainties, such as load levels and wind power outputs. They would be less effective and even impractical when the operational uncertainties change significantly.

On the other hand, considering the increasing number of economic and technical objectives, such as steady-state voltage stability, short-term voltage stability, investment cost, and the robustness of planning decisions, the prevalent multi-objective VAR resources planning model, with no more than three objectives, is no longer sufficient. Although the weighted-sum method can be adopted as a prevalent approach to address this many-objective issue, the information will be lost by using the weighted-sum method to combine several objectives into one. Besides, it is also challenging to determine the weight parameters for a practical engineering problem.

To address the inadequacies of the existing works, a robust VAR planning method for the voltage stability enhancement (including steady-state and short-term) of a wind-penetrated power system toward many objectives (i.e. four or more objectives) is proposed in this chapter.

20.2 Robustness Assessment of Planning Decisions

As illustrated in Figure 20.1, the variation of operational uncertainties might compromise the optimality of the solution. Compared with the operational status considered in VAR planning problems, the actual operational status, due to the inevitable variation of operational uncertainties, might lead to a lower steady-state voltage in post-contingency, a delayed voltage recovery, or more severe voltage

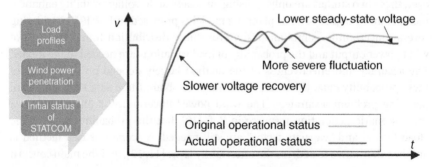

Figure 20.1 Impact of variations of uncertainties on the voltage dynamics.
Source: Adapted from Chi et al. [13] / With permission of IEEE.

fluctuation. To address this inadequacy, the concept of robust optimality is proposed in this chapter. It is used to describe the robustness of the solutions. Furthermore, sensitivity region is employed to describe the sensitivity of these trade-off solutions, and a robust many-objective optimization method is developed with a robustness index (RI), which evaluates the robust optimality of the solutions quantitatively and also serves as one of the objectives. Finally, a many-objective robust optimization problem is formulated.

20.2.1 Acceptable Sensitivity Region (ASR)

As shown in Figure 20.1, random variations of uncertain variables could lead to changes in the values of objective functions (namely, the evaluation results). A robust solution is a solution with objective functions whose values are insensitive to or less sensitive to uncertainty realizations. Given a solution u_0 that is analyzed for its robustness, acceptable sensitivity region (ASR) [14] can be formed in the space of objective function value around the nominal value $f_i(x_0, u_0, \widetilde{w}_0)$, as shown in Figure 20.2a. Subscript "0" of the variables in f_i stands for the original solution. ASR is the maximum acceptable variation of the objective function value

Figure 20.2 Illustration of (a) acceptable sensitivity region (ASR) and (b) its corresponding Δf_i. *Source:* Adapted from Chi et al. [13] / With permission of IEEE.

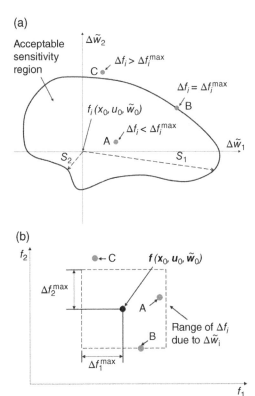

when subjected to the variations of specific uncertain parameters ($\Delta\widetilde{w}$). For a better elaboration, a case of two uncertainties ($\Delta\widetilde{w}_1$ and $\Delta\widetilde{w}_2$ in Figure 20.2a) and two objective functions (f_1 and f_2 in Figure 20.2b) are used as an example. Figure 20.2a shows the shape of ASR when the objective functions are subjected to uncertainties. Point A is the scenario when the variation of f_i, which is represented as Δf_i, is smaller than the maximum acceptable variation Δf_i^{max} when subjected to a specific uncertain parameter. Point B represents the scenario when Δf_i equals Δf_i^{max}; point C represents the scenario when Δf_i is larger than Δf_i^{max}. Correspondingly, Figure 20.2b illustrates the range of Δf_i (marked as a dashed rectangle) due to $\Delta\widetilde{w}_i$. The points A, B, and C in Figure 20.2a correspond to points A, B, and C of the Δf_i ranges (Figure 20.2b), respectively. The smaller the range of Δf_i is, the more robust the solution is, because it means that the same variation of uncertainties leads to a smaller change in the objective function.

For a practical engineering problem, the shape of ASR is always asymmetrical and irregular. It is time-consuming to calculate the area of ASR directly. Therefore, the concept of sensitive direction is employed in this chapter to estimate the ASR quantitatively. For instance, as shown in Figure 20.2a, the change rate of Δf_i is the smallest in direction s_1, which indicates that the objective function is the least sensitive to the variations of uncertainty in this direction. The change rate of Δf_i is largest in direction s_2, which indicates that the objective function is most sensitive to the variations of uncertainty in this direction.

20.2.2 Robust Optimality and Robust Pareto Front

In a conventional VAR planning model, the optimality of a solution is only achieved (or validated) under one or a group of specified operational conditions.

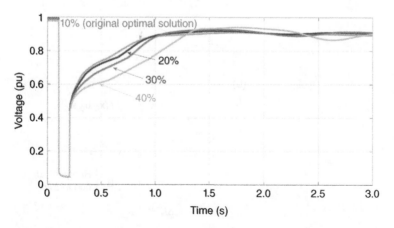

Figure 20.3 Impact of wind power penetration level on the voltage dynamics. *Source:* Adapted from Chi et al. [13]/With permission of IEEE.

Figure 20.3 illustrates the impact of varying wind power output on the post-contingency voltage response. First, the voltage response of an optimal solution with a 10% wind power penetration level is considered as an original case. As the penetration level increases, the dynamic voltage response is getting worse. In this chapter, the concept of *robust optimality* is proposed to describe the phenomena and is incorporated into the VAR planning model to improve the robustness of the solutions.

The Pareto front based on a fixed $\tilde{w}_0$ is defined as a nominal Pareto front. Generally, a robust Pareto front is inferior to the nominal Pareto front [15], in terms of the objective functions. Figure 20.4 illustrates three typical scenarios:

a) the nominal Pareto front is completely superior to the robust Pareto front;
b) there are one or several overlaps between the two Pareto fronts. The rest of the solutions from the robust Pareto front are suboptimal, compared with the nominal Pareto front.
c) the robust Pareto front is a subset of the nominal Pareto front.

Ideally, a perfect Pareto front is the one that does not change when subjected to variations of uncertainties (to be the same as the nominal Pareto front). However, in practice, it is not always the case. The robust Pareto set is generally inferior to the nominal Pareto set, but at the same time, with higher robustness. The purpose of introducing the idea of robust Pareto front is for the selection of solutions more robust to variations of uncertainties. Robust optimality refers to the insensitivity of the

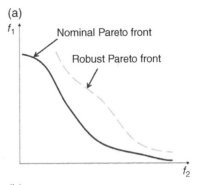

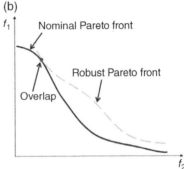

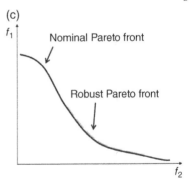

Figure 20.4 Comparison between nominal Pareto front and Robust Pareto front. (a) the nominal Pareto front is completely superior to the robust Pareto front, (b) there are one or several overlaps between the two Pareto fronts, (c) the robust Pareto front is a sub-set of the nominal Pareto front. *Source:* Adapted from Chi et al. [13] / With permission of IEEE.

optimal solutions when subjected to uncertainties. Although a robust solution might be a suboptimal Pareto solution, the robust optimality of the robust Pareto solution is improved when subjected to the variations of $\tilde{w}$. Besides, there is no probability distribution needed in the proposed robust VAR planning model. Given the difficulty in reducing the detrimental impact of uncertainties on a wind-penetrated power system, the robust Pareto front is an acceptable and practical trade-off.

20.2.3 Worst-case Sensitivity Region (WCSR)

Generally, the larger the area of ASR is, the more robust the original solution is. However, for a practical engineering problem, the shape of ASR might be extremely irregular. Calculating the area of ASR directly is time-consuming. On the other hand, the worst case is not fully reflected in the robustness analysis under this scenario. Therefore, based on the sensitive direction discussed in Section 20.2.1 and Figure 20.2, a concept of worst-case sensitivity region (WCSR) is used to quantitatively estimate the robustness of the solution under multiple uncertainties, with an acceptable computation complexity.

WCSR is defined as the largest n-dimensional sphere within the ASR (n is the number of uncertainties). A 2-dimensional case is illustrated in Figure 20.5. Instead of calculating the area of ASR, the radius, R^w, of the symmetric WCSR is employed as an indicator of robustness against uncertainties. Specifically, R^w is calculated based on L_p-norm in (20.1), of which p can be 1, 2, or ∞. Different L_p-norms are illustrated in Figure 20.6.

$$\min_{\Delta \tilde{w}} R^f(\Delta \tilde{w}) = \left[\sum_n^{N^u} |\Delta \tilde{w}_n|^p \right]^{1/p} \tag{20.1}$$

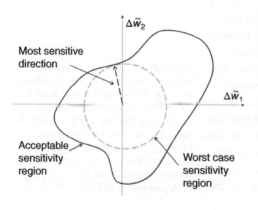

Figure 20.5 Illustrations of WCSR. *Source:* Adapted from Chi et al. [13] / With permission of IEEE.

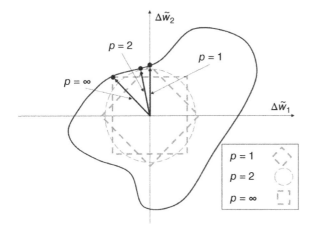

Figure 20.6 Illustrations of different L_p-norms. *Source:* Adapted from Chi et al. [13] / With permission of IEEE.

$$\text{s.t.} \quad \max_{i = 1, 2, \ldots, N^f} \left(\left| \Delta f_i - \Delta f_{0,i} \right| \right) \leq \tau_i^f \tag{20.2}$$

where $\Delta f_{0,i}$ indicates an acceptable deviation of objective function caused by $\Delta \tilde{w}$. Δf_i is defined as the difference between $f_i(x_0, u_0, \tilde{w}_0 + \Delta \tilde{w})$ and $f_i(x_0, u_0, \tilde{w}_0)$. Since the actual area of ASR can be estimated through the radius, the robustness of a solution can be estimated by R^w as (20.1).

20.2.4 Feasibility Sensitivity Region (FSR)

Feasibility sensitivity region (FSR) is employed to evaluate the feasibility of the solution, by checking the constraints. As illustrated in Figure 20.7, any $\Delta \tilde{w}$ within

Figure 20.7 Illustrations of FWCSR. *Source:* Adapted from Chi et al. [13] / With permission of IEEE.

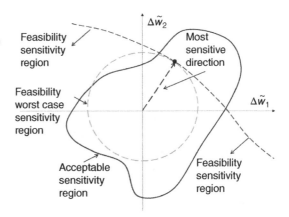

feasibility worst-case sensitivity region (FWCSR) will not compromise the feasibility of the solutions. R^f, calculated as L_p-norms in (20.3), is used to evaluate the solution's feasibility. The boundary of FWCSR is determined by (20.4) and (20.5).

$$\min_{\Delta \tilde{w}} R^f(\Delta \tilde{w}) = \left[\sum_n^{N^u} |\Delta \tilde{w}_n|^p \right]^{1/p} \tag{20.3}$$

$$\text{s.t.} \quad \max_{i = 1, 2, \dots, N^h} [h_i(x_0, u_0, \tilde{w}_0 + \Delta \tilde{w})] \leq 0 \tag{20.4}$$

$$\max_{i = 1, 2, \dots, N^g} [|g_i(x_0, u_0, \tilde{w}_0 + \Delta \tilde{w})|] \leq \tau_i^g \tag{20.5}$$

20.3 Many-objective Dynamic VAR Planning Model

Considering the increasing number of objectives when steady-state voltage stability, short-term voltage stability, investment cost, and the new robust metrics proposed in Section 20.1 are incorporated in VAR planning model, a compact mathematical model of a many-objective optimization problem is presented as (20.6–20.8), wherein (20.6) F is the overall objective function and $f(\cdot)$ is the sub-objective function. x and $\tilde{w}$ indicate the power system's state and the uncertain variables, respectively. u denotes the decision variables. Constraints (20.7) are the steady-state power balance constraints and post-contingency power balance constraints. Constraints (20.8) represent the steady-state operational limits, the post-contingency rotor angle constraint, and LVRT/HVRT constraint. This compact mathematical model will be expanded equation by equation in the following subchapters.

$$\min_u F\left[f_1(x, u, \tilde{w}), f_2(x, u, \tilde{w}), \dots, f_{Nf}(x, u, \tilde{w}) \right] \tag{20.6}$$

$$\text{s.t.} \quad G(x, u, \tilde{w}) = 0 \tag{20.7}$$

$$H(x, u, \tilde{w}) \leq 0 \tag{20.8}$$

In this chapter, the capacity of STATCOM at each installation bus is the decision variable. Two operational uncertainties are incorporated for the robust VAR planning model: (i) wind power generation, and (ii) the difference between the actual output of STATCOMs before a contingency and their rated capacity. It should be noted that other operational uncertainties, such as load levels, can also be incorporated if needed.

20.3.1 Optimization Objectives

The many-objective robust optimization problem consists of four sub-objectives:
(1) total planning cost, (2) $TVSI^r$, (3) $VCPI^p$, and (4) Robustness Index (RI).

20.3.1.1 Total Planning Cost (f_1)

The first objective function is the total planning cost. It is calculated as (20.9).

$$f_1 = C^{invest} = \sum_{i \in \mathscr{B}_s} C_i^{invest} \tag{20.9}$$

$$C_i^{invest} = \begin{cases} c_i^{install} + c^{cap} \cdot cap_i, & \text{if } cap_i \neq 0 \\ 0, & \text{if } cap_i = 0 \end{cases} \tag{20.10}$$

20.3.1.2 Time-Constrained TVSI (f_2)

The second objective function is for short-term voltage stability evaluation. It is
calculated as (20.11). Like previous chapters, a risk-based approach is used in this
chapter for the contingency analysis, considering their probabilities and diverse
impacts on the voltage performance of the system. The time-constrained short-
term voltage stability that can accurately reflect the severity of a large disturbance,
namely $TVSI_k^r$, is calculated based on (13.29), as presented in Chapter 14.

$$f_2 = \sum_{k \in \mathscr{K}} TVSI_k^r \cdot p_k \tag{20.11}$$

20.3.1.3 Prioritized VCPI (f_3)

The original VCPI neglects the priority and impact of different lines on the power
system since it simply calculates the sum of VCPIs of all the lines as the final result.
In this chapter, a prioritized index based on the original VCPI is developed as
(20.12), where the $VCPI_l$ is calculated based on (13.14). $VCPI_k^p$ takes the priorities
of lines into evaluation. In practice, the weighting factors for each line (q_l) are
determined by three criteria: (i) the ratio of power flow over line capacity,
(ii) the geographical location of lines, and (iii) sensitivity analysis when subjected
to different STATCOM capacities.

$$VCPI_k^p = \sum_{l \in \mathscr{L}} \left(q_l \cdot VCPI_l - \sum_{l \in \mathscr{L}} VCPI / N^l \right)^2 / N^l \tag{20.12}$$

The third objective function, calculated as (20.13), is for steady-state voltage stability assessment. A risk-based criterion is also used here, like (20.11).

$$f_3 = \sum_{k \in \mathscr{K}} VCPI_k^p \cdot p_k \tag{20.13}$$

20.3.1.4 Robustness Index (f_4)

For a practical engineering problem, R^w and R^f can be calculated for each solution and the solutions can be compared. However, with R^w and R^f alone, the decision-makers still cannot tell if this decision is robust or not. To address this inadequacy, a RI based on WCSR and FWCSR is proposed as (20.14). R^{ref} is a reference robust radius that is determined by an acceptable uncertainty vector $\Delta \widetilde{w}$. By introducing the reference, not only the robustness of a solution can be optimized with other objectives, but also it can be judged without comparison with other solutions: it is considered as a robust solution if f_4 is smaller than one and it is not robust if f_4 is larger than one.

$$f_4 = R^{ref} / \min\left(R^w, R^f\right) \tag{20.14}$$

20.3.2 Constraints

Constraints (20.15) are active and reactive power flow balances. The steady-state operational constraints, including voltage magnitude, line capacity, and generation output capacity, are expressed in constraints (20.16).

$$\begin{cases} P_g + \tilde{P}_w - P_d - V_i \sum_{j \in \mathscr{B}} V_j \left(G_{ij} \cos \theta_{ij} + B_{ij} \sin \theta_{ij}\right) = 0 \\ Q_g + \tilde{Q}_w + Q^{stat} - Q_d - V_i \sum_{j \in \mathscr{B}} V_j \left(G_{ij} \sin \theta_{ij} - B_{ij} \cos \theta_{ij}\right) = 0 \end{cases} \tag{20.15}$$

$$\begin{cases} V_i^{\min} \leq V_i \leq V_i^{\max} \\ L_l^{\min} \leq L_l \leq L_l^{\max} \\ P_g^{\min} \leq P_g \leq P_g^{\max} \\ Q_g^{\max} \leq Q_g \leq Q_g^{\max} \end{cases} , \forall i \in \mathscr{B}, \forall l \in \mathscr{L}, \forall g \in \mathscr{G} \tag{20.16}$$

Constraint (20.17) serves as a transient stability constraint, which means that the maximum allowed rotor angle difference between any two generators ($\Delta \delta_{gg'}^T$) during a transient period of T of a contingency k should be smaller than or equal to ω. This constraint is used to check the transient rotor angle stability for each contingency. Constraints (20.18) are for LVRT and HVRT profiles. $V_{i,t}$ presents the time-varying voltage trajectory,

$$\left[\max\left(\Delta \delta_{gg'}^T\right)\right]_k \leq \omega \quad \forall g, g' \in \mathscr{G}_s, \forall k \in \mathscr{K} \tag{20.17}$$

$$\begin{cases} V_{i,t} - LVRT(t) \geq 0, \forall t \in T, \forall i \in \mathscr{B}_w \\ V_{i,t} - HVRT(t) \leq 0, \forall t \in T, \forall i \in \mathscr{B}_w \end{cases} \tag{20.18}$$

20.4 Many-objective Optimization Algorithm

20.4.1 Non-dominated Sorting Genetic Algorithm-III (NSGA-III)

Non-dominated sorting genetic algorithm- II (NSGA-II) has been demonstrated to be one of the most effective algorithms on many benchmark problems [16], and it is also one of the most prevalent algorithms to solve the multi-objective VAR planning problem [2, 3], as well as other power system planning problems [17–19]. Nonetheless, a direct implementation of the original NSGA-II on a VAR planning problem [3] has some limitations due to the inherent drawbacks of NSGA-II when applied to a complex problem, including premature convergence, poor solution diversity, and the large distance between the solutions. Besides, considering the increasing number of objectives when steady-state voltage stability, short-term voltage stability, investment cost, and the new robust metric proposed in this chapter are incorporated in VAR planning model, the algorithms, such as NSGA-II and MOEA/D [2, 6], used to solve the multi-objective optimization problem (two or three objectives) are no longer sufficient. As reported in Ref. [20], for many-objective (four or more objectives) optimization problems, the performance of NSGA-II is not satisfactory. The weighted-sum method is a prevalent approach to address this issue. However, the information will be lost by using the weighted-sum method to combine several objectives into one, and it is also very difficult to determine the weight parameters for a practical engineering problem.

As an advanced alternative for NSGA-II, a reference point-based many-objective evolutionary algorithm, called NSGA-III, is proposed by Deb and Jain in [21]. Generally, the framework of NSGA-III is quite similar to that of NSGA-II [16]. However, NSGA-III has a better performance in solving many-objective problems because of an improved selection mechanism of the offspring. The diversity of individuals in a population is maintained by adaptively supplying and updating well-distributed reference points. The main body of NSGA-III is described as follows:

Algorithm 20.1   NSGA-III

1) Classify the population.
2) Calculate the number of reference points (H) to place on the normalized hyperplane.
3) Realize the non-dominated sorting.
4) **For** $t = 1$ *Stopping criteria* do.
5) Select P_1 and P_2 through tournament method.

(Continued)

Algorithm 20.1 (Continued)

6) Perform crossover between P_1 and P_2 (with probability P_c).
7) Realize the non-dominated sorting.
8) Normalize the population members.
9) Associate population members with the reference points.
10) Apply the niche preservation.
11) Keep the niche-obtained solution for the next generation.
12) **End for**

20.4.1.1 Classification of Population

For a parent population (of size N) at the tth generation (P_t), the offspring population (Q_t) with N members is obtained by the recombination and the mutation of P_t. The combination of the offspring population and parent population, namely C_t, are sorted to different nondominant levels $(F_1, F_2, ..., F_l)$ in order to preserve elite members. So, a new population S_t is constructed by selecting individuals of each nondominant level until the size of $S_t \geq N$ (starting from the first nondominant level). For $S_t = N$, no further action is needed, and the next generation begins with $P_{t+1} = S_t$. For $S_t > N$, since members from one to the second last front are already selected, the remaining population members are selected from the last front.

20.4.1.2 Determination of Reference Points

Das's systematic approach [22] is employed to place points on a normalized hyperplane that is equally inclined to all objective axes and has an intercept of one on each axis. If p divisions are considered with each objective, the total number of reference points (H) in a G-objective problem is defined by (20.19).

$$H = \binom{G + p - 1}{p} \tag{20.19}$$

20.4.1.3 Normalization of the Population Members

The normalization of objective values is carried out according to (20.20–20.22). The ideal point of S_t is determined by the identification of the minimum value (z_m^{min}) for each objective function $(m = 1, 2, ..., M)$, and by the construction of the ideal point $\bar{z} = (z_1^{min}, z_2^{min}, ..., z_m^{min})$. $f'_m(x)$ is the translated objective function. $ASF(x, \omega)$ is the extreme point value in each objective axis and $f''_m(x)$ is the normalized objective function. ω_m denotes the weight vector of each objective, and a_m denotes the intercept of the mth objective axis.

$$f'_m(x) = f_m(x) - z_m^{min} \qquad (20.20)$$

$$ASF(x, \omega) = \max_{m-1}^{M} f'_m(x)/\omega_m \qquad (20.21)$$

$$f_m^n(x) = f'_m(x)/\left(a_m - z_m^{min}\right) \qquad (20.22)$$

20.4.1.4 Association Among Reference Points and Solutions

After normalization, a reference point will be associated with each population member. A reference line corresponding to each reference point on the hyperplane is defined by joining the reference point with the origin. The shortest perpendicular distance between each individual of S_t and each reference line is calculated. If the reference line of a reference point is closest to a population member in the normalized objective space, it is considered to be associated with the population member.

20.4.1.5 Niche Preservation Operation

After the association operation, "niche preservation" operation is carried out to select individuals (of ith level) that are associated with each reference point. Details of the niche preservation procedures can be found in Ref. [21].

20.4.1.6 Generation of Offspring

After the formation of P_{t+1}, Q_{t+1} can be generated by applying the usual crossover and mutation operator by randomly choosing parents from P_{t+1}.

20.4.2 Adaptive NSGA-III

Although NSGA-III has been implemented to solve engineering problems successfully [23, 24], there are still some shortcomings when NSGA-III is employed to solve complex many-objective engineering problems.

20.4.2.1 Fixed Mutation Rate

Although a smaller mutation rate is favorable for the 1st generation because of the relatively good diversity, a larger mutation rate is better to improve the diversity when the diversity of later generations becomes poor. Since NSGA-III uses a fixed mutation rate for all generations, it would be difficult to select a suitable mutation rate.

20.4.2.2 Random Sampling

The sampling method adopted in NSGA-III is simple random sampling. When it generates a new sample point, it does not take previously generated sample points into consideration. This pure randomness might lead to an inferior generation: some sampling points clustered closely, while some intervals in the space had no samples at all.

20.4.2.3 Simulated Binary Crossovers (SBX)

Simulated binary crossovers (SBX) used in the original NSGA-III have a strong local search ability. However, it prefers offspring near the parents to offspring away from parents. This preference increases the possibility of local optimality, especially for a complex practical engineering problem.

To address the aforementioned disadvantages and improve the original NSGA-IIII for a practical engineering application, an adaptive NSGA-III-LHS is proposed in this chapter with the following three improvements.

20.4.2.4 An Adaptive Mutation Rate

An adaptive strategy is developed as (20.25) to adaptively adjust the mutation rate based on not only the generation number but also the crowdedness of a generation. Firstly, a crowdedness index (CI) is defined in (20.23). Then, it is unified in (20.24) to make sure that the CIs of the nth generation are in the range of [0, 1].

$$CI_n^* = \sum_{d=1}^{N^{ind}} \left(u_{d,n} - u_n^{avg}\right) / \left(u_n^{max} - u_n^{min}\right) \tag{20.23}$$

$$CI_n = \frac{CI_n^* - CI_n^{min}}{CI_n^{max} - CI_n^{min}} \tag{20.24}$$

Based on CI, an adaptive mutation rate p_m can be calculated as (20.25).

$$P_m = p_0 + \frac{CI_n \cdot (1 - p_0) \cdot (n - 1)}{n^{max} - 1} \tag{20.25}$$

where n and n^{max} are the current and maximum generations, respectively.

The improvements in the adaptive mutation rate are twofold: (i) for the first few generations, the result of $(1 - p_0) \cdot (n - 1)/(n^{max} - 1)$ is relatively small, and as the optimization proceeds, the result of $(1 - p_0) \cdot (n - 1)/(n^{max} - 1)$ increases gradually; (ii) since CI_n entirely depends on the crowdedness of a population, p_m will also be modified as the crowdedness of the population changes. By applying (20.25), instead of being a fixed value like the original NSGA-III or blindly increasing proportionally as the number of generations increases, the mutation rate becomes more adaptive.

20.4.2.5 Latin Hypercube Sampling

NSGA-III and most other genetic algorithms use a simple random sampling approach to generate the 1st generation. Since the simple random sampling does not consider the previous samples when generating new ones, it will bring a nonuniform sampling problem, as discussed at the beginning of this subchapter.

So, dozens of repeated experiments are required to obtain an average result to cancel out this nonuniform problem. However, considering the high computation burden of a practical engineering problem, these repeated experiments are time-consuming and unacceptable, especially for a VAR planning problem considering short-term voltage stability and various operational uncertainties. Therefore, Latin hypercube sampling (LHS), an experimental design method first proposed in Ref. [25], is adopted in this chapter to generate the first generation. Generally, the samples generated by LHS can represent the real variability in a sampling space (in this chapter, it is the range of potential STATCOM capacity) more accurately than simple random sampling.

If a random input vector X has m dimensions $X = (X_1, ..., X_m)$ and each X_j is independent of the others ($j = 1, ..., m$). An LHS set can be generated as (20.26):

$$X_{i,j} = F^{-1}\left(\frac{\pi_j(i) - U_{i,j}}{n}\right), i = 1, 2, ..., n; j = 1, 2, ..., m \tag{20.26}$$

where n is the total number of samples and $\pi_j(1), ..., \pi_j(n)$ are random permutations of $[1, ..., n]$. $U_{i,j}$ is a random variable from a standard uniform distribution. The LHS approach is characterized by a segmentation of the assumed probability distribution into a number of randomly sampled and non-overlapping intervals with equal probability.

20.4.2.6 Differential Evolution (DE) Operator

Differential evolution (DE) operator, which has a good global search ability [26], is used in this chapter instead of the original SBX. It utilizes the differences between randomly selected individuals as evolutionary dynamics. Unlike the preference of SBX over nearer offspring, DE operator uses a "jump in neighborhood search" approach (adding weighted vectors to the target vector properly) to control the evolutionary variation. Procedures of DE operator are as follows:

a) **Input**: Population size and the population P_n.
b) **Random selection**: Select three individuals $x_{r_1,n} \neq x_{r_2,n} \neq x_{r_3,n}$. They do not dominate each other from P_n.
c) **Mutation operator**: Generate mutant vector $v_{i,n+1} = [v_{i,1}, v_{i,2}, ..., v_{i,j}, ..., v_{i,D}]$ (D represents the dimension of a problem and K is the scale factor).

$$v_{i,n+1} = x_{r_1,n} + K \cdot (x_{r_2,n} - x_{r_3,n}) \tag{20.27}$$

d) **Cross operator**: Generate a trial vector $u_{i,n+1} = [u_{i,1}, u_{i,2}, ..., u_{i,j}, ..., u_{i,D}]$ by exchanging the components between the x_i and v_i. Usually, there are two crossover methods used in DE: binomial and exponential. In this chapter, the binomial crossover method is employed as (20.28).

$$u_{i,n+1} = \begin{cases} v_{i,n+1}, \text{if } rand(0,1) \leq CRorj = j^{rand} \\ x_{i,n}, \text{if } rand(0,1) > CRorj \neq j^{rand} \end{cases} \tag{20.28}$$

where CR is a crossover rate, and j^{rand} is a number randomly selected from $[1, 2, ..., D]$.

20.4.3 Computation Steps

The computational flowchart is illustrated in Figure 20.8, and the computation steps are described as follows:

1) **Initialization**: After the algorithm parameters setting, the candidate bus is selected. The individuals of the first generation are generated based on LHS.
2) **Steady-state voltage stability and robustness evaluation**: Steady-state voltage stability is evaluated with the different priorities of buses. The evaluation result for the base case, $f_3(x_0, u_0, \widetilde{w}_0)$, is used to calculate the uncertainty sensitivity region and feasibility sensitivity region.
3) **Short-term voltage stability and robustness evaluation**: The voltage performance of the system in the post-contingency stage is evaluated based on the voltage trajectories obtained from the time-domain simulations. The evaluation result for the base case, $f_2(x_0, u_0, \widetilde{w}_0)$, is used to calculate the uncertainty sensitivity region and feasibility sensitivity region.
4) **Calculations of objectives**: Four objectives are calculated in a risk-based approach, including total planning cost, evaluation result for steady-state voltage stability, evaluation result for short-term voltage stability, and RI.
5) **Individuals updating**: Find the best individual of the current generation and then update it as the best individual of the generation. The update further contains two steps: (i) comparison between each individual, and (ii) selection of the best Pareto solution.
6) **Termination**: Two stopping criteria are adopted in this chapter: (i) change in fitness function is smaller than a specific tolerance, and (ii) the maximum iteration number is reached. If one of them is satisfied, the computation procedure will be terminated, and the final optimization result will be presented as a Pareto front.
7) **Mutation update and offspring generation**: If the stopping criteria are not satisfied, the mutation rate will be modified as (20.25). Then, the computation procedure will go to Step 2 after new offspring are generated.
8) **Break criteria**: If any post-contingency constraints are violated, the assessment of the individual will be terminated. The objective functions will be calculated accordingly with high penalty parameters. Then, the computation procedure will go to Step 4.

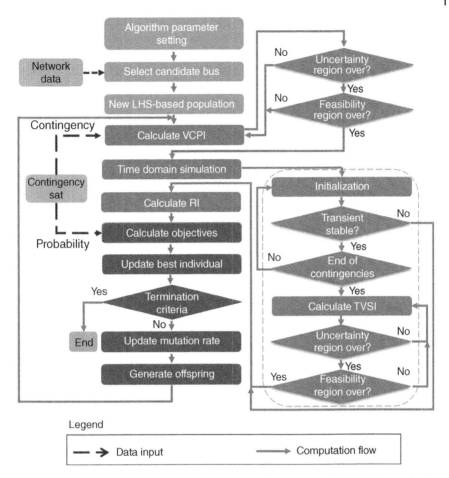

Figure 20.8 Illustrations of FWCSR. *Source:* Adapted from Chi et al. [13] / With permission of IEEE.

20.5 Case Studies

20.5.1 Test System and Parameters

The effectiveness and efficiency of the proposed method are validated on the New England 39-bus test system with several modifications, as illustrated in Figure 20.9. Compared with the original test system, synchronous generators at bus 30, bus 32, and bus 34 are replaced by wind farms. All the wind turbines are modeled as double-fed induction generators in PSS®E 33.4 [27]. Specifically, the generator model, the mechanical model, the electrical model, and the pitch control model are the same ones used in previous chapters. The STATCOM model is SVSMO3 as illustrated in Figure 15.4, Section 15.2. An industry-standard complex load model "CLOD" [27], used in Chapter 17, is also adopted to simulate the time-varying load dynamics. The main parameters for case studies are listed in Table 20.1. In a real-world application, the parameters for the CLOD model can be determined based on historical load data.

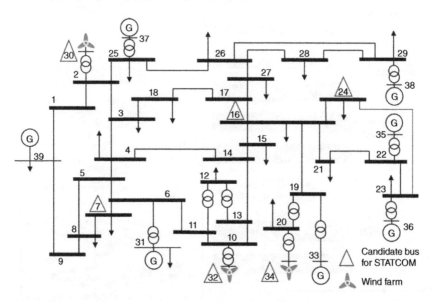

Figure 20.9 New England 39-bus test system. *Source:* Adapted from Chi et al. [13] / With permission of IEEE.

Table 20.1 Parameters for simulation setting.

Parameters	Value
$c^{install}$	1.5 million/unit
c^{cap}	0.05 million/MVAR
p_0	0.15
ω	π
t^{dl}/t^{du}	0.5/0.5 s
v^{dl}/v^{du}	0.9/1.1 pu
Penalty parameter (α^l and α^u)	2.0
LHS levels	20 (0 ∼ 100 MVAR)
LM, SM, DL, TS, CP, K_p	25%, 15%, 10%, 10%, 10%, 2
Fault setup	3-phase short-circuit (cleared after 0.1 s)

Source: Adapted from Chi et al. [13] / With permission of IEEE.

Contingency analysis is based on the method used in Ref. [19]. Faults on line 2–3, line 6–7, and line 16–24 are selected as three critical contingencies. Two operational uncertainties are considered in the case studies: (i) the initial state of STATCOMs (0.6 \widetilde 1.0 pu) before the contingency, (ii) and wind power generation (20 \widetilde 30% penetration level). The peak load level is selected in the case studies because the dynamics of loads have a more profound impact on the short-term voltage stability of the system than load levels [28]. $p = 2$ (namely, *Euclidean* norm) is used in (20.1) and (20.3) for the calculation of RI. The steady-state analysis is carried out in MATLAB 2018b with MATPOWER 6.0, and the time-domain simulation is carried out in PSS®E 33.4.

20.5.2 Numerical Simulation Results and Comparisons

The overall simulation time is 220 790 seconds. The computation burden is acceptable for a long-term VAR planning problem. Figure 20.10 shows a Pareto front with 56 robust Pareto optimal solutions (as the planning decisions). As a four-dimensional illustration, RI (f_4) is shown in different colors.

Like in previous chapters, the fuzzy membership function [29] is used to select the final compromise solution from the robust Pareto front for further comparison and analysis. The mathematical details of the fuzzy membership function can be found in Section 20.3. A compromise solution is also marked in Figure 20.10. The installation plans of the solution for each candidate bus are listed in Table 20.2.

Two additional methods (method B and method C) are compared in Table 20.3 to show the advantages of the proposed method.

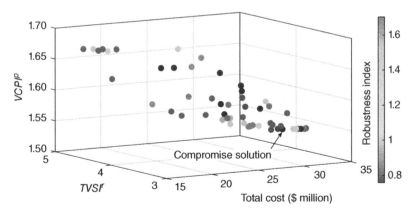

Figure 20.10 Pareto optimal solutions of the proposed method. *Source:* Adapted from Chi et al. [13] / With permission of IEEE.

Table 20.2 Installation decisions (in MVAR).

Bus No.	7	16	24	30	32	34
Capacity	96.3	97.2	100	29.7	98.5	47.7

Source: Adapted from Chi et al. [13] / With permission of IEEE.

Table 20.3 Simulation results.

Methods	Total cost (f_1)	Improved TVSI (f_2)	Improved VCPI (f_3)	Robustness index (f_4)	Computation time
A	**32.47**	**3.825**	1.539	**0.7589**	**220 790 s**
B	32.64	3.834	**1.492**	1.0707	25 036 s
C (avg.)	32.93	3.845	1.568	0.8289	2 322 498 s
C (best)	33.03	3.831	1.545	0.7956	227 696 s

Source: Adapted from Chi et al. [13] / With permission of IEEE.

Method A: It is the proposed method (with four objectives) based on the voltage stability indices introduced in this chapter. The proposed adaptive NSGA-III-LHS is used to solve the problem. The solution used for the comparison and analysis is a compromise solution selected from the robust Pareto front based on the fuzzy membership function [29].

Method B: It is the conventional method (with only three objectives, robust optimality is not considered) based on the voltage stability indices introduced in this chapter. NSGA-II is employed to solve the problem. The solution used for the comparison and analysis is a compromise solution selected from the normal Pareto front based on the fuzzy membership function.

Method C: It is the proposed method (with four objectives) based on the voltage stability indices introduced in this chapter. The original NSGA-III is utilized to solve the problem. Two solutions are selected from the robust Pareto front based on two different criteria, respectively. The first solution (marked as "avg." in Table 20.3) is a mean value of compromise solutions from 10 simulation runs (in Table 20.3, computation time is the total time for 10 runs). Repeated simulations are conducted to offset the problem caused by the simple random sampling, as discussed in Section 20.3. The second solution (marked as "best" in Table 20.3) is a compromise solution selected from the robust Pareto front based on the fuzzy membership function.

For all the evaluation results, a smaller value indicates better voltage stability or higher robustness, as listed in Table 20.3. The voltage stability performance of method A (the proposed method) is close to that of method B (conventional method). Method B even has better steady-state voltage stability (VCPI) compared with method A. However, the robust optimality of the solution from method B is much worse than that of the solution from method A. Although method B does not consider the robustness of the solution, the RI can still be calculated for method B based on Section 20.2, which is 41.1% worse than that of method A. Figure 20.11 compares the voltage trajectories of bus 16 between method A and method B under the worst scenario: the wind power penetration level is 30%, and the output of STATCOMs before the contingency is at the rated capacity. Once the fault is cleared, method A has a fast voltage recovery, while method B experiences a relatively slow voltage recovery.

Figure 20.12 shows two Pareto fronts of method A and method B. The RI of all Pareto optimal solutions of method B is calculated based on Section 20.2. Most of the solutions of method B have worse robustness (marked as diamonds) compared with the solutions of method A (marked as circles). On the other hand, some solutions of method B are better than those from method A, in terms of voltage stability and total cost. This is exactly the scenario presented in the robust analysis in Section 20.1, Figure 20.4b.

Compared with solutions of method C (mean value of 10 simulation runs), solutions of method A not only have better evaluation results for steady-state and short-term voltage stability, but also cost less. Besides, the computation burden

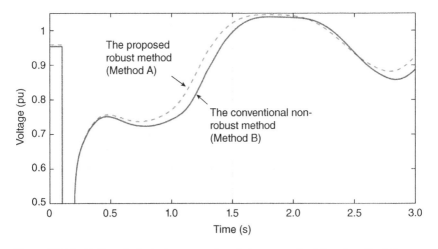

Figure 20.11 Voltage trajectories comparison between method A and method B.

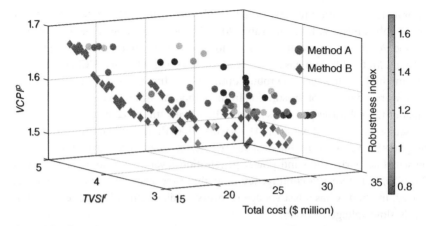

Figure 20.12 Pareto optimal solutions comparison of method A and method B.

of method A is much lighter (90.5% reduction) than that of method B. When compared with the best compromise solution from all the solutions generated by the 10 simulation runs, namely method C (best) in Table 20.3, method A still outperforms method C in all aspects, including voltage stability, robust optimality, total cost, and computation time. It demonstrates that the proposed adaptive NSGA-III-LHS effectively avoids the drawbacks of the original NSGA-III.

Figure 20.13 shows the advantage of the time-constrained TVSI. It is the post-contingency voltage trajectory of bus 34. For a clearer illustration of the voltage

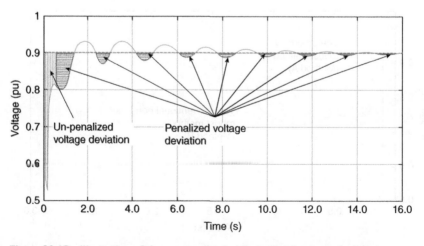

Figure 20.13 Illustration of the un-penalized and penalized voltage deviations.

deviation, the installation plan for STATCOMs used here is not the compromise solution of the proposed method. Based on the definition of the original TVSI, which does not differentiate between the common voltage deviations (within the required time window) and the delayed voltage recovery, the evaluation result (0.2828 after normalization) for short-term voltage stability is calculated by integrating all the voltage deviations. However, since the prolonged voltage deviation after six seconds contributes little to the final evaluation result, this minor but frequent voltage violation is not effectively reflected in the evaluation result. On the other hand, given the protection mechanism of wind turbines, they will be in danger of automatically triggering because of minor but frequent voltage deviations (after the four seconds). These dangers cannot be fully revealed in the original TVSI. By using the proposed time-constrained TVSI, these prolonged voltage deviations are penalized. So, a significant 64.63% increase in the evaluation result (0.4656 after normalization) can clearly reveal the potential danger imposed on wind turbines. Furthermore, a larger variation can also make the adaptive NSGA-III-LHS more efficient in finding optimal solutions.

20.5.3 Scalability and Computation Efficiency Discussion

When the short-term voltage stability is considered, it is time-consuming to solve the VAR planning problem due to its highly nonconvex and nonlinear properties, as well as the complicated time-domain simulations. Specifically, for this robust VAR planning problem, the total time cost T is generally proportional to the following parameters.

$$T \propto \left(N^c, N^{gen}, N^{pop}, N^{dec}, N^{unc}, T^{dts}, T^{OPF} \right) \tag{20.29}$$

where N^c is the number of critical contingencies verified in time-domain simulations. N^{gen}, N^{pop}, and N^{dec} are the number of generations, population size, and the number of decision variables for the evolutionary algorithm, respectively. N^{unc} is the number of uncertainties considered in the problem. T^{dts} denotes computation time for a single time-domain simulation. T^{OPF} is the computation time for a single OPF run.

As the studied network becomes larger, the corresponding increase in computation cost is relatively moderate when compared with the increase in total buses. This is partly because the number of candidate buses needed for a larger system, which determines the number of decision variables of the problem, is not necessary to increase proportionally with the number of the total buses. Another case study carried out on a Nordic 74-bus system [30] is used for comparison, as in Table 20.4. Although the number of total buses increases from 39 to 74 (89.7% increase), the computation time only slightly increases from 220 790 to 246 583 seconds (10.5%).

Table 20.4 Computation costs for different test systems.

Test system	Total computation time (s)
New England 39-bus	220 790
Nordic 74-bus	246 583

Source: Adapted from Chi et al. [13] / With permission of IEEE.

On the other hand, VAR planning of a large-scale power system usually follows an area-based approach rather than taking the whole system as an undivided entity [31] in practice. Studying the VAR planning problem in this area-based way can minimize the reactive power transmission between different areas and maximize the self-sufficiency of each area. Besides, it is also a long-standing industrial standard and engineering practice in countries like China [32, 33]. Therefore, it is practical to divide a large power system into areas and then study the VAR planning problem within each area, while other areas can be modeled as one or several equivalent subsystems. On the other hand, a distributed computing platform [34] can also be developed for PSS®E-based power system analysis. For a control center of a utility grid, parallel computing is prevalent. Although there is a concern over the communication efficiency between different processors and control centers, the computational burden of the proposed method can be significantly reduced by assigning tasks of contingency verifications and operation scenario verification using parallel computation techniques [34]. So, the computation burden is manageable for a real-world application.

Nomenclature

Indices

d index for individuals in a generation
g, g' index for generators
i, j indices of buses/objectives/constraints
k index for contingency
l index for lines
n index for uncertainties/generations in NSGA-III
o index for an original state
p index for Pareto solutions
t index for time

Sets

$\mathcal{B}$	set of all buses
$\mathcal{B}_s$	set of STATCOM buses
$\mathcal{B}_w$	set of wind turbine buses
$\mathcal{G}$	set of all generators
$\mathcal{G}_s$	set of synchronous generators
$\mathcal{L}$	set of lines
$\mathcal{K}$	set of contingencies
$\mathcal{P}$	set of Pareto optimal solution
$\mathcal{T}$	set of time
$\boldsymbol{u}$	set of control variables
$\widetilde{\boldsymbol{w}}$	set of uncertainties
$\boldsymbol{x}$	set of state variables

Variables

B_{ij}	susceptance between bus i and j
cap_i	reactive power source capacities at bus i
CI_n	crowdedness of the nth generation (normalized)
CI_n^*	crowdedness of the nth generation (un-normalized)
G_{ij}	conductance between bus i and j
L_l	apparent power flow on line l
p_k	probability of contingency k
p_m	adaptive mutation rate
P_g	active power output of generator g
P_l^r	active power transferred to the receiving end through the line l
P_n	parent population
$\widetilde{P_w}$	active output of wind turbine w
Q_g	reactive power output of generator g
Q_l^r	reactive power transferred to the receiving end through the line l
Q^{stat}	reactive power output of STATCOM
$\widetilde{Q_w}$	reactive power output of wind turbine w
R^f	radius of FWCSR
R^w	radius of WCSR
s_1	sensitive direction
s_2	insensitive direction
S_n	offspring population
$TVSI^r$	time-constrained TVSI
$u_{d,n}$	value of the fitness function of individual d in the nth generation

u_n^{avg} average value of fitness functions in n^{th} generation

$V_{i,t}$ voltage magnitude of bus i at time t

$VCPI^P$ prioritized VCPI of the whole system

$VCPI_l$ VCPI of line l

V^s annual maximum load equivalent hours

Z^s wind power curtailment hours

Δf_i variations of objective function i

$\Delta f_{0,i}$ acceptable deviation of objective function i

$\Delta \widetilde{w}$ variations of uncertainty

$\Delta \delta_{gg'}^T$ rotor angle deviation between 2 generators

β_p the pth Pareto solution from a Pareto Front

φ phase angle of load

ψ phase angle of line impedance

θ_{ij} voltage angle difference between bus i and j

μ_i^p satisfactory degree for ith objective function

Parameters

$c^{install}$ installation cost of STATCOM

c^{cap} unit cost of STATCOM

CP percentage of constant power loads in CLOD model

DL percentage of discharging lighting loads in CLOD model

$HVRT(t)$ time-varying voltage trajectory constraints for HVRT

K_p parameter for voltage-dependent real power load in CLOD model

LM percentage of large motor loads in CLOD model

$LVRT(t)$ time-varying voltage trajectory constraints for LVRT

N number of an initial population

N^b number of total buses

N^f number of objective functions

N^g number of equality constraints

N^h number of inequality constraints

N^{ind} number of individual (in a generation)

N^l number of total power lines

N^u number of uncertainties

p parameter for L_p-norm

p_0 mutation rate of the first generation

q_l priority of line l

R^{ref} reference robust radius

SM percentage of small motor loads in CLOD model

t^c	fault clearing time
t^{dl}	maximum voltage deviation time (lower)
t^{du}	maximum voltage deviation time (upper)
t^{end}	end of time-domain simulation
t^f	fault time
TS	percentage of transformer saturation loads in CLOD model
v^{dl}	lower voltage threshold
v^{du}	upper voltage threshold
α^l	penalty parameter for voltage violation (lower)
α^u	penalty parameter for voltage violation (upper)
τ_i^f	tolerance for an acceptable deviation $\Delta f_{0,\,i}$
τ_i^g	tolerance for constraint g_i
ω	rotor deviation in an extreme condition

Functions

C^{invest}	total investment cost
$f(\cdot)$	sub-objective functions
F	overall objective function
g_i	equality constraint i
G	equality constraints
h_i	inequality constraint i
H	inequality constraints

References

1 Sapkota, B. and Vittal, V. (2009). Dynamic VAr planning in a large power system using trajectory sensitivities. *IEEE Transactions on Power Systems* 25 (1): 461–469.

2 Xu, Y., Dong, Z.Y., Xiao, C. et al. (2015). Optimal placement of static compensators for multi-objective voltage stability enhancement of power systems. *IET Generation Transmission and Distribution* 9 (15): 2144–2151.

3 Liu, J., Xu, Y., Dong, Z.Y., and Wong, K.P. (2017). Retirement-driven dynamic VAR planning for voltage stability enhancement of power systems with high-level wind power. *IEEE Transactions on Power Systems* 33 (2): 2282–2291.

4 Eghbal, M., Yorino, N., El-Araby, E.E., and Zoka, Y. (2008). Multi-load level reactive power planning considering slow and fast VAR devices by means of particle swarm optimisation. *IET Generation Transmission and Distribution* 2 (5): 743–751.

5 Farsangi, M.M., Nezamabadi-pour, H., Song, Y.H., and Lee, K.Y. (2007). Placement of SVCs and selection of stabilizing signals in power systems. *IEEE Transactions on Power Systems* 22 (3): 1061–1071.

6 Xu, Y., Dong, Z.Y., Meng, K. et al. (2014). Multi-objective dynamic VAR planning against short-term voltage instability using a decomposition-based evolutionary algorithm. *IEEE Transactions on Power Systems* 29 (6): 2813–2822.

7 López, J., Pozo, D., Contreras, J., and Mantovani, J.R. (2016). A multiobjective minimax regret robust VAr planning model. *IEEE Transactions on Power Systems* 32 (3): 1761–1771.

8 Faried, S.O., Billinton, R., and Aboreshaid, S. (2009). Probabilistic technique for sizing FACTS devices for steady-state voltage profile enhancement. *IET Generation Transmission and Distribution* 3 (4): 385–392.

9 Han, T., Chen, Y., and Ma, J. (2018). Multi-objective robust dynamic VAR planning in power transmission girds for improving short-term voltage stability under uncertainties. *IET Generation Transmission and Distribution* 12 (8): 1929–1940.

10 Fang, X., Li, F., Wei, Y. et al. (2015). Reactive power planning under high penetration of wind energy using Benders decomposition. *IET Generation Transmission and Distribution* 9 (14): 1835–1844.

11 Tahboub, A.M., El Moursi, M.S., Woon, W.L., and Kirtley, J.L. (2017). Multiobjective dynamic VAR planning strategy with different shunt compensation technologies. *IEEE Transactions on Power Systems* 33 (3): 2429–2439.

12 Wu, L. and Guan, L. (2019). Integer quadratic programming model for dynamic VAR compensation considering short-term voltage stability. *IET Generation Transmission and Distribution* 13 (5): 652–661.

13 Chi, Y., Xu, Y., and Zhang, R. (2021). Many-objective robust optimization for dynamic var planning to enhance voltage stability of a wind-energy power system. *IEEE Transactions on Power Delivery* 36 (1): 30–42.

14 Gunawan, S. and Azarm, S. (2004). Non-gradient based parameter sensitivity estimation for single objective robust design optimization. *Journal of Mechanical Design* 126 (3): 395–402.

15 Miettinen, K. (2012). *Nonlinear Multiobjective Optimization*. Springer Science & Business Media.

16 Deb, K., Pratap, A., Agarwal, S., and Meyarivan, T.A. (2002). A fast and elitist multiobjective genetic algorithm: NSGA-II. *IEEE Transactions on Evolutionary Computation* 6 (2): 182–197.

17 Mazhari, S.M., Monsef, H., and Romero, R. (2015). A multi-objective distribution system expansion planning incorporating customer choices on reliability. *IEEE Transactions on Power Systems* 31 (2): 1330–1340.

18 Arabali, A., Ghofrani, M., Etezadi-Amoli, M. et al. (2014). A multi-objective transmission expansion planning framework in deregulated power systems with wind generation. *IEEE Transactions on Power Systems* 29 (6): 3003–3011.

19 Ghatak, S.R., Sannigrahi, S., and Acharjee, P. (2018). Optimised planning of distribution network with photovoltaic system, battery storage, and DSTATCOM. *IET Renewable Power Generation* 12 (15): 1823–1832.

20 Köppen, M. and Yoshida, K. (2007). Substitute distance assignments in NSGA-II for handling many-objective optimization problems. In: *International Conference on Evolutionary Multi-Criterion Optimization*, 727–741. Berlin, Heidelberg: Springer.

21 Deb, K. and Jain, H. (2013). An evolutionary many-objective optimization algorithm using reference-point-based nondominated sorting approach, part I: solving problems with box constraints. *IEEE Transactions on Evolutionary Computation* 18 (4): 577–601.

22 Das, I. and Dennis, J.E. (1998). Normal-boundary intersection: a new method for generating the Pareto surface in nonlinear multicriteria optimization problems. *SIAM Journal on Optimization* 8 (3): 631–657.

23 Chen, F., Zhou, J., Wang, C. et al. (2017). A modified gravitational search algorithm based on a non-dominated sorting genetic approach for hydro-thermal-wind economic emission dispatching. *Energy* 121: 276–291.

24 Wu, X., Li, J., Shen, X., and Zhao, N. (2020). NSGA-III for solving dynamic flexible job shop scheduling problem considering deterioration effect. *IET Collaborative Intelligent Manufacturing* 2 (1): 22–33.

25 McKay, M.D., Beckman, R.J., and Conover, W.J. (2000). A comparison of three methods for selecting values of input variables in the analysis of output from a computer code. *Technometrics* 42 (1): 55–61.

26 Ali, M., Siarry, P., and Pant, M. (2012). An efficient differential evolution based algorithm for solving multi-objective optimization problems. *European Journal of Operational Research* 217 (2): 404–416.

27 Siemens Power Technologies International (2013). PSS®E 33.4 Model Library. New York: Siemens.

28 Zhang, R., Xu, Y., Dong, Z.Y., and Wong, K.P. (2016). Measurement-based dynamic load modelling using time-domain simulation and parallel-evolutionary search. *IET Generation Transmission and Distribution* 10 (15): 3893–3900.

29 Agrawal, S., Panigrahi, B.K., and Tiwari, M.K. (2008). Multiobjective particle swarm algorithm with fuzzy clustering for electrical power dispatch. *IEEE Transactions on Evolutionary Computation* 12 (5): 529–541.

30 Cutsem, T.V., Glavic, M., Rosehart, W. et al. (2015). Test Systems for Voltage Stability Analysis and Security Assessment. PES-TR19. IEEE.

31 Lee, Y. and Song, H. (2019). A reactive power compensation strategy for voltage stability challenges in the Korean power system with dynamic loads. *Sustainability* 11 (2): 326.

32 Technical Guidelines on Voltage and Reactive Power of the Power System, SD325-1989. Beijing: National Energy Administration of China; 1989.

33 Miao, Z.M., Liu, W., Xue, J. et al. (2008). Technical Principles on Reactive Power Compensation of the Power System, State Grid Corporation of China, Q/GDW 212-2008, Beijing: State Grid Corporation of China.

34 Meng, K., Dong, Z.Y., Wong, K.P. et al. (2010). Speed-up the computing efficiency of power system simulator for engineering-based power system transient stability simulations. *IET Generation Transmission and Distribution* 4 (5): 652–661.

Index

Stability-Constrained Optimization for Modern Power System Operation and Planning,
First Edition. Yan Xu, Yuan Chi, and Heling Yuan.
© 2023 The Institute of Electrical and Electronics Engineers, Inc.
Published 2023 by John Wiley & Sons, Inc.

IEEE Press Series on Power and Energy Systems

Series Editor: Ganesh Kumar Venayagamoorthy, Clemson University, Clemson, South Carolina, USA.

The mission of the IEEE Press Series on Power and Energy Systems is to publish leading-edge books that cover a broad spectrum of current and forward-looking technologies in the fast-moving area of power and energy systems including smart grid, renewable energy systems, electric vehicles and related areas. Our target audience includes power and energy systems professionals from academia, industry and government who are interested in enhancing their knowledge and perspectives in their areas of interest.

1. *Electric Power Systems: Design and Analysis, Revised Printing*
   Mohamed E. El-Hawary

2. *Power System Stability*
   Edward W. Kimbark

3. *Analysis of Faulted Power Systems*
   Paul M. Anderson

4. *Inspection of Large Synchronous Machines: Checklists, Failure Identification, and Troubleshooting*
   Isidor Kerszenbaum

5. *Electric Power Applications of Fuzzy Systems*
   Mohamed E. El-Hawary

6. *Power System Protection*
   Paul M. Anderson

7. *Subsynchronous Resonance in Power Systems*
   Paul M. Anderson, B.L. Agrawal, J.E. Van Ness

8. *Understanding Power Quality Problems: Voltage Sags and Interruptions*
   Math H. Bollen

9. *Analysis of Electric Machinery*
   Paul C. Krause, Oleg Wasynczuk, and S.D. Sudhoff

10. *Power System Control and Stability, Revised Printing*
    Paul M. Anderson, A.A. Fouad

11. *Principles of Electric Machines with Power Electronic Applications, Second Edition*
    Mohamed E. El-Hawary

Printed and bound by CPI Group (UK) Ltd, Croydon, CR0 4YY

16/04/2025